Nanocrystal Technologies in Biomedical Sciences

Neelesh Kumar Mehra
Kuldeep Kumar Bansal
Editors

Nanocrystal Technologies in Biomedical Sciences

Advances in crystal development, characterization and targeted drug delivery

Springer

Editors
Neelesh Kumar Mehra
Department of Pharmaceutics
National Institute of Pharmaceutical
Education and Research (NIPER)
Hyderabad, Telangana, India

Kuldeep Kumar Bansal
Pharmaceutical Sciences Laboratory
Faculty of Science and Engineering
Åbo Akademi University, Biocity
Turku, Finland

ISBN 978-981-95-2967-4 ISBN 978-981-95-2968-1 (eBook)
https://doi.org/10.1007/978-981-95-2968-1

This Springer imprint is published by the registered company Springer Nature Singapore Pte Ltd.
The registered company address is: 152 Beach Road, #21-01/04 Gateway East, Singapore 189721, Singapore

Preface

In pharmaceutical and biomedical sciences, addressing poor bioavailability is a paramount concern during drug development. Innovative strategies are being employed to enhance the solubility, dissolution rate, and physiochemical properties of Biopharmaceutical Classification System (BCS) II and IV drugs. Drug nanocrystals, utilizing nanotechnology, emerge as a solution by leveraging the crystalline nature of drugs to address critical challenges. This carrier-free colloidal drug delivery system operates at the nano-size range, where the crystalline drug particle core is enveloped by a stabilizing agent (surfactant/polymer). However, selecting a stabilizer remains a persistent challenge for certain drug molecules.

It provides a thorough exploration of critical aspects related to nanocrystal preparation for drug delivery. It covers classification, stabilizer selection criteria, bottom-up and bottom-down preparation, characterization methods, recent advancements, toxicological considerations, regulatory insights, and clinical status. The book advocates for implementing quality by design (QbD) and a design of experimentation approach to minimize errors in product optimization and validation for nanocrystal formulation development. Encompassing crystal technologies, nanocrystals, translational aspects, and the theory of crystal formation, the book delves into novel advanced approaches with regulatory considerations for product development. Contributions from global experts, including renowned formulators, researchers, and academicians, ensure diverse insights and expertise across various specializations.

Encompassing crystal technologies, nanocrystals, translational aspects, and the theory of crystal formation, the book delves into novel advanced approaches with regulatory considerations for product development. Contributions from global experts, including renowned formulators, researchers, and academicians, ensure diverse insights and expertise across various specializations.

This book will be based on the basic applied fundamentals, concepts, and theory of multifunctional nanocarriers with an emphasis on targeted drug delivery in pharmaceutical and biomedical sciences. We believe that there is a vacuum of scientific and industrial understanding, as no comprehensive book is available which comprises the abovementioned table of contents in pharmaceutical and research applications. The prime objective of the authors is to exhaustively cover the targeting, receptor-mediated targeting, and recent advancements using these multifunctional nanocarriers with their regulatory aspects. The proposed book would be edited through the contribution of multi-authored book chapter across the globe from eminent scientists/experts.

Hyderabad, Telangana, India — Neelesh Kumar Mehra
Turku, Finland — Kuldeep Kumar Bansal

Contents

Editors and Contributors

About the Editors

Neelesh Kumar Mehra is working in the Department of Pharmaceutics, National Institute of Pharmaceutical Education & Research (NIPER), Hyderabad, India. He earned his PhD from Dr. H.S. Gour University, Sagar, and PostDoc from Irma Lerma Rangel College of Pharmacy, Texas A & M Health Science Centre, Kingsville, TX, USA. He served as Manager in Product Development, Sentiss Research Centre, Sentiss Pharma Pvt Ltd., Gurgaon, for development, scale-up, and technology transfer of complex ophthalmic, inhalation, and optic pharmaceutical products. He received "TEAM AWARD" for successful commercialization of Ophthalmic Suspension product. He has authored more than 140 peer-reviewed publications in high repute international journals, and more than 50 book chapters. He has four granted and four filed patent on manufacturing process and composition to improve therapeutic efficacy for topical delivery. Five PhD students and 36 MS students have completed their research work under his supervision, and he is currently mentoring 10 PhD and 7 MS students for dissertations/research projects. He has received numerous outstanding awards including Young Scientist and Team Awards for his research output. He is a peer reviewer of various international journals and publishers.

He edited six books on multifunctional nanocarriers, dendrimers, carbon nanotubes, ophthalmic drug delivery, biopharmaceutical, and nanotechnology. He has rich research and teaching experience in the formulation and development of complex, innovative, ophthalmic and injectable biopharmaceutical products including micro- and nanotechnologies for regulated market and so on.

According to announcement by the Stanford University Scientists based on the database and standardized citation indicators, Dr. Mehra is included in top 2% scientists in the world in the area of Pharmacology and Pharmacy, consecutively for 3 years.

Kuldeep Kumar Bansal, an adjunct professor, has been affiliated with the Department of Pharmacy at Åbo Akademi University in Turku, Finland, since 2017. He commenced his tenure at Åbo Akademi University as a Postdoctoral researcher, focusing on the development of polymer-based nano-formulations. During his tenure, he produced a novel polymer named poly (jasmine lactone) (PJL) and pushing technology forward to identify commercial pathways for drug delivery applications. He obtained his PhD from the University of Nottingham, United Kingdom, in 2015. His study primarily focuses on the production of novel polymers for enhanced delivery of toxic and hydrophobic pharmaceuticals. He possesses experience in working with pharmaceutical companies in the research and development and quality assurance sectors. He has produced over 40 peer-reviewed publications in reputable international journals along with more than 4 book chapters. He serves as a peer reviewer for numerous international journals and publishers.

He possesses one awarded patent on polymer and composition from renewable sources. He is presently supervising four PhD students and has overseen more than ten MS students in their dissertation and research projects.

Contributors

Kuldeep Kumar Bansal Pharmaceutical Sciences Laboratory, Faculty of Science and Engineering Åbo Akademi University, Biocity, Turku, Finland

Sonali Batra Department of Pharmaceutical Sciences, Indira Gandhi University, Meerpur, Rewari, Haryana, India

Nitisha Beniwal Department of Biomedical Engineering, Indian Institute of Technology Hyderabad, Sangareddy, Telangana, India

Padakanti Sandeep Chary College of Pharmacy and Pharmaceutical Sciences, Florida A&M University, Tallahassee, FL, USA

Vaibhav Dhumal Department of Pharmaceutics, National Institute of Pharmaceutical Education and Research (NIPER), Hyderabad, Telangana, India

Anish Dhuri Department of Pharmaceutics, National Institute of Pharmaceutical Education and Research (NIPER), Hyderabad, Telangana, India

Xiaoyun Du School of Pharmaceutical Sciences, Fudan University, Key Laboratory of Smart Drug Delivery of MOE and National Key Laboratory of Advanced Drug Formulations for Overcoming Delivery Barriers, Shanghai, China

Anuja Dubey Government Bilasa Girls PG Autonomous College, Bilaspur, Chhattisgarh, India

Maha F. Emam Division of Pharmaceutical Chemistry and Technology, University of Helsinki, Helsinki, Finland

Erika Espo Division of Pharmaceutical Chemistry and Technology, University of Helsinki, Helsinki, Finland

Bhavana Gajare Department of Pharmaceutics, National Institute of Pharmaceutical Education and Research (NIPER), Hyderabad, Telangana, India

Shyam Sudhakar Gomte Department of Pharmaceutics, National Institute of Pharmaceutical Education and Research (NIPER)-Ahmedabad, Palaj, Gandhinagar, Gujarat, India

Aakanchha Jain Department of Pharmaceutics, National Institute of Pharmaceutical Education and Research (NIPER)-Ahmedabad, Palaj, Gandhinagar, Gujarat, India

Chirag Jain Centre for Industrial Pharmacy and Pharmaceutical Regulatory Affairs, Amity Institute of Pharmacy, Amity University, Noida, Uttar Pradesh, India

Neha Jain Centre for Pharmaceutics, Amity Institute of Pharmacy, Amity University, Noida, Uttar Pradesh, India

Krishna Kant Jangde Department of Pharmacy, Guru Ghasidas Vishwavidyalaya (A Central University), Bilaspur, Chhattisgarh, India

Atte Junnila Division of Pharmaceutical Chemistry and Technology, University of Helsinki, Helsinki, Finland

Anssi-Pekka Karttunen Division of Pharmaceutical Chemistry and Technology, University of Helsinki, Helsinki, Finland

Karthikeyan Kesavan Department of Pharmacy, Guru Ghasidas Vishwavidyalaya, Bilaspur, Chhattisgarh, India

Pooja Khade Department of Pharmaceutics, National Institute of Pharmaceutical Education and Research (NIPER), Hyderabad, Telangana, India

Pradeep Kumar Department of Botany, Swami Shraddhanand College, University of Delhi, Delhi, India

Sandeep Kumar School of Pharmaceutical Sciences, Delhi Pharmaceutical Sciences and Research University, New Delhi, India

Vishal Labhade Department of Pharmaceutics, National Institute of Pharmaceutical Education and Research (NIPER), Hyderabad, Telangana, India

Yi Lu School of Pharmaceutical Sciences, Fudan University, Key Laboratory of Smart Drug Delivery of MOE and National Key Laboratory of Advanced Drug Formulations for Overcoming Delivery Barriers, Shanghai, China

Akanksha Malaiya Nanomedicine and Bioengineering Research Laboratory, Department of Pharmacy, Indira Gandhi National Tribal University, Amarkantak, Madhya Pradesh, India

Bhagyoday Tirth Pharmacy College, Sagar, Madhya Pradesh, India

Niveditha Manne Department of Pharmaceutics, National Institute of Pharmaceutical Education and Research (NIPER), Hyderabad, Telangana, India

Neelesh Kumar Mehra Department of Pharmaceutics, National Institute of Pharmaceutical Education and Research (NIPER), Hyderabad, Telangana, India

Dinesh Kumar Mishra Department of Pharmacy, Guru Ghasidas Vishwavidyalaya (A Central University), Bilaspur, Chhattisgarh, India

Parasuraman Mohan Department of Pharmaceutics, Vellalar College of Pharmacy, Erode, Tamil Nadu, India

Sunny Mukherjee Department of Biomedical Engineering, Indian Institute of Technology Hyderabad, Sangareddy, Telangana, India

Eetu Myöhänen Division of Pharmaceutical Chemistry and Technology, University of Helsinki, Helsinki, Finland

Irem Namli Division of Pharmaceutical Chemistry and Technology, University of Helsinki, Helsinki, Finland

Puja Nayak Department of Pharmacy, Guru Ghasidas Vishwavidyalaya (A Central University), Bilaspur, Chhattisgarh, India

Rishi Paliwal Nanomedicine and Bioengineering Research Laboratory, Department of Pharmacy, Indira Gandhi National Tribal University, Amarkantak, Madhya Pradesh, India

Shivani Rai Paliwal Department of Pharmacy, Guru Ghasidas Vishwavidyalaya (A Central University), Bilaspur, Chhattisgarh, India

Ekta R. Pardhi Pharmaceutical Nanotechnology Research Laboratory, Department of Pharmaceutics, National Institute of Pharmaceutical Education and Research (NIPER), Hyderabad, Telangana, India

Ruchita Patil Pharmaceutical Nanotechnology Research Laboratory, Department of Pharmaceutics, National Institute of Pharmaceutical Education and Research (NIPER), Hyderabad, Telangana, India

Shruti Patil Department of Pharmaceutics, National Institute of Pharmaceutical Education and Research (NIPER)-Ahmedabad, Palaj, Gandhinagar, Gujarat, India

Vasu Peddinti Department of Pharmaceutics, National Institute of Pharmaceutical Education and Research (NIPER)-Ahmedabad, Palaj, Gandhinagar, Gujarat, India

Leena Peltonen Division of Pharmaceutical Chemistry and Technology, University of Helsinki, Helsinki, Finland

Jothimani Rajeswari Department of Pharmacy, Guru Ghasidas Vishwavidyalaya, Bilaspur, Chhattisgarh, India

Aravind Kumar Rengan Department of Biomedical Engineering, Indian Institute of Technology Hyderabad, Sangareddy, Telangana, India

Ankur Sharma Hyalo Technologies, Somerest, NJ, USA

Kritika Sharma School of Pharmaceutical Sciences, Delhi Pharmaceutical Sciences and Research University, New Delhi, India

Priya Sharma Department of Pharmacy, Guru Ghasidas Vishwavidyalaya, Bilaspur, Chhattisgarh, India

Sumit Sharma School of Pharmaceutical Sciences, Delhi Pharmaceutical Sciences and Research University, New Delhi, India

Pankaj Kumar Singh Department of Pharmaceutics, National Institute of Pharmaceutical Education and Research (NIPER), Hyderabad, Telangana, India

Anitha Sriram Department of Pharmaceutics, National Institute of Pharmaceutical Education and Research (NIPER), Hyderabad, Telangana, India

Sourabh Tapekar Department of Pharmaceutics, National Institute of Pharmaceutical Education and Research (NIPER), Hyderabad, Telangana, India

Tapani Viitala Division of Pharmaceutical Chemistry and Technology, University of Helsinki, Helsinki, Finland
Pharmaceutical Sciences Laboratory, Åbo Akademi University, Turku, Finland

Wei Wu School of Pharmaceutical Sciences, Fudan University, Key Laboratory of Smart Drug Delivery of MOE and National Key Laboratory of Advanced Drug Formulations for Overcoming Delivery Barriers, Shanghai, China

Rati Yadav Pharmaceutical Nanotechnology Research Laboratory, Department of Pharmaceutics, National Institute of Pharmaceutical Education and Research (NIPER), Ministry of Chemical and Fertilizer, Hyderabad, Telangana, India

Devaraj Yallappa Department of Pharmaceutics, National Institute of Pharmaceutical Education and Research (NIPER)-Ahmedabad, Palaj, Gandhinagar, Gujarat, India

Abbreviations

3D	3 dimensional
3TC	Lamivudine
ACQ	Aggregation-caused quenching
ADME	Absorption, distribution, metabolism, and excretion
AgNPs	Silver nanoparticles
AIE	Aggregation-induced emission
API	Active pharmaceutical ingredient
ARA-UR	Cytarabine uracil
AUC	Area under curve
AZT	Zidovudine
BBB	Blood-brain barrier
BCS	Biopharmaceutics classification system
BE-NCT	Baicalein-nicotinamide
BODIPY	Boron-dipyrromethene
BR-NCs	Brinzolamide nanocrystals
BS	Bovine serum albumin
BZM	Brinzolamide
CA	Cabotegravir
CAP	Cellulose acetate phthalate
CBD-NC	Cannabidiol nanocrystals
CC	Cocrystals
CL	Cholecalciferol
CLS	Confocal laser scanning microscopy
CM	Residues of traditional Chinese medicine
C_{max}	Maximum plasma concentration
CNCs	Cellulose nanocrystals
CPFX/NAC/Cur DPIs	Ciprofloxacin-curcumin-N-acetyl cysteine dry powder inhaler
Cs	Saturation solubility
CT	Computed tomography
CUR	Curcumin
CUR-NS	Curcumin nanosuspension
DDS	Drug delivery services

DIW	Direct ink writing
DMNs	Dissolving microneedles
DSC	Differential scanning calorimetry
FITC	Fluorescein 5(6)-isothiocyanate
FNB	Fenofibrate
FPR	Flamma Fluro®
FRET	Fluorescence resonance energy transfer
FRET	Forster resonance energy transfer
GBC	Glibenclamide
Gd	Gadolinium
GI	Gastrointestinal
HAND	HIV-associated neurocognitive disorder
HNCs	Hybrid nanocrystals
HPH	High-pressure homogenizer
HPMC	Hydroxy propyl methyl cellulose
HRTEM	High-resolution transmission electron microscopy
HyNASDs	Hybrid nanocrystal-amorphous solid dispersions
IDD-P	Insoluble drug delivery particle
ISO	International Organization for Standardization
ITZ NS	Itraconazole nanosuspension
IVIVC	In vitro-in vivo correlation
LD	Laser diffractometry
MDCK	Madin-Darby canine kidney cells
MMPs	Matrix metalloproteinases
MRI	Magnetic resonance imaging
MRT	Mean residence time
MTX-NC-MN	Methotrexate nanocrystals loaded microneedle array
mV	Milli volt
MYR-NIC	Myricetin-nicotinamide
Na	Data not available (na)
NAC	N-acetyl cysteine
Nano GA	18β-Glycyrrhetinic acid nanocrystals
NCCs	Nano-cocrystals
NCs	Nanocrystal
NIR	Near infrared
NPs	Nanoparticles
NRG	Naringenin
NSSPE	Nanocrystals self-stabilized Pickering emulsion
ODD	Ocular drug delivery
OECs	Olfactory ensheathing cells
O-PT-NC-Folate	Paclitaxel-loaded folate receptor-targeted NC
PA	Paeoniflorin
PAP-HCl	Phenazopyridine hydrochloride
PAP–PI	Phenazopyridine phthalimide
PBPK	Physiologically based pharmacokinetics;

PBS	Phosphate-buffered saline
PCS	Photon correlation spectroscopy
PDI	Polydispersity index
PTX/NC-PSS	Paclitaxel nanocrystals stabilized by sodium poly styrene sulfonate
PTX-DSF	Paclitaxel-disulfiram
PVA	Poly vinyl alcohol
PVP-NiS2/FeS2	Polyvinylpyrrolidone-nickel disulfide/iron disulfide
QT	Quercetin
QT-HNCs	Quercetin hybrid nanocrystals
RES	Reticuloendothelial system
RPV	Rilpirivine
RT	Room temperature 25 °C
SDS	Sodium dodecyl sulfate
SEM	Scanning electron microscope
SLS	Sodium lauryl sulfate
TEM	Transmission electron microscopy
TPA	12-O-tetradecanoyl phorbol-13-acetate
TPGS 1000	D-α-Tocopheryl polyethylene glycol 1000 succinate
UCNCs	Up-conversion nanocrystals
USFDA	US Food and Drug Administration
VAP peptides	Vesicle-associated membrane protein-associated protein peptides
WBM	Wet ball milling
XRD	X-ray diffraction

1 Nanocrystal Technologies: Concepts, Fundamentals, and Strategies

Priya Sharma, Jothimani Rajeswari, Parasuraman Mohan, and Karthikeyan Kesavan

Abstract

The advent of contemporary nanoscience with nanocrystals (NCs) occurred in the 1980s and continues to this day. NCs, or nanometer-sized solid drug particles, have crystalline features and unique electrical, optical, and chemical capabilities due to their nanoscale dimensions. Nanocrystals, unlike other nanoformulations, do not require a carrier, reducing carrier-related toxicity. Nanocrystal formulations offer the advantage of dose reduction as they do not require the use of a carrier system for preparation, thereby decreasing the overall drug dosage. NCs can be developed by using both top-down and bottom-up techniques. They appear in a crystalline state, as the name implies, but may be indeterminate depending on the method of manufacture. NCs must be uniform in size, shape, internal structure, and surface chemistry. Nanocrystals have the potential to address challenges across various industries, including drug delivery for poorly soluble compounds, cosmetic research and products, biotechnology, and catalysis. It is found in some research that NCs help enhance the bioavailability of weakly water-soluble APIs, that is, BCS class II and IV drugs having low bioavailability due to their less hydrophilic nature. Primary uses in the biomedical sciences include increased solubility, enhanced bioavailability, targeted drug delivery, controlled release, regenerative medicine, herbal formulations, and bioimaging. This chapter presents a detailed overview of nanocrystal principles, basics, and tactics.

P. Sharma · J. Rajeswari · K. Kesavan (✉)
Department of Pharmacy, Guru Ghasidas Vishwavidyalaya, Bilaspur, Chhattisgarh, India

P. Mohan
Department of Pharmaceutics, Vellalar College of Pharmacy, Erode, Tamil Nadu, India

N. K. Mehra, K. K. Bansal (eds.), *Nanocrystal Technologies in Biomedical Sciences*,
https://doi.org/10.1007/978-981-95-2968-1_1

Keywords

Nanocrystals · Drug nanocrystals · Nanoparticles · Nano-formulations · Drug delivery · Biomedical sciences

Abbreviations

3D	3 Dimensional
API	Active pharmaceutical ingredient
AUC	Area under curve
BCS	Biopharmaceutics classification system
CAB	Cabotegravir
CBD-NC	Cannabidiol nanocrystals
C_{max}	Maximum plasma concentration
CMR	Residues of traditional Chinese medicine
CNCs	Cellulose nanocrystals
C_s	Saturation solubility
DDS	Drug delivery services
DIW	Direct ink writing
DPPH	2,2-Diphenyl-1-picryl-hydrazyl-hydrate
DSC	Differential scanning calorimetry
Gd	Gadolinium
GI	Gastrointestinal
HAND	HIV-associated neurocognitive disorder
HPH	High-pressure homogenization
ISO	International Organization for Standardization
MRI	Magnetic resonance imaging
mV	Milli volt
NAC	N-acetylcysteine
Nano GA	18β-Glycyrrhetinic acid nanocrystals
NCs	Nanocrystals
NRG	Naringenin
NSSPE	Nanocrystals self-stabilized Pickering emulsion
PCLA	Poly(ε-caprolactone-*co*-lactide)-*b*-poly(ethylene glycol)-*b*-poly (ε-caprolactone-*co*-lactide
PCS	Photon correlation spectroscopy
PDI	Polydispersity index
PTX	Paclitaxel
PTX-NCs	Paclitaxel nanocrystals
QT	Quercetin
QT-HNCs	Quercetin hybrid nanocrystals
RPV	Rilpirivine
TEM	Transmission Electron Microscopy
TPA	12-O-tetradecanoyl phorbol-13-acetate

UCNC	Upconversion nanocrystals
USFDA	US Food and Drug Administration
WBM	Wet ball milling
XRD	X-ray diffraction

1.1 Introduction

Pure solid particles having a mean diameter less than 1 μm and a crystalline nature are known as drug nanocrystals. The technology presents a remarkable chance to administer hydrophobic medications. Its distinctiveness stems from the fact that nanocrystals are made of medicament or the payload, removing the carrier's auxiliary function. Furthermore, crystalline dispersions in liquid media are frequently stabilized by the use of stabilizers or surfactants [1].

Nanocrystals contain minute crystals of parent drugs in the size range (100–1000 nm) and make the effective transport of drugs inside the cells and among the cells [2].

The concept of nanocrystals is straightforward yet ingenious. As the size decreases, the physicochemical properties undergo modifications, increasing kinetic solubility. This phenomenon is explained by the nanocrystals' greater curvature, which raises the velocity of dissolution and, as a result, increases solubility. The dissolving velocity (Noyes–Whitney equation) of weakly soluble actives is improved by the increased solubility and surface area. Passive diffusion via biological membranes is encouraged by the greater concentration gradient produced by the higher solubility. Furthermore, because of the "gecko effect," which is the increased surface-to-volume ratio that occurs with decreasing size, nanosized materials have higher membrane adherence. The nanocrystal technology is beneficial for drug delivery because it increases retention time at the absorption site, which makes it easier for active ingredients to penetrate. These beneficial qualities are utilized in a number of pharmaceutical therapeutic products intended for oral use, including Tricor®, Emend® (Aprepitant, MSD), and Rapamune® (Sirolimus, Pfizer) [3].

Drug nanocrystals that are dispersed in a liquid media are referred to as "nanosuspensions." They can be made by employing intense particle size reduction techniques in the presence of a surfactant or stabilizer to break down larger drug dispersions or by precipitating drug particles from an organic solvent and then adding an aqueous surfactant/stabilizer solution [4]. According to Stern's definition, a formulation must have a 90% volume diameter (D90) $<$ 2.5 mm and a median diameter capacity (D50) $<$ 1 mm in order to be classified as nanosuspension [5].

Drug nanocrystals are an interesting way to administer medications that are not very soluble in water. When distributed in a liquid medium, they need to be stabilized using surfactants or polymeric steric stabilizers to avoid aggregation and/or Ostwald ripening during storage. Water, other aqueous solutions, or nonaqueous solutions can be used as the dispersion media in the formulation of drug nanocrystal suspensions. The intrinsic attributes of the raw material, such as its kinetic and thermodynamic properties, are changed as the particle size is reduced to the nanoscale range [6]. Utilization of drug nanocrystals is a versatile strategy that can be applied to all drugs having low solubility since they can be directly converted into pharmacological particles of a size of nanometers [7].

Enhancing the solubility of saturation, dissolution rate, and potentially the mucoadhesive property of the drug can improve oral bioavailability. Nanocrystal formulations for oral administration offer several benefits, including the following.

1.1.1 Faster Absorption

Since NCs greatly affect the design and modification of drug nanocrystals demonstrating improved performance both in vitro and in vivo, the nanocrystals play a critical role in accelerating absorption and enhancing overall bioavailability [8].

1.1.2 Improved Oral Bioavailability

Particle surface area, curvature, saturation solubility, dissolving rate, and bioavailability all improve as particle size is reduced to the nanoscale. The drug NCs offer substantial benefits in terms of increased bioavailability of drug when administered orally [9]. Nanocrystals are remarkably effective in increasing bioavailability; they can increase the bioavailability of drugs that are poorly soluble by 2–30 times [10].

1.1.3 Rapid Onset of Action

Through a quick onset of pharmacological effects for the treatment, the nanocrystals method may be a viable formulation strategy for increasing the bioavailability of active components [11].

1.1.4 Enhanced Dose Proportionality

Some nanocrystal formulations (danazol, megestrol) have shown improved dosage proportionality or linear pharmacokinetics, most likely as a result of their faster rate of dissolution in the gastrointestinal tract's absorption region. Therefore, using a nanocrystal formulation technique may potentially result in enhanced bioavailability and the additional benefit of linear pharmacokinetics and dose proportionality [12].

1.1.5 Required Dose Reduction

Due to their great solubility, oral nanosuspensions are especially used to increase the rate and extent of drug absorption. A lower dosage may be necessary as a result of improved bioavailability, which might reduce the likelihood of adverse effects and possibly increase patient compliance [13].

1.1.6 Versatile Dosage Form

Nanocrystals are adaptable and suitable for various dosage forms due to their nanoscale size, improved solubility, and distinctive physical properties. Pills, capsules, sachets, or powder can be used for oral intake. Because of their small particle sizes, nanosuspensions can also be administered intravenously, resulting in 100% bioavailability [14].

1.2 History

The birth of modern nanoscience with NCs is, however, attributed to a much later period: beginning in the early 1980s and extending to the present [15]. In the early 1990s, drug nanocrystals were developed, as indicated by the initial patent applications from Nanosystems; this has since been acquired by Elan Drug Delivery. Liversidge et al. conducted pioneering research on male beagle dogs and prepared danazole nanosuspension by wet milling technique to compare the bioavailability with other formulations [16]. After oral administration, the nanocrystals are dissolving in the intestinal tract. There is no or negligible uptake of particles from the gut. Pharmaceutical scientists have endeavored to address this issue cautiously by utilizing existing, proven technologies instead [17].

However, it is believed that the emergence of contemporary nanoscience with NCs occurred much later, starting in the early 1980s and continuing to the present [18]. According to "Nanosystems," early patent applications, drug NCs were created in the early 1990s.

Since then, Elan Drug Delivery has bought it. Liversidge et al. conducted pioneering research on male beagle dogs and prepared danazole nanosuspension by wet milling technique to compare the bioavailability with other formulations [16]. The nanocrystals are dissolving in the gastrointestinal tract following oral ingestion. Particles from the intestines are either completely absent or barely absorbed. Pharmaceutical scientists have endeavored to address this issue cautiously by utilizing existing, proven technologies instead [17].

Drug nanocrystal technology has been a primary emphasis in the pharmaceutical industry for the previous two decades. One of its significant advantages lies in its ability to formulate poorly soluble drugs. In the last 10 years of the twentieth century, numerous studies were conducted to explore the processes used to formulate drug nanocrystals and their physical and chemical characteristics in vitro, including process parameters, formulation composition, physical and chemical stability, crystalline structure, enhanced rate of dissolution and saturation solubility, bioadhesion, and more. Over the last decade, there has been increasing attention on the effectiveness of nanocrystals in vivo on animals and clinical studies as well, leading to many exciting discoveries [7].

Humans have been aware of and utilizing inorganic nanocrystals, that is, those that contain elements typically considered inorganic for more than 2000 years. The Lycurgus cup, for instance, is a very elaborate glass cup that was created during the fifth to fourth centuries BC and gold nanocrystals were used to color its glass. Since the publication of *Panacea Aurea-Auro Potabile* in 1616, which implied that drinking solutions of gold colloids could be beneficial for human health in the majority of diseases, biomedical uses for inorganic nanocrystals have been claimed for decades. However, inorganic nanocrystals did not start to find valid biomedical applications until the twentieth century and the introduction of medical imaging methods. One example of this is the use of a thorium colloid called Thorotrast in X-ray imaging. Iron oxide nanoparticle formulations became available for use as contrast agents in MRI throughout the 1990s; Feridex, for instance, received FDA approval in 1996 [19].

Nanocrystals were initially developed to improve the oral absorption of poorly soluble medications. For the development of first-generation NCs, pearl milling technique was used for breaking down the components. This technology was patented by Liversidge et al. in 1990 [20] and was owned by Nanosystems, which Elan Drug Delivery has subsequently purchased [21]. Subsequently, Muller et al. then created NCs by dissolving the drug powder in a surfactant solution, which were then done milling by pearl milling method for a few hours and extended to a few days [22]. Furthermore, DissoCubes, a second-generation NC made with high-pressure homogenization (HPH), was released. Previously owned by DDS (Drug Delivery Services) in Germany, the patent for NCs is currently held by Skye Pharma. This company introduced various size-reduction techniques, including insoluble drug delivery particle technology. Between 2000 and 2005, the US FDA approved five different NCs. Additionally, Johnson & Johnson's injectable NCs of paliperidone palmitate for the treatment of schizophrenia were approved by the US FDA in 2009. NCs have been utilized extensively in both industry and academic studies to administer hydrophobic medications via a variety of routes. As a result, more than 20 products have been approved, and several clinical trials are presently underway at various levels. There are already more than a dozen NC oral formulations that have received clinical approval, indicating their efficacy in increasing bioavailability through enhanced solubility [23].

1.3 Selection Criteria of Stabilizer for the Preparation of NCs

A significant amount of additives may be needed to ensure adequate stabilization since the surface area of the particles in a nanosuspension is considerably higher than that of particles in a typical suspension. Thus, it is essential to carefully evaluate the nature and degree of concentration of the stabilizing agent utilized in order to ensure successful formulation, regardless of the process chosen to formulate nanosuspensions. Polymeric stabilizers and surfactant stabilizers may both serve this particular function [24].

The successful production of nanocrystals relies on the selection of the right stabilizers. Stabilizers that are frequently employed include polymers or surfactants, including TPGS, cellulose derivatives, poloxamers, and polysorbates. Furthermore, other excipient types, including the use of superdisintegrants during milling, have been investigated for their potential as stabilizers. Stabilizers often also improve the solubility of drug substances to a certain degree. Due to the lack of efficient methods to evaluate stabilizers and a lack of knowledge about the interactions in colloidal systems, selecting a stabilizer can be challenging [25].

The formulation scientist's prior experiences and empirical data serve as the main foundation for the stabilizer selection. Although the exact factors influencing the stabilization effect are unknown, it is crucial to take into account more than just the interaction forces between the drug surface and the stabilizer molecule. Those forces can be hydrophilic/hydrophobic, hydrogen bonding, ionic, or other weak forces. Additionally, it is necessary to have a high zetapotential greater than $\pm$ 30 mV for the development of an electric barrier or a chain length that is sufficient to produce a steric barrier [26]. Different parameters for the selection of stabilizers for the preparation of NCs are mentioned in Fig. 1.1 & Table 1.1 describe the different classes of stabilizers commonly employed to stabilize the nanosuspensions.

1.4 Methods of Preparation of Nanocrystals

There are two fundamental methods for creating drug nanocrystals: bottom-up technology, top-down technology, and the combination thereof. All of these methods for the preparation of NCs are shown in Fig. 1.2.

1.4.1 Bottom-Up Technique

Bottom-up technology is also named as hydrosol technology, it was invented by Sucker, and Sandoz (Novartis in the present) acquired the intellectual property in 1988. This method involves dissolving the drug in a solvent, which is introduced to a nonsolvent afterward, leading the finely distributed drug nanocrystals to precipitate. The precipitation method is easy to use and requires inexpensive tools. One of the primary methods is the employment of static mixers or micro-mixers, which replicate lab-scale settings by imitating precipitation circumstances in a tiny container [28].

In this technique, one begins with the molecule in solution, which is made up of either crystalline or amorphous components. The fundamental strategy of this technique is to dissolve the medication in a solvent, then add this to a nonsolvent,

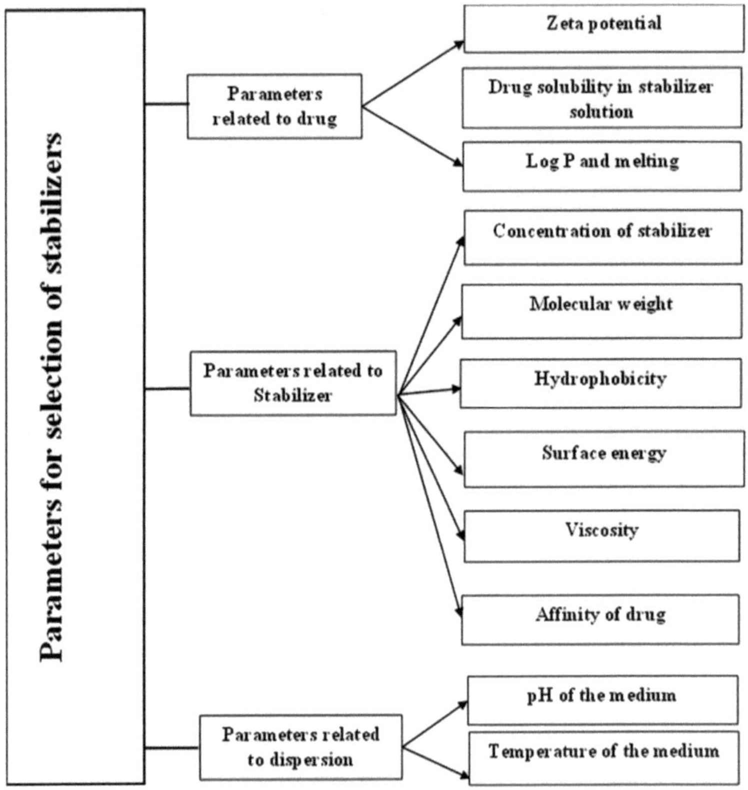

Fig. 1.1 Parameters for the selection of stabilizers

Table 1.1 Classes for stabilizers frequently employed to stabilize nanosuspensions [27]

Category		Stabilizers
Surfactants	Ionic	SLS, docusate sodium, polyethylene imine, cetrimonium chloride, Dowfax 2A1
	Nonionic	Poloxamer 188, Poloxamer 407, TPGS, Tween 80, Plantacare 2000, Saponins, block copolymers of PEO-PPO-PEO
	Amphoteric	Lecithin
Polymers	Synthetic	PVP, PVA, PVA-PEG, graft copolymer
	Semisynthetic	HPMC, HPC, HEC, CMC-Na, PEGs
	Natural	Sodium alginate, chitosan
Food proteins		Whey protein isolate, soybean protein isolate, b-lactoglobulin
Miscellaneous		Amino acids, copolymers

which causes the drug to precipitate. Controlling the particle structure (amorphous vs. crystalline) and preventing the crystals from growing to the micrometer-size range are essential steps in this procedure [17]. What follows are a few methods that fall under the category of bottom-up techniques.

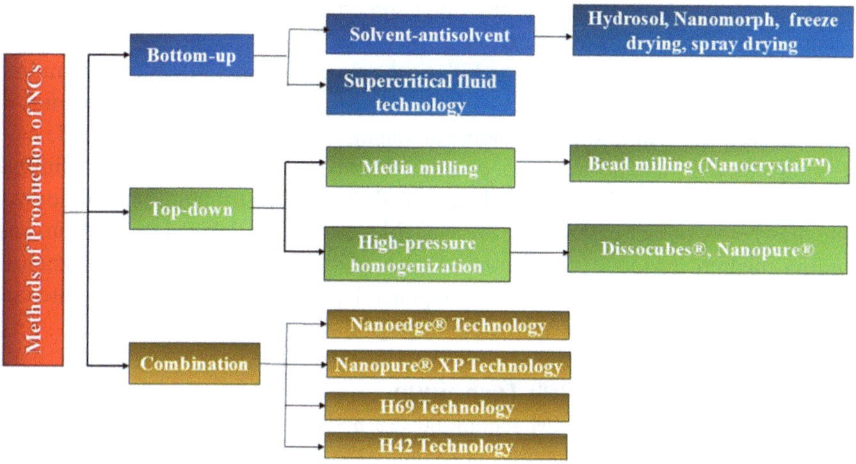

Fig. 1.2 Different methods of production of NCs

1.4.1.1 Precipitation Method (Solvent–Antisolvent)

The basic concept of precipitation method is the formation of drug NCs using continuous mixing of organic solvent and a mixture of drug and solvent that is compatible with water into an antisolvent usually water or another aqueous-based vehicle. A supersaturated environment is generated in water medium that advances of nucleation and precipitate formation of drug components, and arising the buildup of NCs [6].

1.4.1.2 Supercritical Fluid Technology (SCF)

SCF technology implies the use of supercritical fluids, for example, supercritical CO_2, having 31.1 °C as the critical temperature and 72.9 atm pressure. Due to the lesser polarity of supercritical carbon dioxide, lipophilic APIs easily dissolve and form uniform solution mixture and are extended using capillary tube, ultimately creating tiny drug fragments. This formulation is influenced by the expansion process and the solution's increased diffusivity [29].

1.4.2 Top-Down Technique

The decreasing particle size in top-down techniques is dependent on high-pressure collisions or mechanical attrition, which can lead to contamination because of instrumental wear and tear.

The suspension is homogenized under high pressure and milling method. The medium for suspension is usually aqueous, though polyethylene glycol and oils may as well be utilized for medium purposes. This method is quite environmentally sustainable due to the less use of organic solvents. At the end of the milling process, the drug is found in the crystal form while some polymorphic alterations may also occur. Crystallinity may decrease as a result of high-pressure homogenization, although, in this case, water also stabilizes the drug crystals [30]. Wet ball milling (WBM) process may be used to prepare nanosuspensions of weakly soluble drugs in water as well as organic solvents. Some advantages of wet ball milling include ease to setup, ease to handle the drug quantity between 1 and 400 mg/mL, minimal discrepancy in different batches, and a relatively limited distribution of particle size. The drawbacks of wet ball milling are unstable on storage, delayed production, and milling medium residues. Compared to other methods, the top-down approach is more practical from an industrial and general perspective. Several commercial formulations and patents are accessible through the top-down method [31].

1.4.2.1 Media Milling (MM) Technique

The main components of MM technique are a redistributing section for milling process, one cooling component for controlling the rise in heat, and milling medium to reduce the particles till nanosizing. Fundamentally milling medium is constructed of substances like zirconia, glass beads, and derived products of stiff polystyrene. It may be operated as per the requirement of finished good either a continuous mode, called circulation type, or a discontinuous mode, called batch type. For the production of small uniform NCs, continuous mode is most frequently used [32].

1.4.2.2 High-Pressure Homogenization (HPH)

There are several established technologies for using the homogenization procedure to formulate nanocrystals. Microfluidization technique, Dissocubes®, and Nanopure® are one of those techniques belonging to those classes. They diminish the particles by liquid efflux from front impaction in force till 1700 bar. Collision and cavitation take place when a very high-pressure impact applies. A major drawback of this method is that it requires at least 50–75 cycles to achieve the desired nanoscale-size range. In comparison to milling, this makes the process more labor-intensive and comparatively time-consuming. HPH has a reduced chance of producing contaminants than wet medium ball milling [31].

1.4.3 Combination Technique

The shortcomings of each technique can be addressed by combining several techniques. A combined technology is known as Nanoedge, where precipitation followed by a second high-energy step was created by Baxter International Inc. (Deerfield, IL) [33].

1.4.3.1 Nanoedge® Technology

Nanoedge is a combination strategy in which a high shear and/or heat transferring procedure are used after drug microprecipitation. In the first stage, the precipitation process produces crystalline or amorphous drug nanocrystals by using a particular solvent–antisolvent mixture. The drug particles may be completely crystalline, somewhat crystalline, or completely amorphous at this phase. By applying only once or continue energy (direct heat/mechanical stress) and thermal rest, annealing transforms thermodynamically unstable particles into a more stable state. In the addition of further energy, additional nucleation and clustering of crystals stops and conversion of amorphous as well as unstable crystal structure into its stable crystal forms. Because the crystals are friable after the first step, the next energy-addition step is faster and more effective [34].

1.4.3.2 Nanopure® XP Technology

In 2005, Moschwitzer and Lemke created the Nanopure® XP technology, which modifies the starting material by evaporating it and then homogenizing it under high pressure. The main advantage of this technology involves its potential to specifically reduce homogenization duration that is essential; nevertheless, the material is a very tough crystal in nature. The process permits using a diverse range of solvents since the whole solvent is evaporated completely to eliminate any worries about the toxic effects of the solvent [35].

1.4.3.3 H69 Technology

This technology combines cavitation, precipitation, and HPH methods. The timing of the process is a crucial difference between NanoEdge™ and H69 technology. In this process, precipitation, along with HPH steps, occurs simultaneously. When the crystals of drugs appear in the time of precipitation phase, the crystals are subjected to cavitation withstanding the homogenizer's impact and shear forces. This procedure improves the homogeneity of the final nanocrystals and finely controls crystal development [36].

1.4.3.4 H42 Technology

The concepts of HPH and spray drying are combined in H42 technology. The H42 process employs a spray drying approach to first uniformly deliver the hydrophobic drug among stabilizers. A coarse suspension is generated afterward when this drug–stabilizer mixture is allowed to dissolve in an aqueous solution. The resulting suspension is exposed to HPH, ultimately regulating the particle size. Despite being widely used in commercial production, H42 technology has a low yield and might not be appropriate for medications that are heat-sensitive [36].

1.5 Characteristics of Drug Nanocrystals

1.5.1 Increased Dissolution Rate

Increased dissolution rate is the most significant benefit of drug nanocrystals compared to bulk drugs. This is primarily because of the high interfacial surface resulting from small sizes of particles. For instance, regarding the use of spherical particles, the surface area Vs. vol. (S/V) ratio is $3/r$. Therefore, reducing the size of particles from 50 μm (usual for bulk pharmaceuticals) to 500 nm triggers the Noyes–Whitney equation to increase the rate of dissolution by about 100-fold [30]. The solvent and temperature are the only factors affecting the saturation solubility (C_s), a drug-specific constant. As per the Nernst–Brunner/Noyes–Whitney theory, the available surface area for dissolution is closely correlated with the solid active pharmaceutical ingredient's (API) rate of dissolution.

$$\frac{dx}{dt} = \frac{AD(C_s - X_d)}{Vh} \tag{1.1}$$

where dx/dt is the dissolution rate, X_d is the amount of dissolved substance, A represents the particle surface area, D denotes the dissolution rate constant, V is the volume of fluid that is available for dissolution, C_s indicates the saturation solubility, and h is the effective layer thickness.

The rate of dissolution will enhance with reduction of size of the particles till sub-micron range because of the larger surface area. Prandtl equation demonstrates that an improved pace of dissolution results from a decrease in the thickness of the diffusion layer. Additionally, the Freundlich–Ostwald equation shows that the API's saturation solubility increases in nanosize form.

$$S = S_\infty \ \exp \frac{2\gamma m}{\gamma \rho RT} \tag{1.2}$$

where S is the saturation solubility of a drug having size in nanoscale, S_∞ represents the saturation solubility of a large infinite drug crystal, γ shows interface tension of crystal and medium, m represents the molecular weight of substance, r is the radius of particle, ρ is the density, R is the gas constant, and T is the temperature.

Ultimately, adding surfactants to nanosuspension formulations increases surface wetting, leading to improved dissolution rates compared to traditional micronized formulations. This results in greater therapeutic drug efficacy and prevents the elimination of new drug compounds due to low solubility. As a result, innovative and patented technologies are developing within the pharmaceutical sector [37].

1.5.2 Higher Saturation Solubility

The thickness of the layer of diffusion and the level of saturation are affected by the presence of particles of nanosize; when it is less than about 50 μm, the reduction in diffusion layer starts, leading to an increase in solubility. The Ostwald–Freundlich theory states the enhancement of dissolution by enhancement of saturation concentration. This was initially formulated for droplets of liquid in the gaseous phase, though since then it has been demonstrated for the particles of solids in solvent when the particle size is roughly less than 1 μm:

$$S_{NP} = S_0 \left(\frac{2V_{my}}{RT_r} \right) \tag{1.3}$$

were V_m is the molar volume, γ stands for the tension at the interface, R denotes the gas constant, S_{NP} is the solubility of nanoparticles having r radius, S_0 stands for the solubility of material, and T is the temperature. The impact of size of particles on saturation concentration is more noticeable when the particle size is reduced; below 100 nm, the increase is even exponential. The effect is first noticeable, with particle sizes less than 1 μm.

Increased intake and raised saturation levels lead to a situation of supersaturation, which has been shown to enhance drug penetration. The difficulty in vivo lies in sustaining supersaturation until permeation occurs and preventing uncontrolled precipitation or crystallization.

1.5.3 Supersaturated State

Drug nanocrystals lead to the formation of a supersaturated state, which is thermodynamically unstable, as a result of their increased solubility in comparison to thermodynamic solubility. For instance, the concentration levels of the dissolved drug adjacent to the sample surface with nanocrystals measuring 580 nm were over five times higher than the concentration levels achieved with bulk indomethacin when drug nanocrystals were compressed onto a plane surface [31]. In an additional investigation, to determine the water dissolution capability for oral nimodipine NCs of different size, conventional shake flask method was utilized for evaluating them as well as determining the correlation of solubility and pharmacokinetic parameters in regard to both the nimodipine NCs and Nimotop®. The oral NCs drug delivery system has emerged as a potential strategy for improving the oral bioavailability of drugs that are insoluble or inadequately soluble [38].

1.5.4 Nanocrystals for Solubility Enhancement

Adding a small amount of surfactant to nanocrystals gives them all the benefits of nanoformulation, including increased saturation solubility and dissolution velocity [39]. The APIs from Biopharmaceutical classification systems class II and IV have solubility issues; the most effective method for resolving these issues is nanotechnology. For the BCS class II and IV medications, nanocrystal technologies have been presented as favorable multipurpose formulation techniques. The rate of dissolution and saturation solubility of active substances can be efficiently increased by nanocrystals due to their higher surface-to-volume ratio [40].

Rossier et al. formulated Ricobendazole NCs stabilized by Poloxamer 188 using spray drying and wet milling techniques for the treatment of helminthiasis. This instance contrasts with the physical combination of Poloxamer 188 and ricobendazole. Their redispersed NCs, though, demonstrated a higher oral bioavailability in canines. $AUC0 - \infty$ was 1.9 times higher, and the maximum plasma concentration (C_{max}) was higher than that of the micronized drug. The authors proposed that the increased specific surface area of the medication, which encourages higher saturation concentrations, was the cause of the NCs' enhanced in vivo performance [41]. In order to increase curcumin's solubility, Gye Hwa Shin et al. effectively produced an amorphous curcumin-TPGS nanocrystal by ultrasonic homogenization. When compared to crude curcumin, this product demonstrated a remarkable 433-fold increase in water solubility [42]. Resveratrol, an anti-parkinsonism drug, has achieved greater bioavailability, a 3.1-fold rise in C_{max}, and a 3.5-fold increase in dissolution rate for its nanocrystal formulation compared to the initial coarse suspension [43].

1.6 Advantages of Drug Nanocrystals

Drug nanocrystals offer distinct advantages compared to drug nanocarriers. Unlike nanocarriers, drug nanocrystals do not constrain the rate of encapsulation and may accommodate a broad and adaptable range of dosages. This flexibility enables easy fulfillment of clinical requirements, allowing for the administration of high dosages (therapeutic dose >500 mg) in the form of nano-preparations. The merits of nanocrystals are also depicted in Fig. 1.3 and Table 1.2.

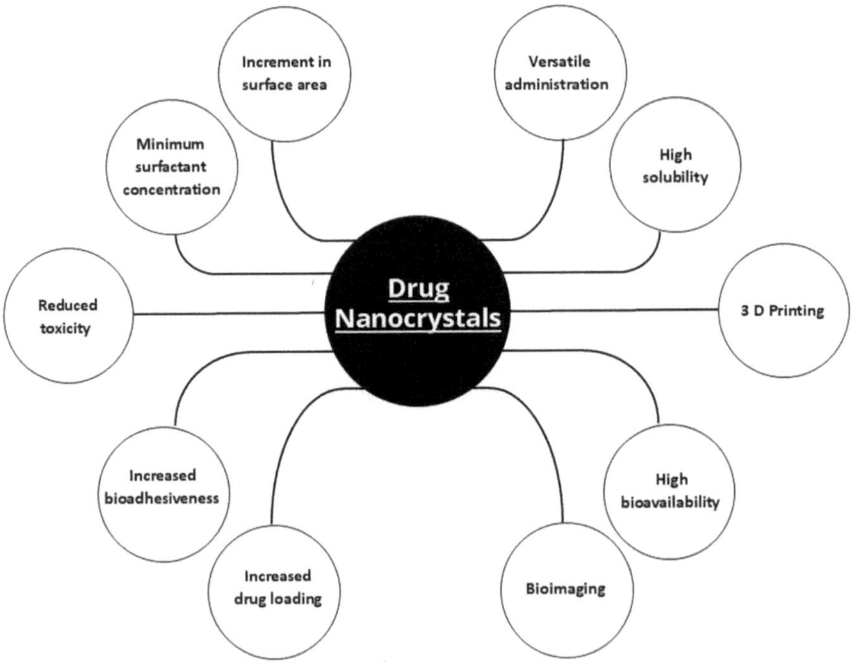

Fig. 1.3 Advantages of nanocrystals

Table 1.2 Merits of nanocrystals

S. no.	Physicochemical attribute	Possible advantages
1	Solid phase: enhanced drug loading	Reduced dose concentration, that is, essential for ocular SC and IM instillation.
2	Solid phase: increased stability	Increased stability for extended period (long term) in chemical and physical aspects, leading to increased resistance to oxidation and hydrolysis.
3	Smaller particle size: Enhanced rate of drug dissolution	For oral delivery: The rate of absorption and concentration enhanced, leading to increased drug bioavailability, as indicated by the AUC in area under the plasma concentration–time curve, onset of action, and peak drug plasma concentration. This reduces fluctuations as well as minimizes the differences between fed and fasted states. Pulmonary: Enhances drug delivery to the deeper areas of the lung.
4	Higher dosage of the API without the use of strong vehicles	Enhanced pharmacological effectiveness and decreased unwanted effects and fewer side effects (particularly for IV administration).
5	Particulate dosage form	The mucoadhesion is increased, leading to longer retention in the gastrointestinal (GI) tract and improved bioavailability; surface properties are altered for achieving site-specific drug delivery [45].
6	Sustained release	Nanostructured materials, because of its elevated surface and porosity volume, can manage and maintain drug release as carriers. Controlled and sustained drug release often prolongs the therapeutic effect of the formulation [46].
0.7.	High solubility	The very fast expansion of the surface for contact with liquid medium resulting in an enhancement in drug solubility. Solubility limit of the drug and the concentration difference in the intestinal wall along with blood flow rises [46].
8.	Improved permeability	Boosts the absorption of medications with low water solubility and those categorized as BCS class IV. Rababimide nanocrystals exhibit enhanced physico-mechanical traits and are evaluated using a bio-mimicking 3D intestinal permeability model [47].
9.	Pharmacokinetic and biodistribution	The pharmacokinetic and biodistribution characteristics of the drug formulation might be altered by manipulating the interaction between the nanocrystals and biological tissues [48].

SC subcutaneous, *IM* intramuscular, *IV* intravenous

- Nanocrystals can be processed into various usable forms. Techniques like fluidized bed drying, freeze-drying, and spray drying may solidify nanocrystal suspensions, forming solid agents like tablets, capsules, and other injectables, for example, lyophilized powder.

- Nanoparticle size may be precisely controlled. Particle size affects the solubilizing effects and bioavailability of medications taken orally, making it an essential factor in the efficacy of nanopreparations. As the medication is nanocrystallized, the size that was measured is a precise representation of the particle size.
- The techniques for creating nanocrystals are diverse, straightforward, and easy for mass manufacturing and industrial uses. Drug nanocrystals can be produced using standard technology such as wet grinding machines, high-pressure homogenizers, and high-pressure microjets [44].

1.7 Limitations of Nanocrystal Technologies

1.7.1 Temperature and Environmental Adjustment

Various studies have demonstrated that the structure of metal nanocrystals can alter dynamically and occasionally reversibly in response to changes in temperature and environmental factors. These patterns of behavior have been explained by the fact that the interfacial free energy density is highly influenced by temperature as well as the adsorption of molecules on the NC's surface [49].

Comprehensive research has demonstrated that the form of a metal nanocrystal can experience dynamic (and occasionally reversible) alterations due to changes in temperature and the environment it is in. This phenomenon was explained by the catalysis community; also, it is associated with the interfacial free energy density that is strongly impacted by both temperature and adsorption of molecules onto the surface of NCs.

1.7.2 Toxicity and Effect on the Environment

Certain nanocrystals, especially those that include heavy metals like cadmium found in CdSe quantum dots, can be toxic and may present environmental hazards. This limitation obstructs their use in biomedical applications and consumer goods [50].

1.7.3 Aggregate Formation and Self-Assembling Challenges

Nanocrystals typically cluster together because of van der Waals interactions, which can affect how well they disperse in solvents or matrices, resulting in inconsistent properties in composites. This clustering diminishes their efficiency in applications that necessitate uniform dispersion, such as drug delivery, photovoltaics, or catalysis [31].

1.7.4 Cost and Economic Viability

The cost of high-quality nanocrystals, particularly for large-scale uses, can be extremely high because of the costs of raw materials, intricate synthesis methods, and purification processes. This affects the economic viability of incorporating nanocrystals into products such as consumer electronics, solar energy devices, and medical diagnostic tools [51].

1.7.5 Limited Long-Term Stability of Optical and Electronic Properties

Nanocrystals can undergo alterations in their optical and electronic characteristics over time, influenced by factors like photobleaching, oxidation, or thermal instability. This presents difficulties for their use in displays, solar cells, and sensors, where it is essential to ensure long-term reliability [52].

1.7.6 Complexity in Synthesis and Scale-Up

These have a consistent size, shape, and composition that necessitates careful management of the synthesis conditions. The intricate nature of these methods, along with the requirement for highly pure reagents, raises expenses and complicates large-scale production, thereby restricting their industrial use [18].

1.8 Characterization of Nanocrystals

1.8.1 Particle Size, Size Distribution, and Zeta Potential

Photon correlation spectroscopy (PCS) is a commonly used method that relies on dynamic light scattering principles for the assessment of NCs for average particle size in relation to z values, zeta potential (the density of NC's surface charge), and particle size distribution. The polydispersity measure can range between 0 (indicating monodisperse particles) and 0.500 (indicating polydisperse particles); thus, PDI measure is crucial for nanoformulation's physical stability. A lesser PDI measure value (<0.3) for a nanoformulation indicates longer stability compared to formulations with a high PDI [23].

For the determination of particle size of nanosuspensions for using via IV route, Coulter counter technique is crucial. The zeta potential of a formulation aids in predicting storage stability. Generally, the aggregation of particles is very less if they contain enough zeta potential, availing sufficient inter-particulate repulsion. Formulations qualify as electrostatically stable if their zeta potential falls within a limit of ± 30 mV; they represent that the system is quite stable. If the zeta potential of a system is ± 20, it denotes a structurally stabilized systems having short stability. A zeta potential of ± 60 mV is the usual rule for efficient stabilization of the nanosystem, whereas values in the ±5 mV range imply quick aggregation [23].

1.8.2 Morphological Characterization

Scanning electron microscopy and transmission electron microscopy are primary techniques to examine and identify morphological characteristics of nanocrystals. By analyzing the relative intensity and clarity of the diffraction peaks corresponding to specific crystallographic directions, comprehensive morphological details can be obtained through SEM. These methods are useful for estimating the average dimensions and forms of the nanocrystals [53]. In addition to other features of the nanocrystal film structure, the SEM pictures of the film fracture surface clearly show the pitch, handedness and direction, and director. Thus, SEM is helpful to analyze the nanocrystals [54].

Atomic force microscopy is utilized for visualization of the surface topography and assessing rough texture variations in thin layers made up of alternating layers of xyloglucan and cellulose nanocrystals, along with those comprising flexible poly (allylamine hydrochloride) and stiff cellulose nanocrystals arranged in alternating layers. The surficial characteristics and rough texture of cellulose films coated with spin were analyzed using atomic force microscopy. AFM is used to analyze the topography, elastic characteristics, and adhesive capabilities of distinct wood-derived cellulose nanocrystals (CNCs) [55].

The hot stage X-ray powder diffraction technique is used to study solid-state forms in relation to temperature, operating on the principle of X-ray diffraction from lattice planes, which provides insights into polymorphic forms (characterized by distinct diffraction peaks) and amorphous forms (indicated by the absence of peaks). The results obtained are presented in a diffractogram that offers both qualitative and quantitative data (such as the degree of crystallinity). This method can utilize samples in powder, paste, or slurry forms [56].

Perhaps the best technique for examining NCs and their assemblies is transmission electron microscopy (TEM). Standard low-efficiency TEM examinations capture images of the groups of nanocrystals, allowing the statistical analysis for the size and shape of nanocrystals of a sample. High-resolution TEM imaging exposes specific shape and internal structure for nanocrystals [15].

X-ray diffraction (XRD): The arrangement and symmetry of the nanocrystals within the assembly and its extension can be investigated using X-rays across various scales, from atomic structures to the nanoscale, and up to the mesoscale (hundreds of nanometers) through XRD techniques. Varieties of X-ray powder diffraction patterns, obtained from nanocrystals exhibiting different lattice symmetries, are presented: cubic copper, monoclinic tenorite, tetragonal anatase, rutile, and hexagonal hydroxyapatite [57]. Nanocrystal drug sample preparation for X-ray spectroscopy analysis often involves freeze-drying. Liquid nanocrystal dispersions designed for dosing in their solid form must be freeze-dried as the role of freeze-drying in determining nanocrystal drug stability is critical during translation [1].

1.8.3 Solid-State Properties

Raman spectroscopy: Raman spectroscopy is a spectroscopic technique employed to analyze and assess cellulose nanomaterials. This spectroscopic method provides crucial data regarding crystallinity and the identification of sulphate esters on the surfaces of CNCs. The characterization of various types of CNCs using the 380-Raman technique has been extensively applied. Because of its capability to gather polymorphic information, Raman spectroscopy has successfully characterized and quantified materials with different crystalline structures and aggregation states. This makes it possible to comprehend the physicochemical behavior of CNCs, enabling enhancement of these properties for diverse applications [58].

The infrared (IR) technique serves as an efficient analytical tool for examining ligand interactions on the surface of up-conversion nanocrystals (UCNCs). Several chemical functional groups found in the molecule can be identified within the infrared region of the spectrum. Ligands exhibit continuous vibrations, with the energy levels associated with these movements primarily residing in the IR range, specifically from wavenumber 4000 to 200 cm^{-1}. Infrared analysis could be employed to investigate unusual visual characteristics of UCNCs by observing the ligand's involvement in chemical processes [59].

Nuclear magnetic resonance (NMR): A new NMR technique has been introduced as an effective method for characterizing organic compounds that are confined on inorganic nanoparticles, including UCNCs and quantum dots [60]. Shifting of NMR signals is a key feature indicating the ligands attached to the surface of UCNCs; thus, the magnetic nucleus can be protected from the external magnetic field by the nanoparticle. For example, the chemical shift observed in case of oleylamine, which changes from 2.64 to 2.68 ppm as a result of the nanoparticle's presence [61].

1.8.4 Thermal Analysis

Differential scanning calorimetry (DSC) is a commonly employed technique for investigating the thermal properties of drugs and drug nanocrystals. DSC analysis is conducted to assess the crystalline state of the drug and the interaction between excipients and the drug following the production of nanocrystals. This is particularly crucial for drugs that exist in various polymorphic forms [62].

Thermogravimetric analysis (TGA): The thermogravimetric analysis is carried out to predict the rate of change of weight with respect to temperature (°C) under the nitrogen (flow rate between 20 and 60 mL/min) environment [63]. Here is how TGA is applied to nanocrystals and what insights it offers.

TGA serves as an important method for assessing nanocrystals, offering valuable information regarding their thermal stability, composition, and decomposition characteristics. TGA evaluates the mass of a sample as it is gradually heated, making it especially effective for studying the thermal stability of nanocrystals. Often, nanocrystals necessitate organic ligands or surfactants on their surfaces to uphold stability and dispersion. TGA can determine the existence and thermal stability of these ligands by monitoring mass loss at specific temperatures linked to ligand decomposition. TGA is capable of differentiating between various phases or components in composite nanocrystals by identifying unique weight-loss events. For instance, in core–shell nanocrystals, the outer layer may decompose at a different temperature compared to the core. By performing TGA at different heating rates, researchers can acquire kinetic information regarding the thermal decomposition of nanocrystals, uncovering details about the reaction order, activation energy, and other kinetic parameters [64, 65].

1.8.5 Solidification of Nanocrystal Suspensions

Solidification is used for stabilization, and solid preparations tend to be more stable than liquid ones. Solidifying nanocrystal suspensions can reduce the formation of unstable factors like aggregation and Ostwald ripening. Therefore, nanocrystal suspensions are often transformed into a solid state. Subsequently, the solid powders are converted into various dosage forms, such as sterile powder for injection, oral tablets, and capsules [27].

1.9 Biomedical Applications of Nanocrystals

Nanocrystals hold significant importance in biomedical applications because of their distinctive physical, chemical, and biological characteristics. Figure 1.4 summarizes the biomedical application of nanocrystals that are also explained. What follows are some of the main domains where nanocrystals are utilized in the biomedical sector.

1.9.1 Improved Bioavailability

Nanocrystal drugs refer to the direct micronization of bulk drugs into nanoscale drug particles, which are submicron colloidal dispersive systems without support. Nanocrystals with a particle size of 100–1000 nm are usually seen as a binary system, containing APIs and selected stabilizers. Compared to the traditional preparations, nanocrystals present many novel and promising pharmacodynamic and metabolic characteristics, such as flexibility for fitting in various dosage forms, enhanced targeting capability, prolonged systemic circulation, accessibility for hybrid platform construction, sustained and controlled release, high bioadhesion, special intracellular mechanism, and multicomponent co-delivery [66].

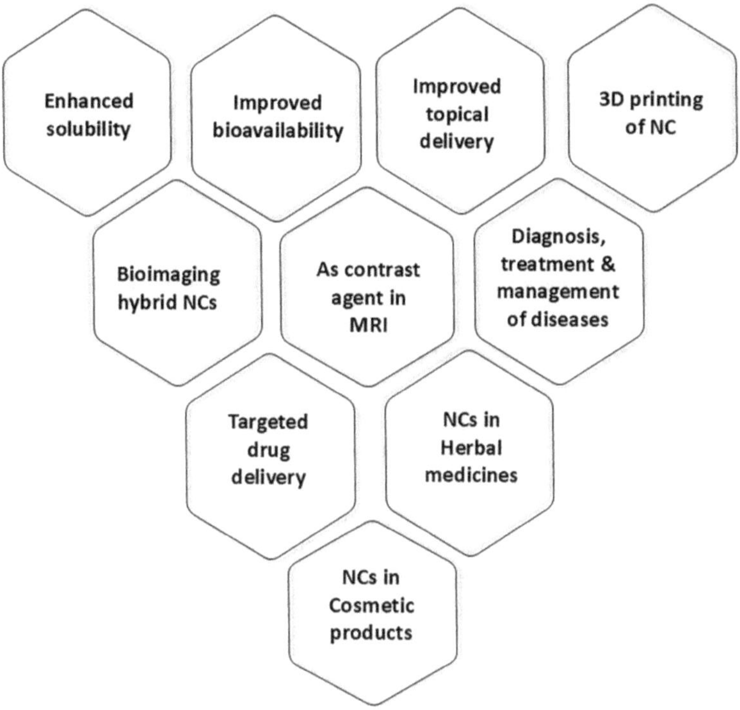

Fig. 1.4 Biomedical applications of nanocrystals

Shen et al. employed a self-discriminating hybrid nanocrystal to investigate the fate of quercetin hybrid nanocrystals (QT-HNCs) in the body after they were administered orally and intravenously. They also assessed the role of integral nanocrystals in enhancing the oral bioavailability of quercetin (QT) by comparing the total exposure of integral QT-HNCs and total QT in the liver. The findings indicated that QT-HNCs could remain in the body as complete nanocrystals for up to 48 hours after oral and intravenous administration. Furthermore, a greater accumulation of integral QT-HNCs was observed in the liver and lungs following both oral and intravenous administration of QT-HNCs [8].

One of the emerging platforms for formulating flavonoids is nanocrystals, described for improving the aforesaid physicochemical and therapeutic issues. Nanocrystals enhance the bioavailability of flavonoids due to decreased particle size and increased dissolution rate, owing to increased curvature available for the medium to remain in contact with drug crystals [67]. The diagrammatic representation of the improvement in bioavailability of poorly soluble drugs is shown in Fig. 1.5.

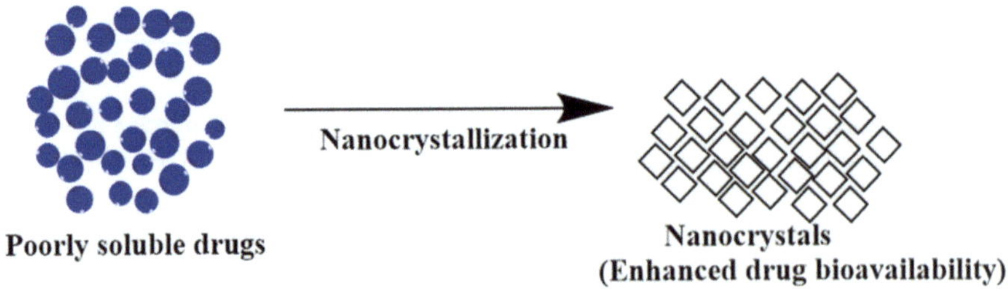

Fig. 1.5 Poorly soluble drug's nanocrystallization improves drug bioavailability

1.9.2 3D Printing of Nanocrystals

3D printing is a fast prototyping and manufacturing technique that creates three-dimensional objects by layering materials on top of one another. Additionally, 3D printing offers the ability to create formulations in a variety of sizes, shapes, and compositions, which can be utilized to optimize the drug release kinetics and dosage. Compared to labor-intensive traditional production techniques, all of these can be achieved with straightforward design modifications utilizing software. 3D technology can be utilized to create release-tailored medication in addition to facilitating the production of patient-specific personalized medicine [68]. Three-dimensional (3D) printing can rapidly generate customized solid 3D objects through the progressive addition of materials using computer-aided design/computer-aided manufacturing (CAD/CAM) [69]. In principle, personalized formulations can be produced by adjusting the size, shape, and dose based on patient specifications [70]. FDA's approval of the first 3D-printed drug called "Spritam" (levetiracetam) has exemplified the success of this approach [71]. Academic and industrial sectors have become more interested in cellulose nanocrystals (CNCs) due to their specific properties such as their exceptional mechanical properties, biodegradability, nontoxicity, and renewability. The prepared nanocrystals showed desirable properties, including great flexibility, high strength, high stiffness, and a high surface-to-volume ratio. The authors concluded that cellulose nanocrystal is a better alternative additive or excipients in the manufacturing of various pharmaceutical dosage forms [72]. The use of 3D printing to create cellulose-based composites for different purposes, including targeted drug delivery, intricate designs, tissue engineering, dental applications, automotive components, and customized food products, has been successfully demonstrated. The utilization of cellulose nanocrystals (CNCs) in combination with different synthetic and natural polymers has been investigated using 3D printing technology. The potential applications of 3D-printed cellulose nanocrystal structures in enormous biomedical fields, such as self-folding tubes, patches, bioimplants, regenerated tissues, and structures with or without cells, are significant.

Sultan and colleagues created biomedical hydrogel scaffolds using CNC/gelatine/sodium alginate via direct ink writing (DIW) printing. The findings showed that the 3D-printed hydrogel nanostructured scaffolds with nanosized pore wall thickness and orientation demonstrated exceptional support for proliferation, interaction, and growth of cells in tissue regeneration [73].

Cui and colleagues utilized cellulose nanocrystals (CNCs) PEG-aliphatic block copolymers of polyester for creating thermo-responsive three-dimensional hydrogel structures for bioprinting using printing method based on extrusion. The hydrogel developed from these CNC-enhanced copolymers displayed outstanding mechanical properties, allowing it to maintain its form with exceptional accuracy and precision. Furthermore, the addition of CNCs notably enhanced the printability of hydrogel-based scaffolds [72].

Lopez et al. first developed and reported the incorporation of NCs into oral solid dosage forms that are 3D-printed using the melting solidification printing method. Here, to create 3D-printed dosage forms, albendazole (ABZ) nanocrystals incorporated into PEG 1500/PG were utilized as a carrier ink. The prepared formulation does not show any physicochemical interactions and no crystalline alterations were visible because of the inclusion of albendazole NCs in the ink. Specifically, 97.8% albendazole crystallinity was observed after the conversion of 3D state when its comparison was done with the physical mixture of compounds. The nanocrystals' particle size, crystallinity, and chemical stability were all preserved both before and after the 180-day storage period. Therefore, by combining 3D printing and nanocrystals, these findings show relevant pharmacological potential for creating stable, fast-release, oral, solid dosage forms of poorly soluble drugs [74].

An in situ-forming robust injectable and CNCs based on 3D-printed hydrogel network loaded with amphiphilic copolymers PLCA [poly(ε-caprolactone-co-lactide)-b-poly(ethylene glycol)-b-poly(ε-caprolactone-co-lactide)] was developed by Phan et al.. By physically combining CNCs with amphiphilic PCLA copolymers, the mechanical and physicochemical characteristics of injectable hydrogels were regulated. The prepared DOX-loaded hydrogel formulation showed prolong delivery along with distinctive release kinetics and exhibited antitumor effect on the in vivo model of MDA-MB-231 tumor xenograft [75].

Santos et al. developed S-nitrosoglutathione loaded into hydrogels consisting of poly (acrylic acid) and Pluronic F127 micelles with thermal response. The innovative process for 3D printing CNC incorporated into partially permeated poly (acrylic acid)/Pluronic F127 hydrogels using poly (acrylic acid) photopolymerization through a digital light processing technique, without the use of organic solvents. S-nitrosoglutathione-charged hydrogels of poly (acrylic acid)/Pluronic F127/CNC exhibit continuous release of nitric oxide by cell proliferation after hydrating them, and the rate of hydration depends on charges on S-nitrosoglutathione, along with the presence of the cellulose nanocrystals [76].

1.9.3 Bioimaging Hybrid Nanocrystals

The process of capturing pictures of living organisms using fluorescent dyes or markers and nanomaterials is commonly known as bioimaging and constitutes a substantial area of study. The utilization of nanoparticles as substances that enhance the visibility of internal body structures for bioimaging with MRI is likely the most extensive application within this field [77]. The bioimaging hybrid nanocrystals have been created to offer additional bioimaging capabilities in living organisms by incorporating small amounts of fluorophores and contrast metals into the structure. As expected, these nanocrystals can directly show how drug crystals behave in living organisms, without the need for fluorescent labeling materials. Hybrid nanocrystals can be used to both treat tumors and track drug particles in living organisms, presenting a new approach to clinical tumor imaging. Generally, metal and fluorescence materials can be used to create bioimaging hybrid nanocrystals [78].

Cellulose nanocrystals are thought to be a new class of bioimaging probes due to the size range 200–300 nm length and diameter 5–50 nm, along the specific surface area of up to 150 m^2/g. The anionic charge of cellulose nanocrystals causes the cell injury while the appropriate alterations in cellulose nanocrystals restrict this side effect [79].

1.9.4 Nanocrystal as a Contrast Agent in MRI

Nanocrystals that are active in MRI can exist in a pair of varieties. Certain core materials, such as iron oxide, manganese oxide, or gadolinium oxide, may be MRI-active nanocrystals. As an alternative, an MRI-active substance can be added to a non-MRI-active nanocrystal, such as phospholipid coatings with gadolinium (Gd) chelates inserted in their headgroups or in a carrier matrix that contains (super)paramagnetic substances. These nanoparticles influence how quickly water in tissues relaxes, creating MR image contrast [80].

Torkashvand and Sarlak developed a double T_1–T_2 contrast medium for MRI dependent on superparamagnetic magnetic oxide nanoparticles. Thermal decomposition was used to create the magnetite nanoparticles, which were subsequently placed onto a hydrophilic cellulose nanocrystal-poly citric acid nanoplatform. The nanocomposite's sufficient cellular uptake, excellent biocompatibility, and low toxicity were demonstrated by inductively coupled plasma optical emission spectroscopy, in vitro cytotoxicity up to 200 μg/mL Fe with HeLa cell lines, and colloidal stability investigations. These findings are encouraging for their potential use in biomedical applications [81].

1.9.5 Nanocrystals for Diagnosis, Treatment, and Management of Diseases

1.9.5.1 Nanocrystals in Cancer Diagnostics and Treatment

Nanocrystals are currently being actively researched as drug delivery materials, especially for cancer treatment. They are considered promising future nanomaterials due to their small size and spindle-shaped structure, which provides numerous hydroxyl groups on the surface and allows for degradation in living cells. The hydroxyl groups on cellulose nanocrystals enable chemical functionalization with antitumor agents, as well as prolonged blood circulation and low cytotoxicity due to their hydrophilic surfaces and favorable elongated shape. Cellulose nanocrystals are also known as nanocrystalline cellulose, crystallites, cellulose nano-whiskers, rod--like cellulose microcrystals, whiskers, or needles, as per the International Organization for Standardization (ISO) and common usage [82].

Carboxylated cellulose nanocrystals (CNCs) were produced from residues of traditional Chinese medicine (CMR) using a sequential periodate-chlorite oxidation method without mechanical treatment. Various spectroscopic and analytical techniques were employed to characterize the resulting nanocelluloses, which indicated that the cellulose structure remained intact after the sequential oxidation. Additionally, it was observed that the thermal properties decreased as the nanocellulose particle size and crystallinity index decreased. This process enhances the potential use of traditional Chinese medicine residues as a viable source of cellulose for the efficient production of nanocellulose with a high carboxyl content, suitable for applications in nanomaterials [83].

One of the most effective antitumor drugs, paclitaxel (PTX), has a poor solubility (<0.1 μg/mL), which has greatly restricted its clinical use. A couple of researchers used a high-pressure homogenizer to create stable paclitaxel nanocrystals (PTX-NCs) without the use of an organic solvent in order to resolve this problem. According to the in vitro findings, PTX-NCs possessed a lower IC 50 value in MCF7 and MDA-MB breast cancer cells than PTX solution. The AUC of PTX-NCs was nearly 9–10 times greater than that of typical PTX crystals, based on the oral pharmacokinetic investigations [84].

Bexarotene, a member of the BSC class II drugs, has had limited practical use in the treatment of leukemia due to its low bioavailability and poor solubility. Bexarotene belongs to BSC class II compounds, whose clinical application in the treatment of leukemia has been limited by its poor solubility and low bioavailability. In order to create the best bexarotene nanocrystals, Chen and colleagues used the precipitation-combined microfluidization approach and screened the optimal stabilizers (F68 and lecithin). When the particle size was decreased, the in vitro dissolution rate of bexarotene-NCs was considerably higher. The in vitro dissolution rate of bexarotene-NCs was significantly higher when the particle size was reduced [85].

1.9.5.2 Nanocrystals in Inflammatory Disease

Nanosuspensions can be delivered to the local inflamed skin or muscle by various routes of administration, the most common and developed being the oral route. Its actual ability to localize the drug in the site of inflammation helps increase the contact time of the drug nanocrystals to the inflamed site, which makes it convenient to use for anti-inflammatory purposes [86]. Nanocrystal formulations are used in herbal medicines for inflammatory diseases such as bronchial asthma (curcumin), anti-osteoporotic (naringenin), treat impotence, and prevent osteoporosis (lipophilic aglycone icaritin) [36].

Naringenin, a plant-based active constituent, when formulated as nanocrystals using techniques like antisolvent, so no precipitation, shows promising results in enhancing its therapeutic effects. For instance, Gera et al. demonstrated that

NRG-NS with a particle size of approximately 117 nm exhibited superior bone repair capabilities and increased ALP levels compared to standard NRG. Furthermore, NRG nanocrystals produced via wet media milling with TPGS. These nanocrystals were three times more effective in reducing cough frequency compared to a placebo and 1.6 times more effective than untreated NRG [87].

Pireddua et al. formulated diclofenac acid nanosuspension for the management of inflammatory skin conditions. Wet media milling was used to formulate the drug nanocrystals, and Poloxamer 188 was used to stabilize those drug NCs. When applied to mouse skin, the drug nanosuspension resulted in a greater amount of diclofenac in the skin than both the commercial formulation and the coarse suspensions. Additionally, the nanosuspension inhibited edema in vivo in a way that was statistically identical to that of the commercial formulation. In contrast, the myeloperoxidase activity in the injured tissue was more inhibited by nanosuspension than by commercial formulation [86].

1.9.5.3 Nanocrystals in Antiretroviral Drug Delivery

Nanocrystals (NCs) are frequently utilized to improve transportation through the nasal-to-brain pathway. NCs are a type of nanoparticles consisting of pure, drug crystals at the nanometer scale, offering various benefits such as high drug capacity and minimal excipient-associated toxicity [88].

Treatment of HIV-associated neurocognitive disorder (HAND): Abbate et al. used a hollow microneedle system to deliver nanocrystal formulations of two antiretroviral drugs, rilpirivine (RPV) and (cabotegravir) CAB, to the mystical pad area of rats for application to either side of the face for treating HIV-associated neurocognitive disorders [89].

Viral infections significantly threaten global health. In 2015, two NC formulations that showed long-term, sustained drug release were made: cabotegravir and rivpivirine. NanoART, a potent nanocrystal product, was developed for site-specific combination method for encapsulation of atazanavir, ritonavir, and efavirenz collectively using high-pressure homogenization technique. The outcomes showed enhanced antiviral activity, effective cellular absorption, and enhanced physical stability. Furthermore, by altering the surface of NCs with polysaccharides, polyethylene glycol, and albumin, their targeting capacity could be enhanced [90].

1.9.5.4 Nanocrystals as Antibacterial Agents

Silver nanoparticles exhibit remarkable antimicrobial activity as inhibiting both gram-positive organisms (*S. aureus*) and gram-negative bacteria (*E. coli*). The incorporation of silver nanoparticles or silver zinc oxide nanoparticles into cellulose nanocrystal substrate is furnished with antimicrobial activity. Comparing cellulose nanocrystals-supported silver zinc oxide nanoparticles to cellulose nanocrystals-free ZnO–Ag heterostructure nanoparticles of the same particle size, the former demonstrated higher antibacterial activity due to their good dispersion.

Shi et al. prepared cellulose nanocrystals (CNCs), where a coating strategy inspired by nature was created, and the combination of nanoparticles showed outstanding stability in solution, significantly enhancing their ability to fight bacteria when using silver nanoparticles as the model particles. To achieve this, CNC underwent modification with dopamine, followed by the generation and anchoring of silver nanoparticles on the CNC surface through the reduction of silver ions by polydopamine-coated CNC. The findings revealed that the stability of silver nanoparticles was notably improved by CNC, leading to a more than fourfold increase in antibacterial activity based on studies with *Escherichia coli* and *Bacillus subtilis* [91].

1.9.5.5 NC Nanocrystals for Ophthalmic Diseases

Nanocrystals have higher saturation solubility, dissolution rate, and mucoadhesive properties. When applied to the eye, they quickly start to dissolve (burst release), and the increased saturation solubility results in a higher concentration gradient, leading to increased diffusive flux into the eye. Nanocrystals that are not completely dissolved remain attached to the eye surface for a longer duration, serving as a depot from which a constant supply of the drug continues to dissolve [92].

Brinzolamide (BRA)-loaded NCs were prepared to reduce the intraocular pressure (IOP) while increasing the drug dissolution. The prepared brinzolamide-loaded NCs exhibited enhanced solubility, dissolution, substantially lowered the IOP in an ocular hypertensive simulation model, and decreased cell toxicity into the corneal epithelial cell of human in comparison to commercial formulation [29].

Romero et al. developed cationic nanocrystals containing 0.05% dexamethasone acetate and 0.10% polymyxin B as an ocular preparation; it is prepared by wet bead milling method, that is, a low-key self-developed technique. This prepared formula offered increased residence time due to the small particle size and enhanced mucoadhesion through positive ions in the eye, eventually resulting in increased bioavailability [90].

1.9.5.6 Nanocrystals for Dermatological Diseases

Dermal nanosuspension is mainly of interest if conventional formulation approaches fail the use of drug nanocrystals, leading to an increased concentration gradient between the formulation and the skin. Using drug nanocrystals creates a higher concentration gradient between the formulation and the skin, leading to increased saturation solubility and supersaturated formulations, which enhance drug absorption through the skin. The use of positively charged polymers as stabilizers for the drug nanocrystals can further enhance this effect. The opposite charge increases the affinity of the drug nanocrystals to the negatively charged stratum corneum [93].

Different sizes of Hesperetin nanocrystals were produced through small-scale bead milling and high-pressure homogenization, and then assessed for their effectiveness in penetrating the skin. The findings indicate that nanocrystals can enhance the skin penetration of insoluble active ingredients due to the small size of the NCs playing a crucial role; smaller nanocrystals were found to enhance the penetration of poorly water-soluble actives and facilitate deeper penetration into the skin's deeper layers [3].

1.9.6 Nanocrystal-Targeted Drug Delivery/Versatile Routes for Delivery

1.9.6.1 Improve Topical Delivery

Improved topical delivery of medium-soluble drug: The NC's application has alleviated the delivery of moderately soluble drugs topically by addressing issues such as poor skin penetration and low topical bioavailability. Nanocrystals are effective in transporting drugs to or into the cells due to their high drug loading and low concentration of surfactants for stabilization. They enhance the saturation solubility and dissolution rate of the drug, leading to an increase in the concentration gradient between the applied formulation and the skin membrane. This, in turn, increases the passive diffusion of the drug through the skin. Nanocrystals also have a large surface area, which improves their adhesiveness to the skin surface, thereby increasing the residence time of the formulation on the skin [94].

Mupirocin nanocrystals (MP-NCs) were prepared using a modified bottom-up technique employing a D-optimal response surface design and a nanoprecipitation approach for optimization. The results of optimized MPNC generated nanosized uniform spherical particles. According to the in vitro data, MP-NC's saturation solubility increased tenfold, and its rate of dissolution increased ninefold. Ex vivo permeation tests utilizing the skin of pig ears revealed a twofold increase in MP-NC dermal deposition, with the maximum drug deposition taking place at a skin depth of 500 μm. Following lyophilization, the best MP-NC formulations were added to a 2% w/w cream. Microbiological investigations demonstrated that MP-NC's minimum bactericidal and inhibitory concentrations had decreased by an attribute of 16. Rats treated with MP-NC cream showed more rapid and effective healing of infected burn wounds than rats treated with commercially available Avoban ointment, according to in vivo experiments employing a rat excision burn wound model.

According to our findings, nanocrystallization of MP may offer a way for larger concentrations of a topically applied MP to penetrate the skin, reach pertinent infected regions, and maybe have a local antibacterial effect [85].

The majority of the studies that have been published so far have been on evaluating nanocrystals for better topical distribution in vitro and/or ex vivo. Only a few researchers have studied topical nanocrystals in vivo. In skin permeation study, the nanosuspension of resveratrol containing Poloxamer 188 and Tween 80 exhibited better skin permeation compared to conventional nanosuspension. In contrast, the in vitro DPPH (2,2-diphenyl-1-picryl-hydrazyl-hydrate) test revealed greater antioxidant activity (>90%) for both the coarse suspension and the nanosuspension. This investigation proved that the drug's antioxidant activity was not increased by the formulation content or the particle size reduction procedure [95].

1.9.6.2 Hair Follicle Targeting

Targeting pharmacological molecules into the hair follicle is another intriguing application for nanocrystals. Once the nanocrystals have penetrated the hair follicle, they may create a depot from which the active ingredient is released, consequently resulting in a highly efficient and prolonged cutaneous medication administration. The main formulation characteristics that were considered to potentially impact the nanocrystals' tendency to penetrate were the vehicle's viscosity, lipophilicity, and polarity (charge), in that order. It is possible that variations in the formulation's polarity and/or viscosity might affect how easily the nanocrystals are released from the vehicle, which may modify how well they penetrate the follicles [96].

Curcumin nanocrystals were formulated and submerged into gels having a variety of properties. The ability of penetration of these particles was evaluated using the pig ear model (ex vivo model), and the result demonstrated the efficient penetration of formula into the infundibulum, that is, the lower part of hair follicles that ultimately shows less obstructive passive diffusion of drugs. The permeation of curcumin into the dermal layer of skin shows that there are no other qualifications required for the hair follicle targeting for NCs' formulations. So, allowing the formulation of media that is loaded with NCs may be independently modified as per the requirements of an individual with no hindrance in the permeation of drug NCs inside the hair follicles [97].

1.9.6.3 Nanocrystals for the Pulmonary Route

Nanocrystals are excellent carriers for poorly soluble drugs when delivered through the pulmonary route. They can be suspended for nebulization or incorporated into powders for inhalation. While nanocrystal-based pulmonary drug delivery is still in its early stages, it has demonstrated the possibility of increasing the bioavailability of drugs in the lungs. However, currently, there is not enough concrete proof regarding the NC's fate in the lungs. It is important to acknowledge the differences in both the chemical and physical characteristics of different drugs. Variations in solubility may play a significant role in producing conflicting outcomes. Modifications in the particle size, shape, and surface characteristics could also influence the fate of nanocrystals in the lungs [98].

Guan et al. assessed the viability of dry powder inhalations of ciprofloxacin NCs jointly delivered along with N-acetylcysteine for improving the penetrability and enhancing the bioavailability into the lungs. The ciprofloxacin nanocrystals (NCs) were produced using the antisolvent precipitation method. The resulting dry powder is composed of ciprofloxacin NCs. While the procurement of N-acetyl-cysteine was done by freeze-drying afterward using spray drying methods. To formulate dry powder inhalations of ciprofloxacin NCs/N-acetylcysteine joint delivery system, the inhaled lactose was blended with this powder. This research illustrates that the spray-dried powder inhalation exhibiting a synergistic action and improved flow property is highly required drug delivery in the lungs while minimizing the risk of severe lung injury [99].

1.9.6.4 NCs for Oral Administration

Nanocrystalline dispersions, tablets, and other solid formulations are frequently difficult to formulate and administer via the GI tract, despite the many success stories. Despite the numerous success stories, formulation and delivery of nanocrystalline dispersions, tablets, and other solid formulations via the GI tract are often not straightforward processes. The effects of physiological variables, such as pH fluctuations and peristalsis, on the nanocrystals are difficult to predict. The rate and amount of dissolution of NCs in the gastrointestinal tract are mainly affected by the acidic pH of drug [100].

Aceclofenac-loaded NCs were prepared by Rahim et al. in order to increase the rate of dissolution, drug bioavailability of oral Aceclofenac, and ultimately its analgesic property in Swiss albino rabbits. The production of nanosuspensions was done by precipitation ultrasonication process. Stabilizers or excipients used were HPMC, PVP, and sodium dodecyl sulphate. The result showed that Aceclofenac NCs enhanced the saturation solubility to 2.6- to 4.5-fold in comparison to raw drug in a solution of stabilizer and also with the only raw drug [101].

1.9.6.5 Nanocrystals for the Intravenous Route

Nanocrystals for the intravenous route: Intravenous drug delivery is a crucial area of development for nanocrystals. The lack of carrier materials reduces the risk of excipient-related toxicity. Nanocrystals, being suspended for injection rather than dissolved, enable a small volume to carry a sufficient drug dosage. The intravenous use of nanocrystals of dantrolene sodium (Ryanodex) and meloxicam (Anjeso) has been approved by the FDA. Ryanodex was found to be equivalent to the dantrium solution based on the 95% confidence interval of major pharmacokinetic parameters. As a result, it was assumed that Ryanodex dissolves immediately in the blood [102].

1.9.6.6 NCs for Enhanced Topical Drug Delivery

The low bioavailability for topically applied drugs is influenced by several factors, with the primary factor of barrier function of the stratum corneum layer of the skin. When the drug is applied topically, it can permeate the skin through the trans-appendageal route. The drug NCs may attain the target area via the trans-appendageal route. Drug absorption is improved because of the concentration gradient that is produced between the formulation and the skin by the higher saturation solubility of drug nanocrystals.

The use of NCs' drug delivery is a modern approach in dermal delivery of drug, with limited number of patents and published works. NC-based products could aid in treating both superficial and deep tissue-level skin infections by providing immediate and controlled drug release. Nanocrystal technology has been effectively utilized in cosmeceuticals for products such as sunscreens, antiaging treatments, moisturizers, and makeup foundation. Nanocrystal-based drug products have shown promise not only in treating skin diseases; moreover, in enhancing the drug formulation stability and decreasing the unwanted effects, such as irritant effect and toxic effect [103].

Quan et al. prepared 18β-glycyrrhetinic acid NCs (Nano GA) using the HPH method. The dermal permeability-enhancing potential of Nano GA was studied using ex vivo experiments with mouse skin. The in vivo evaluation of its topical anti-inflammatory effects was conducted using a mouse ear model induced by 12-O-tetradecanoyl phorbol-13-acetate (TPA). Following nanocrystallization, Nano GA exhibited greater dermal permeability compared to coarse GA. The Nano GA hydrogel demonstrated superior edema relief properties and more effectively suppressed inflammation progression compared to the coarse 18β-glycyrrhetinic acid hydrogel and hydrogel of indomethacin. The findings recommend that for the treatment of skin inflammation the Nano GA might be an effective topical formula [104].

1.9.7 Nanocrystals in Herbal Medicine

Herbal medicine nanocrystals have been explored for various therapeutic applications, including cancer, inflammatory diseases, cardiovascular diseases, mental/nervous diseases, and antimicrobial infections. The application of nanocrystals in herbal medicines for cancer therapy is mentioned as antitumor (annonaceous acetogenins, isoliquiritigenin, gambogenic acid), prostatic carcinoma (oridonin, silibinin), breast cancer (celastrol), colorectal cancer (andrographolide analogue), and anticancer (oleanolic acid) [36].

An oral drug nanocrystal self-stabilized Pickering emulsion (NSSPE), was developed using nanocrystals of purarin, a poorly soluble active component from *Puerariae Radix,* as solid particle stabilizers, with *Ligusticum chuanxiong* essential oil serving as the main oil phase. The NSSPE was designed to improve the oral bioavailability of puerarin. The results revealed that the puerarin nanocrystals could stabilize Pickering emulsion of *Ligusticum chuanxiong* essential oil without additional stabilizers, and that the pickering emulsion could improve the oral bioavailability of puerarin. These findings suggest that the drug nanocrystal self-stabilized pickering emulsions are a promising oral drug delivery system for traditional Chinese medicine containing poorly soluble ingredients and volatile oils [105].

1.9.8 Nanocrystals in Cosmetic Products

The saturation solubility of active pharmaceutical ingredients is enhanced by converting them into NCs because of their particle size ranging into nanometer range, along with the increased rate of dissolution due to the greater surface area. A larger concentration gradient is produced between the epidermal membrane and the topically applied formulation due to the enhanced saturation solubility, resulting in longer contact time of the drug formula into the skin. The hair follicle targeting is achieved by the small particles (nanometer-sized). Furthermore, they demonstrated enhanced adherence to cell membranes or surfaces, leading to prolonging the formulation's residence time in the skin.

Nanocrystal-based cosmetic products that contain poorly soluble antioxidants such as cosmetics based on nanocrystals that include poorly soluble antioxidants, such Juneva's "Juvedicalage-decoder cream" with rutin from St. Margrethen, Switzerland, and La Prairie's "Cellular Serum Platinum Rare" with hesperidin from Zurich, Switzerland, that provide skin protection and anti-aging qualities were introduced [106].

Table 1.3 Nanocrystal cosmetic marketed products

Product name	Active ingredient	Use of product
La Roche-Posay Anthelios Melt-In Sunscreen Milk	Zinc oxide nanocrystals	Skin production
L'Oréal Paris Age Perfect Intense Hydration Night Mask	Hyaluronic acid nanocrystals	Moisturizer
Olay Regenerist micro Sculpting cream	Hyaluronic acid, aminopeptide, vit B3	Anti-aging moisturizer
Pantene Pro-V Nanorepair	Nano-keratin	Hair care production
L'Oréal Paris EverPure Repair & Protect	Nano-ceramides	Hair care production
Maybelline Dream Pure BB Cream (nano-pigments)	Salicylic acid	Skin production
EltaMD UV Clear Broad-Spectrum SPF 46	Zinc oxide nanocrystals	Sunlight protectant
CeraVe Sunscreen Lotion SPF 50	Nano-zinc oxide	UV light protectant

Taking inspiration from the protective mechanisms found in nature, Mendoza et al. developed an efficient UV-blocking nanomaterial using cellulose nanocrystals grafted with diethyl sinapate CNC-DES. Cytotoxicity tests showed that CNC-DES did not result in serious harm to fibroblast cells in mice after 24 h of exposure. In summary, CNC-DES possesses potent antioxidant and UV-blocking qualities, and it is also lightweight, nongreasy, biocompatible, and water-dispersible. These highlight the remarkable capacity of CNCs that are grafted with diethyl sinapate as UV filters influenced by nature in the upcoming cosmetics generation products, particularly for those that is designed for hyper-reactive skin [107].

In 2007, nanocrystal-based cosmetics were introduced with skin protection and anti-aging capabilities that incorporate poorly soluble antioxidant, and the studies show that formulations based on nanocrystals have a lot of potential for topical administration [108] Different nanocrystal-based cosmetic products available in the market are listed in Table 1.3.

1.10 Recent Studies of Nanocrystals

As research advanced during the past 20 years, a number of innovative upgrading procedures or extraction techniques were created. Examples include enzymatically producing cellulose nanocrystals from recycled pulp, preparing chitin nano-whiskers or cellulose nanocrystals in an ionic liquid, mechanically separating starch nanocrystals using ceramic membrane filtration, preparing cellulose nanocrystals using sonochemical assistance, and extracting cellulose nanocrystals using a combination of mechanical shearing, enzymatic, and acid hydrolysis [109]. Several nanocrystal drug products have been effectively sold, although the majority are meant for oral use, except for Invega Sustenna®, which is administered via intramuscular injection [110].

In the past few decades, there have been numerous new applications discovered in almost every technological field. These include areas such as human health, electronics, optics, energy storage and production, chemical catalysis, and sensing [111].

Peng et al., in 2020, reported a series of red-emitting biodegradable up-conversion nanocrystals (UCNCs). The inorganic potassium heptafluozirconate (K3ZrF7:Yb/Er) doped with Yb3+/Er3+ possess a dynamically "soft" crystal structure with a K+ cation and a water-soluble [ZrF7]3− cluster. When exposed to water, both in vitro and in vivo, the red-emitting K3ZrF7:Yb/Er UCNCs demonstrate a pH-dependent biodegradation capability. The intrinsic red up-conversion luminescence, which measures the rapid biodegradation rate, can be adjusted, especially in a mildly acidic tumor microenvironment (pH ~ 5–6). Also, unlike the nondegradable β-NaYF4:Yb/Er UCNCs, which mostly collect in the primary organs of mice, the final biodegradation products of K3ZrF7:Yb/Er UCNCs may be quickly eliminated from the body of mice without showing any signs of toxicity. These findings clearly would help future clinical translations and biomedical uses of lanthanide-doped inorganic UCNCs [112].

Bonaccorso et al. prepared curcumin-loaded nanocrystals for delivering drugs from the nose to the brain and targeting the central nervous system. The key findings include nasal compatibility. The formulation's pH and osmolarity were carefully adjusted to prevent any harm to nasal cilia, ensuring safe use and improving cellular uptake. The nanocrystals showed significantly greater absorption by olfactory glial cells compared to free curcumin, suggesting enhanced permeability. These results indicate that curcumin NCs could serve as an effective delivery system for brain-targeted therapies via the intranasal route, offering promise for treating CNS conditions [113].

Khan et al., developed nanocrystals (NC) of capsaicin to enhance water solubility and skin adhesion. The optimized NCs had a particle size of 120 nm, a surface charge of −20.7 mV, and a PDI of 0.48, indicating stability. Drug content was 85% with 90% entrapment efficiency. In vitro release showed 89.94% of the drug released in 24 h, and the NCs exhibited superior anti-inflammatory activity compared to Dicloran®. This suggests that capsaicin NCs could lead to more effective and cost-efficient dosage forms [114].

In 2022, Fu and colleagues worked to enhance the bioavailability of cannabidiol (CBD) and CBD nanocrystals (CBD-NC) are injected intramuscularly (IM). The NCs were produced using the antisolvent precipitation process. When freeze-dried with bovine serum albumin as a cryoprotective, the powder displayed better dispersibility, with a mean particle size of 139.1 ± 1.4 nm after reconstitution. The powder that is freeze-dried has also been evaluated for Ph, drug encapsulation, in vitro dissolution, and through in vivo studies using rats. The in vivo results showed a substantial rise in CBD's $C_{\max}$ and AUC_{0-24} hours when it is injected intramuscularly compared to oral administration for CBD oil solution and CBD nanocrystals. This suggests that nanosized CBD is a simple and effective nanosystem for intramuscular delivery with improved bioavailability. These findings also support further development of CBD-NC retentive dosage forms [115].

Seo prepared a new nanocrystal formulation of silybin, called HM40, designed to overcome its low water solubility, which limits its absorption and use as medicine. They created HM40, a nanocrystal form of silybin, to improve its solubility. The tested HM40 exhibited better physical properties, stability, releases of silybin, and its absorption in rats and humans. The results found that HM40 remained stable for 2 years, released silybin more effectively, and significantly improved its absorption (bioavailability) compared to standard silybin capsules. In short, HM40 shows promise as a new oral drug formulation for silybin, potentially leading to better treatments using this beneficial compound [116].

1.11 Recent Patents on Nanocrystals

Petersen invented nanocrystal-based topical nanosuspension, cream cosmetic formulation containing the various active components like hesperidin, rutin, apigenin, ascorbyl, apigenin, ascorbyl palmitate, and resveratrol for the treatment of anti-aging, and skin protection purposes. The nanosuspension contains stabilizers like lecithin, PVP, PVA, lauryl, decyl, cetearyl glucoside, inutec SP1, sodium cocoamphoacetate, and poloxamer grades. The production method is followed by pearl milling method and high-pressure homogenization (HPH). The prepared formulation showed particle size of less than 1000 nm. The prepared cosmetic nanoformulations showed better photoprotective mechanism and reduced susceptibility to ultraviolet light [117].

Vergnault et al. formulated spironolactone-loaded nanosuspension and cream dosage forms for the treatment of acne, alopecia, hirsutism, and rosacea. In this preparation, sodium docusate was used as a stabilizer and the production method adopted high-pressure homogenization (HPH). The nanocrystals showed the particle size within the range of 400–600 nm. The formulated nanoparticulate spironolactone showed improved antimicrobial activity against acne-related bacteria compared to aqueous cream [118].

Gupta et al. patented nanosized silver sulfadiazine loaded into various semisolid forms like cream, lotion, ointment, and gel for the treatment of burn and wound-healing activity. In this preparation, PVP was used as a stabilizer and dyno mill was used as a homogenizer. The prepared formulation of nonocrystal size ranges from 150 to 500 nm. The optimized formulations showed better therapeutic efficacy compared to the commercial product in a burn and model of wound infection [119].

Petersson patented calcipotriol monohydrate loaded into nanosuspension, cream, or ointment with Poloxamer 188 used as a stabilizer. The nanocrystals prepared by high-pressure homogenization (HPH) technique and formed crystals showed the diameter ranges from 200 to 600 nm. The formulated calcipotriol monohydrate displayed better chemical stability. The nanocrystal incorporated cream displayed twofold the stimulation for the specific gene compared to the marketed available daivonex formula for the treatment of psoriasis [120] Some commercially available nanocrystal formulations are shown in Table 1.4.

Table 1.4 Commercially available brands of drug nanocrystals [44, 121]

Marketed preparation	Drug	Targeted disease	Company name
Rapamune	Sirolimus	Preventing organ rejection in patients 13 years and older who have had kidney transplants	Pfizer Inc., USA
Emend	Aprepitant	Preventing nausea and vomiting caused by chemotherapy	MSD Pharmaceuticals Pvt Ltd., New Delhi
Tricor	Fenofibrate	Lowering triglycerides, overall cholesterol, and LDL cholesterol	AbbVie Inc., USA
Megace	Megestreol acetate	Sometimes used in the treatment of breast cancer and prostate cancer and also employed to alleviate hot flushes in women going through menopause	Bausch&Lomb U. K Limited
Triglide	Fenofibrate	Hypercholesterolemia	First Horizon Pharmaceutical, France
Invega, Sustenna	Paliperidone, palmitate	Schizophrenia	Janssen Pharmaceutica Belgium

1.12 Future Perspectives

Treating various diseases using nanocrystal-based drug delivery system is a better option due to the nanometric size, with the typical drug loading capacity of up to 90%. Nanocrystals offering a huge potential for the development of poorly soluble pharmaceutical ingredient and provide better therapeutic efficacy compared with conventional dosage forms. Nonetheless, the current applications of nanocrystal techniques are rather limited. The future is to expand its clinical application to benefit an increased number of patients. The nanocrystal-based preparations loaded into other drug delivery systems may provide sustained drug release pattern through various routes. The future perspective might concentrate on the development of controlled-release nanocrystals of both hydrophilic and lipophilic drugs through various routes of administration. Parenteral nanocrystal dosage forms are used as a future choice for the treatment of cancer, and most of the cancer drugs come under the category of water-insoluble. Many case studies concluded that nanocrystal-based various delivery systems proved significantly improved drug delivery pattern and treatment of diseases but the corresponding area is to be improved for the treatment of diseases.

The high drug loading capacity of nanocrystrals is efficient to release a large quantity in the targeted site compared to nanomedicines. This system is suitable for multidrug delivery system or a combination of therapeutic agents. Nanocrystal-based vaccine delivery: Developing nanocrystal-based vaccines for infectious diseases. Personalized nanomedicine: Tailoring nanocrystal-based delivery systems to individual patient's needs. Nanocrystal-based gene therapy: Using nanocrystals to deliver genetic material for gene therapy. 3D printing-based delivery: Using 3D printing to create customized nanocrystal-based delivery systems. Stimuli-responsive delivery: Developing nanocrystal-based delivery systems that respond to environmental stimuli.

Conflict of Interest The authors declare no conflict of interest, financial or otherwise.

References

1. Jarvis M, Krishnan V, Mitragotri S (2019) Nanocrystals: a perspective on translational research and clinical studies. Bioeng Transl Med 4:5–16. https://doi.org/10.1002/btm2.10122
2. Parmar PK, Wadhawan J, Bansal AK (2021) Pharmaceutical nanocrystals: a promising approach for improved topical drug delivery. Drug Discov Today 26:2329–2349. https://doi.org/10.1016/j.drudis.2021.07.010
3. Pelikh O, Stahr P-L, Huang J et al (2018) Nanocrystals for improved dermal drug delivery. Eur J Pharm Biopharm 128:170–178. https://doi.org/10.1016/j.ejpb.2018.04.020
4. Chogale M, Ghodake V, Patravale V (2016) Performance parameters and characterizations of nanocrystals: a brief review. Pharmaceutics 8:26. https://doi.org/10.3390/pharmaceutics8030026
5. Senthil Rajan D, Subashini R, Murugananthan G (2021) Pharmaceutical nanocrystals. In: Handbook on nanobiomaterials for therapeutics and diagnostic applications. Elsevier, San Diego, pp 409–420
6. Fontana F, Figueiredo P, Zhang P et al (2018) Production of pure drug nanocrystals and nano co-crystals by confinement methods. Adv Drug Deliv Rev 131:3–21. https://doi.org/10.1016/j.addr.2018.05.002
7. Gao L, Zhang D, Chen M (2008) Drug nanocrystals for the formulation of poorly soluble drugs and its application as a potential drug delivery system. J Nanopart Res 10:845–862. https://doi.org/10.1007/s11051-008-9357-4
8. Shen B, Shen C, Zhu W, Yuan H (2021) The contribution of absorption of integral nanocrystals to enhancement of oral bioavailability of quercetin. Acta Pharm Sin B 11:978–988. https://doi.org/10.1016/j.apsb.2021.02.015
9. Junyaprasert VB, Morakul B (2015) Nanocrystals for enhancement of oral bioavailability of poorly water-soluble drugs. Asian J Pharm Sci 10:13–23. https://doi.org/10.1016/j.ajps.2014.08.005
10. Liu J, Tu L, Cheng M et al (2020) Mechanisms for oral absorption enhancement of drugs by nanocrystals. J Drug Deliv Sci Technol 56:101607. https://doi.org/10.1016/j.jddst.2020.101607
11. Ochi M, Kawachi T, Toita E et al (2014) Development of nanocrystal formulation of meloxicam with improved dissolution and pharmacokinetic behaviors. Int J Pharm 474:151–156. https://doi.org/10.1016/j.ijpharm.2014.08.022
12. Chen M-L, John M, Lee SL, Tyner KM (2017) Development considerations for nanocrystal drug products. AAPS J 19:642–651. https://doi.org/10.1208/s12248-017-0064-x
13. Rabinow BE (2004) Nanosuspensions in drug delivery. Nat Rev Drug Discov 3:785–796. https://doi.org/10.1038/nrd1494
14. Bhujbal NS, Cheriyan BV (2020) Recent advances in nanocrystal. Int J Res Pharm Sci 11:5791–5796. https://doi.org/10.26452/ijrps.v11i4.3227
15. Kovalenko MV, Manna L, Cabot A et al (2015) Prospects of nanoscience with nanocrystals. ACS Nano 9:1012–1057. https://doi.org/10.1021/nn506223h

16. Liversidge GG, Cundy KC (1995) Particle size reduction for improvement of oral bioavailability of hydrophobic drugs: I. Absolute oral bioavailability of nanocrystalline danazol in beagle dogs. Int J Pharm 125:91–97. https://doi.org/10.1016/0378-5173(95)00122-Y
17. Müller RH, Gohla S, Keck CM (2011) State of the art of nanocrystals – special features, production, nanotoxicology aspects and intracellular delivery. Eur J Pharm Biopharm 78:1–9. https://doi.org/10.1016/j.ejpb.2011.01.007
18. Talapin DV, Lee J-S, Kovalenko MV, Shevchenko EV (2010) Prospects of colloidal nanocrystals for electronic and optoelectronic applications. Chem Rev 110:389–458. https://doi.org/10.1021/cr900137k
19. Cormode DP, Sanchez-Gaytan BL, Mieszawska AJ et al (2013) Inorganic nanocrystals as contrast agents in MRI: synthesis, coating and introduction of multifunctionality. NMR Biomed 26:766–780. https://doi.org/10.1002/nbm.2909
20. Liversidge GG, Conzentino P (1995) Drug particle size reduction for decreasing gastric irritancy and enhancing absorption of naproxen in rats. Int J Pharm 125:309–313. https://doi.org/10.1016/0378-5173(95)00148-C
21. Banavath H, Raju SK, Tahir Ansari M et al (2010) Nanosuspension: an attempt to enhance bioavailability of poorly soluble drugs. Int J Pharm Sci Res 1:1–11
22. Muller RH, Becker R, Kruss B, Peters K (1999) 54 pharmaceutical nanosuspensions for medcament administrationas systems with increased saturation solubility and rate of solution 75 inventors: 30 foreign application priority data. PCT Pub
23. Pardhi E, Vasave R, Srivastava V et al (2024) Nanocrystal technologies in biomedical science: from the bench to the clinic. Drug Discov Today 29:103913. https://doi.org/10.1016/j.drudis.2024.103913
24. Van Eerdenbrugh B, Vermant J, Martens JA et al (2009) A screening study of surface stabilization during the production of drug nanocrystals. J Pharm Sci 98:2091–2103. https://doi.org/10.1002/jps.21563
25. Shete G, Jain H, Punj D et al (2014) Stabilizers used in nano-crystal based drug delivery systems. J Excip Food Chem 5(4):184–209
26. Tuomela A, Hirvonen J, Peltonen L (2016) Stabilizing agents for drug nanocrystals: effect on bioavailability. Pharmaceutics 8:16. https://doi.org/10.3390/pharmaceutics8020016
27. Li J, Wang Z, Zhang H et al (2021) Progress in the development of stabilization strategies for nanocrystal preparations. Drug Deliv 28:19–36. https://doi.org/10.1080/10717544.2020.1856224
28. Bansal S, College MCP, Kumria R, Bansal M (2012) Nanocrystals: current strategies and trends. Int J Res Pharmaceut Biomed Sci 3:407–419
29. Mohammad IS, Hu H, Yin L, He W (2019) Drug nanocrystals: fabrication methods and promising therapeutic applications. Int J Pharm 562:187–202. https://doi.org/10.1016/j.ijpharm.2019.02.045
30. Peltonen L, Hirvonen J (2018) Drug nanocrystals—versatile option for formulation of poorly soluble materials. Int J Pharm 537:73–83. https://doi.org/10.1016/j.ijpharm.2017.12.005
31. Keck C, Muller R (2006) Drug nanocrystals of poorly soluble drugs produced by high pressure homogenisation. Eur J Pharm Biopharm 62:3–16. https://doi.org/10.1016/j.ejpb.2005.05.009
32. Joseph E, Singhvi G (2019) Multifunctional nanocrystals for cancer therapy: a potential nanocarrier. In: Nanomaterials for drug delivery and therapy. Elsevier, Amsterdam, pp 91–116
33. Wu C, Chen Z, Hu Y et al (2018) Nanocrystals: the preparation, precise control and application toward the pharmaceutics and food industry. Curr Pharm Des 24:2425–2431. https://doi.org/10.2174/1381612824666180515124614
34. Dhaval M, Makwana J, Sakariya E, Dudhat K (2020) Drug nanocrystals: a comprehensive review with current regulatory guidelines. Curr Drug Deliv 17:470–482. https://doi.org/10.2174/1567201817666200512104833
35. Kalhapure RS, Palekar S, Patel K, Monpara J (2022) Nanocrystals for controlled delivery: state of the art and approved drug products. Expert Opin Drug Deliv 19:1303–1316. https://doi.org/10.1080/17425247.2022.2110579

36. Guo M, Qin S, Wang S et al (2023) Herbal medicine nanocrystals: a potential novel therapeutic strategy. Molecules 28:6370
37. Sawant SV, Vilasrao Kadam DJ, Kisan Jadhav DR, Sankpal SV (2011) Drug nanocrystals: novel technique for delivery of poorly soluble drugs. Int J Sci Innov Dis 1:1–15
38. Fu Q, Sun J, Zhang D et al (2013) Nimodipine nanocrystals for oral bioavailability improvement: preparation, characterization and pharmacokinetic studies. Colloids Surf B: Biointerfaces 109:161–166. https://doi.org/10.1016/j.colsurfb.2013.01.066
39. Ahire E, Thakkar S, Darshanwad M, Misra M (2018) Parenteral nanosuspensions: a brief review from solubility enhancement to more novel and specific applications. Acta Pharm Sin B 8:733–755
40. Mirza RM, Ahirrao SP, Kshirsagar SJ (2017) A nanocrystal technology: to enhance solubility of poorly water soluble drugs. J Appl Pharm Res 5:1–13
41. Rossier B, Jordan O, Allémann E, Rodríguez-Nogales C (2024) Nanocrystals and nanosuspensions: an exploration from classic formulations to advanced drug delivery systems. Drug Deliv Transl Res 14:3438–3451. https://doi.org/10.1007/s13346-024-01559-0
42. Shin GH, Li J, Cho JH et al (2016) Enhancement of curcumin solubility by phase change from crystalline to amorphous in cur-TPGS nanosuspension. J Food Sci 81:N494–N501. https://doi.org/10.1111/1750-3841.13208
43. Xiong S, Liu W, Zhou Y et al (2020) Enhancement of oral bioavailability and anti-parkinsonian efficacy of resveratrol through a nanocrystal formulation. Asian J Pharm Sci 15:518–528. https://doi.org/10.1016/j.ajps.2019.04.003
44. Shi J, Guo F, Zheng A et al (2015) Progress in the study of drug nanocrystals. Pharmazie 70: 757–764. https://doi.org/10.1691/ph.2015.5044
45. Zhou Y, Du J, Wang L, Wang Y (2016) State of the art of nanocrystals technology for delivery of poorly soluble drugs. J Nanopart Res 18:257
46. Bai M, Yang M, Gong J et al (2022) Progress and principle of drug nanocrystals for tumor targeted delivery. AAPS Pharm SciTech 23:41
47. Samim Sardar M, Kashinath KP, Kumari M et al (2024) Rebamipide nanocrystal with improved physicomechanical properties and its assessment through bio-mimicking 3D intestinal permeability model. Nanoscale 16:19786–19805. https://doi.org/10.1039/D4NR03137G
48. Fernandes AR, Dias-Ferreira J, Ferreira-da-Silva C et al (2018) Drug nanocrystals. In: Applications of nanocomposite materials in drug delivery. Elsevier, San Diego, pp 239–253
49. Xia Y, Xiong Y, Lim B, Skrabalak SE (2009) Shape-controlled synthesis of metal nanocrystals: simple chemistry meets complex physics? Angew Chem Int Ed 48:60–103. https://doi.org/10.1002/anie.200802248
50. Hardman R (2006) A toxicologic review of quantum dots: toxicity depends on physicochemical and environmental factors. Environ Health Perspect 114:165–172. https://doi.org/10.1289/ehp.8284
51. Dabbousi BO, Rodriguez-Viejo J, Mikulec FV et al (1997) (CdSe)ZnS core−shell quantum dots: synthesis and characterization of a size series of highly luminescent nanocrystallites. J Phys Chem B 101:9463–9475. https://doi.org/10.1021/jp971091y
52. Martynenko IV, Litvin AP, Purcell-Milton F et al (2017) Application of semiconductor quantum dots in bioimaging and biosensing. J Mater Chem B 5:6701–6727. https://doi.org/10.1039/C7TB01425B
53. D'Arienzo M, Scotti R, Di Credico B, Redaelli M (2017) Synthesis and characterization of morphology-controlled TiO_2 nanocrystals. In: Morphological, compositional, and shape control of materials for catalysis, Studies in surface science and catalysis, vol 177. Elsevier, Amsterdam, pp 477–540
54. Majoinen J, Kontturi E, Ikkala O, Gray DG (2012) SEM imaging of chiral nematic films cast from cellulose nanocrystal suspensions. Cellulose 19:1599–1605. https://doi.org/10.1007/s10570-012-9733-1

55. Maghsoudy-Louyeh S, Kropf M, Tittmann BR (2018) Review of progress in atomic force microscopy. Open Neuroimaging J 12:86–104. https://doi.org/10.2174/1874440001812010086
56. Peltonen L, Strachan C (2015) Understanding critical quality attributes for nanocrystals from preparation to delivery. Molecules 20:22286–22300. https://doi.org/10.3390/molecules201219851
57. Giannini C, Ladisa M, Altamura D et al (2016) X-ray diffraction: a powerful technique for the multiple-length-scale structural analysis of nanomaterials. Crystals (Basel) 6:87. https://doi.org/10.3390/cryst6080087
58. Dassanayake RS, Dissanayake N, Fierro JS et al (2023) Characterization of cellulose nanocrystals by current spectroscopic techniques. Appl Spectrosc Rev 58:180–205. https://doi.org/10.1080/05704928.2021.1951283
59. Liu X, Deng R, Zhang Y et al (2015) Probing the nature of upconversion nanocrystals: instrumentation matters. Chem Soc Rev 44:1479–1508. https://doi.org/10.1039/C4CS00356J
60. Boyer J-C, Vetrone F, Cuccia LA, Capobianco JA (2006) Synthesis of colloidal upconverting $NaYF_4$ nanocrystals doped with Er^{3+}, Yb^{3+} and Tm^{3+}, Yb^{3+} via thermal decomposition of lanthanide trifluoroacetate precursors. J Am Chem Soc 128:7444–7445. https://doi.org/10.1021/ja061848b
61. Kádár R, Spirk S, Nypelö T (2021) Cellulose nanocrystal liquid crystal phases: progress and challenges in characterization using rheology coupled to optics, scattering, and spectroscopy. ACS Nano 15:7931–7945. https://doi.org/10.1021/acsnano.0c09829
62. Lin N, Huang J, Chang PR et al (2011) Preparation, modification, and application of starch nanocrystals in nanomaterials: a review. J Nanomater 2011:573687. https://doi.org/10.1155/2011/573687
63. Nagarajan KJ, Ramanujam NR, Sanjay MR et al (2021) A comprehensive review on cellulose nanocrystals and cellulose nanofibers: pretreatment, preparation, and characterization. Polym Compos 42:1588–1630
64. Vivero-Escoto JL, Slowing II, Trewyn BG, Lin VS-Y (2010) Mesoporous silica nanoparticles for intracellular controlled drug delivery. Small 6:1952–1967. https://doi.org/10.1002/smll.200901789
65. Peng HR, Gong MM, Chen YZ, Liu F (2017) Thermal stability of nanocrystalline materials: thermodynamics and kinetics. Int Mater Rev 62:303–333. https://doi.org/10.1080/09506608.2016.1257536
66. Ding Y, Zhao T, Fang J et al (2024) Recent developments in the use of nanocrystals to improve bioavailability of APIs. Wiley Interdiscip Rev Nanomed Nanobiotechnol 16:e1958
67. Gujar K, Wairkar S (2020) Nanocrystal technology for improving therapeutic efficacy of flavonoids. Phytomedicine 71:153240
68. Lamichhane S, Bashyal S, Keum T et al (2019) Complex formulations, simple techniques: can 3D printing technology be the Midas touch in pharmaceutical industry? Asian J Pharm Sci 14:465–479. https://doi.org/10.1016/j.ajps.2018.11.008
69. Li J, Wu C, Chu PK, Gelinsky M (2020) 3D printing of hydrogels: rational design strategies and emerging biomedical applications. Mater Sci Eng R Rep 140:100543. https://doi.org/10.1016/j.mser.2020.100543
70. Xu X, Awwad S, Diaz-Gomez L et al (2021) 3D printed punctal plugs for controlled ocular drug delivery. Pharmaceutics 13:1421. https://doi.org/10.3390/pharmaceutics13091421
71. Jin Q, Li H, Jin Z et al (2018) TPGS modified nanoliposomes as an effective ocular delivery system to treat glaucoma. Int J Pharm 553:21–28. https://doi.org/10.1016/j.ijpharm.2018.10.033
72. Khalid MY, Arif ZU, Noroozi R et al (2023) 3D/4D printing of cellulose nanocrystals-based biomaterials: additives for sustainable applications. Int J Biol Macromol 251:126287. https://doi.org/10.1016/j.ijbiomac.2023.126287
73. Sultan S, Mathew AP (2018) 3D printed scaffolds with gradient porosity based on a cellulose nanocrystal hydrogel. Nanoscale 10:4421–4431. https://doi.org/10.1039/c7nr08966j

74. Lopez-Vidal L, Real JP, Real DA et al (2022) Nanocrystal-based 3D-printed tablets: semi-solid extrusion using melting solidification printing process (MESO-PP) for oral administration of poorly soluble drugs. Int J Pharm 611:121311. https://doi.org/10.1016/j.ijpharm.2021.121311
75. Phan VHG, Murugesan M, Huong H et al (2022) Cellulose nanocrystals-incorporated thermosensitive hydrogel for controlled release, 3D printing, and breast cancer treatment applications. ACS Appl Mater Interfaces 14:42812–42826. https://doi.org/10.1021/acsami.2c05864
76. Santos MI, da Silva LCE, Bomediano MP et al (2021) 3D printed nitric oxide-releasing poly (acrylic acid)/F127/cellulose nanocrystal hydrogels. Soft Matter 17:6352–6361. https://doi.org/10.1039/D1SM00163A
77. Wolfbeis OS (2015) An overview of nanoparticles commonly used in fluorescent bioimaging. Chem Soc Rev 44:4743–4768
78. Zhang X, Guan J, Mao S (2022) Applications of hybrid nanocrystals in drug delivery. In: Hybrid nanomaterials for drug delivery. Elsevier, pp 53–83
79. Xie S, Zhang X, Walcott MP, Lin H (2018) Applications of cellulose nanocrystals: a review. Eng Sci 2:4–16
80. Corti M, Lascialfari A, Micotti E et al (2008) Magnetic properties of novel superparamagnetic MRI contrast agents based on colloidal nanocrystals. J Magn Magn Mater 320:e320–e323. https://doi.org/10.1016/j.jmmm.2008.02.064
81. Torkashvand N, Sarlak N (2019) Fabrication of a dual T1 and T2 contrast agent for magnetic resonance imaging using cellulose nanocrystals/Fe3O4 nanocomposite. Eur Polym J 118:128–136. https://doi.org/10.1016/j.eurpolymj.2019.05.048
82. Lugoloobi I, Maniriho H, Jia L et al (2021) Cellulose nanocrystals in cancer diagnostics and treatment. J Control Release 336:207–232
83. He Q, Bai Y, Lu Y et al (2022) Isolation and characterization of cellulose nanocrystals from Chinese medicine residues. Biomass Convers Biorefinery 129:1081–1089. https://doi.org/10.1007/s13399-022-03380-6
84. Sharma S, Verma A, Teja BV et al (2015) Development of stabilized paclitaxel nanocrystals: in-vitro and in-vivo efficacy studies. Eur J Pharm Sci 69:51–60. https://doi.org/10.1016/j.ejps.2014.11.012
85. Najm MB, Rawas-Qalaji M, Assar NH et al (2022) Optimization, characterization and in vivo evaluation of mupirocin nanocrystals for topical administration. Eur J Pharm Sci 176:106251. https://doi.org/10.1016/j.ejps.2022.106251
86. Pireddu R, Caddeo C, Valenti D et al (2016) Diclofenac acid nanocrystals as an effective strategy to reduce in vivo skin inflammation by improving dermal drug bioavailability. Colloids Surf B: Biointerfaces 143:64–70. https://doi.org/10.1016/j.colsurfb.2016.03.026
87. Gera S, Sampathi S, Maddukuri S et al (2022) Therapeutic potential of naringenin nanosuspension: in vitro and in vivo anti-osteoporotic studies. Pharmaceutics 14:1449. https://doi.org/10.3390/pharmaceutics14071449
88. Formica ML, Real DA, Picchio ML et al (2022) On a highway to the brain: a review on nose-to-brain drug delivery using nanoparticles. Appl Mater Today 29:101631. https://doi.org/10.1016/j.apmt.2022.101631
89. Abbate MTA, Ramöller IK, Sabri AH et al (2023) Formulation of antiretroviral nanocrystals and development into a microneedle delivery system for potential treatment of HIV-associated neurocognitive disorder (HAND). Int J Pharm 640:123005. https://doi.org/10.1016/j.ijpharm.2023.123005
90. Romero GB, Keck CM, Müller RH, Bou-Chacra NA (2016) Development of cationic nanocrystals for ocular delivery. Eur J Pharm Biopharm 107:215–222. https://doi.org/10.1016/j.ejpb.2016.07.005
91. Shi Z, Tang J, Chen L et al (2015) Enhanced colloidal stability and antibacterial performance of silver nanoparticles/cellulose nanocrystal hybrids. J Mater Chem B 3:603–611. https://doi.org/10.1039/C4TB01647E

92. Peters MCC, dos Santos Neto E, Monteiro LM et al (2020) Advances in ophthalmic preparation: the role of drug nanocrystals and lipid-based nanosystems. J Drug Target 28:259–270. https://doi.org/10.1080/1061186X.2019.1663858
93. Kumar P (2017) A review on significance of nanocrystals in drug delivery. World J Pharm Pharm Sci 6:347–358. https://doi.org/10.20959/wjpps201712-10553
94. Liu Y, Zhao J, Wang L et al (2018) Nanocrystals technology for transdermal delivery of water-insoluble drugs. Curr Drug Deliv 15:1221–1229. https://doi.org/10.2174/1567201815666180518124345
95. Sinico C, Pireddu R, Pini E et al (2016) Enhancing topical delivery of resveratrol through a nanosizing approach. Planta Med 83:476–481. https://doi.org/10.1055/s-0042-103688
96. Fang C-L, Aljuffali IA, Li Y-C, Fang J-Y (2014) Delivery and targeting of nanoparticles into hair follicles. Ther Deliv 5:991–1006. https://doi.org/10.4155/tde.14.61
97. Pelikh O, Eckert RW, Pinnapireddy SR, Keck CM (2021) Hair follicle targeting with curcumin nanocrystals: influence of the formulation properties on the penetration efficacy. J Control Release 329:598–613. https://doi.org/10.1016/j.jconrel.2020.09.053
98. Lv Y, Wu W, Corpstein CD et al (2022) Biological and intracellular fates of drug nanocrystals through different delivery routes: recent development enabled by bioimaging and PK modeling. Adv Drug Deliv Rev 188:114466
99. Guan W, Liu Y, Ding S, Wang Y (2021) Ciprofloxacin nanocrystals and N-acetylcysteine co-solidified powders for pulmonary drug delivery: development and in vitro and in vivo characterization. J Nanopart Res 24:41. https://doi.org/10.1007/s11051-022-05427-1
100. de Macedo LO, Masiero JF, Bou-Chacra NA (2024) Drug nanocrystals in oral absorption: factors that influence pharmacokinetics. Pharmaceutics 16:1141. https://doi.org/10.3390/pharmaceutics16091141
101. Rahim H, Sadiq A, Khan S et al (2017) Aceclofenac nanocrystals with enhanced in vitro, in vivo performance: formulation optimization, characterization, analgesic and acute toxicity studies. Drug Des Devel Ther 11:2443–2452. https://doi.org/10.2147/DDDT.S140626
102. Ganta S, Paxton JW, Baguley BC, Garg S (2009) Formulation and pharmacokinetic evaluation of an asulacrine nanocrystalline suspension for intravenous delivery. Int J Pharm 367:179–186. https://doi.org/10.1016/j.ijpharm.2008.09.022
103. Patel V, Sharma OP, Mehta T (2018) Nanocrystal: a novel approach to overcome skin barriers for improved topical drug delivery. Expert Opin Drug Deliv 15:351–368
104. Quan W, Kong S, Ouyang Q et al (2021) Use of 18β-glycyrrhetinic acid nanocrystals to enhance anti-inflammatory activity by improving topical delivery. Colloids Surf B: Biointerfaces 205:111791. https://doi.org/10.1016/j.colsurfb.2021.111791
105. Zhang J, Zhang J, Wang S, Yi T (2018) Development of an oral compound Pickering emulsion composed of nanocrystals of poorly soluble ingredient and volatile oils from traditional Chinese medicine. Pharmaceutics 10:170. https://doi.org/10.3390/pharmaceutics10040170
106. Wadhawan J, Parmar PK, Bansal AK (2021) Nanocrystals for improved topical delivery of medium soluble drug: a case study of acyclovir. J Drug Deliv Sci Technol 65:102662. https://doi.org/10.1016/j.jddst.2021.102662
107. Mendoza DJ, Maliha M, Raghuwanshi VS et al (2021) Diethyl sinapate-grafted cellulose nanocrystals as nature-inspired UV filters in cosmetic formulations. Mater Today Bio 12:100126. https://doi.org/10.1016/j.mtbio.2021.100126
108. Parveen N, Abourehab MAS, Thanikachalam PV et al (2023) Nanocrystals as an emerging nanocarrier for the management of dermatological diseases. Colloids Surf B: Biointerfaces 225:113231
109. Lin N, Huang J, Dufresne A (2012) Preparation, properties and applications of polysaccharide nanocrystals in advanced functional nanomaterials: a review. Nanoscale 4:3274–3294
110. Lu Y, Qi J, Dong X et al (2017) The in vivo fate of nanocrystals. Drug Discov Today 22:744–750
111. Jansons AW, Plummer LK, Hutchison JE (2017) Living nanocrystals. Chem Mater 29:5415–5425

112. Peng P, Wu N, Ye L et al (2020) Biodegradable inorganic upconversion nanocrystals for in vivo applications. ACS Nano 14:16672–16680. https://doi.org/10.1021/acsnano.0c02601
113. Bonaccorso A, Gigliobianco MR, Pellitteri R et al (2020) Optimization of curcumin nanocrystals as promising strategy for nose-to-brain delivery application. Pharmaceutics 12: 476. https://doi.org/10.3390/pharmaceutics12050476
114. Khan BA, Rashid F, Khan MK et al (2021) Fabrication of capsaicin loaded nanocrystals: physical characterizations and in vivo evaluation. Pharmaceutics 13:841. https://doi.org/10.3390/pharmaceutics13060841
115. Fu X, Xu S, Li Z et al (2022) Enhanced intramuscular bioavailability of cannabidiol using nanocrystals: formulation, in vitro appraisal, and pharmacokinetics. AAPS PharmSciTech 23: 85. https://doi.org/10.1208/s12249-022-02239-3
116. Seo SR, Kim GY, Kim MH et al (2024) Nanocrystal formulation to enhance oral absorption of silybin: preparation, in vitro evaluations, and pharmacokinetic evaluations in rats and healthy human subjects. Pharmaceutics 16:1033. https://doi.org/10.3390/pharmaceutics16081033
117. Petersen R (2007) Nanocrystals for use in topical cosmetc formulations and method of production thereof S371. https://doi.org/10.1046/j.1473
118. Vergnault G, Dieter N, Grenier P et al (2011) Topical nanoparticulate spironolactone formulation, pp 1–18
119. Gupta N, Bedi S, Srivastava J et al (2017) Topical pharmaceutical composition comprising nanonized silver sulfadiazine, pp 1–9
120. Petersson K (2014) Calcipotriol monohydrate nanocrystals, pp 1–19
121. Gokce EH, Ozyazici M, Souto EB (2010) Nanoparticulate strategies for effective delivery of poorly soluble therapeutics. Ther Deliv 1:149–167. https://doi.org/10.4155/tde.10.4

Formulation-Based Composites in Nanocrystal Preparation and Advanced Characterization Techniques

2

Krishna Kant Jangde, Chirag Jain, Neha Jain, Anuja Dubey, and Dinesh Kumar Mishra

Abstract

Nanocrystals (NCs) are a key innovation in nanotechnology, particularly for improving drug delivery systems for poorly soluble drugs. There are various methods employed for NC production, including top-down techniques like pearl/ball milling, and high-pressure homogenization; bottom-up methods like precipitation, and cryo-vacuum techniques; and combination approaches like NanoEdge and SmartCrystal technologies. With these methods, NC's size, stability, and solubility can be altered. For the physical and chemical evaluation of developed NCs, advanced characterization techniques such as scanning electron microscopy (SEM), powder X-ray diffraction (PXRD), and dynamic light scattering (DLS) are crucial. These assessments ensure stability, solubility, and efficacy in drug formulations. NCs are widely used in fields like cancer therapy and treatments for neurodegenerative diseases. Despite their potential, challenges like regulatory concerns, toxicity risks, and large-scale production hurdles exist. In this chapter, an attempt has been made to review and compile recent advances in production

K. K. Jangde · D. K. Mishra (✉)
Department of Pharmacy, Guru Ghasidas Vishwavidyalaya (A Central University), Bilaspur, Chhattisgarh, India

C. Jain
Centre for Industrial Pharmacy and Pharmaceutical Regulatory Affairs, Amity Institute of Pharmacy, Amity University, Noida, Uttar Pradesh, India

N. Jain
Centre for Pharmaceutics, Amity Institute of Pharmacy, Amity University, Noida, Uttar Pradesh, India

A. Dubey
Government Bilasa Girls PG Autonomous College, Bilaspur, Chhattisgarh, India

N. K. Mehra, K. K. Bansal (eds.), *Nanocrystal Technologies in Biomedical Sciences*,
https://doi.org/10.1007/978-981-95-2968-1_2

and characterization methods of NCs, emphasizing their future role in overcoming biological barriers and advancing personalized medicine.

Keywords

Nanocrystal · Preparation · Formulation · Characterization

2.1 Introduction

The field of nanotechnology represents a multidisciplinary field that was developed to engineer biological materials, such as atoms, molecules, and supramolecules, at the nanoscale range of approximately 1–100 nm [1, 2]. Collections of particles that aggregate into clusters are called nanocrystals (NCs), and their estimated size ranges from 10 to 400 nm. Basically, NCs' physiochemical features lie between particles and bulk solids [3]. Drug NCs have a straightforward composition, usually consist of only pure drugs, do not require a carrier, and may include trace amounts of stabilizers like surfactants and a filling agent like sucrose, thereby minimizing accessory-related toxicity. This sets them apart from other nano preparations like liposomes, various nanoparticles, as well as solid lipid NCs as the "carrier" [4]. Drug particles aggregate easily because drug NCs frequently have very high surface tensions. NC aggregation can be avoided, and surface tension can be lowered by using an appropriate stabilizer. The hydrophilic portion of the stabilizer forms a double electrical layer and a charge around the drug when it is adsorbed onto the drug surface. When the distance between two drug particles is lowered beyond a certain point, the two layers with the same charges oppose one another and the fragments separate, stopping polymerization [5].

Pharmaceuticals that are poorly soluble in water exhibit a restricted rate of dissolution, which can lead to inadequate biopharmaceutical issues when taken orally. The system used to deliver drugs exhibits a high degree of inter-patient variation, an unstable absorption state, inadequate bioavailability, a lack of dose proportionality, and a delayed first effect [6]. The augmentation of surface area, the dissolution rate, saturating solubility, surface adhesiveness of cell membranes, and oral bioavailability are all brought about by the reduction of particle size to the nanometer range. As a result, NCs exhibit the following characteristics of drug molecules: a limited drug efficiency, several routes of administration, few adverse reactions, high drug load, and quick onset of action. Nimodipine intravenous administration was made possible by a novel strategy utilizing sodium cholic acid-stabilized NCs, which also provided a significant reduction in risk-to-benefit ratio [7].

Because of its higher surface area per unit volume ratio and enhanced dissolution rates (its dissolution speed) associated with nanosizing, nanocrystalline drug technology makes hydrophobic drugs more soluble. The drug crystals are exceptionally well-suited for the rehabilitative use of drugs classified as class II and IV (low solubility) in the Biopharmaceutics Classification System (BCS), which have been unsuccessful in the past. Considered as NC colloidal dispersion (NCD), NC dosage forms have also been demonstrated to be stable in suspensions. The dispersions offer an easy-to-scale-up platform for producing highly stable and commercially viable products. The therapeutic result of drug products containing NCs is influenced by various properties, including their shape, size, and surface charge, stabilizing agent type, and the coating of polymers used during formulation. The effects of rigidity and surface texture on preclinical studies (i.e., in vivo/in vitro) effectiveness assays are studied as additional physicochemical properties [8].

The most common methods of administration are topical, ocular, parenteral, oral, and pulmonary. In the event of consumption, the intestinal tract is where NCs dissolve, leading to a minor cellular absorption of unaltered drug particles through the intestine's cellular, intracellular, or lymphatic routes. By contrast, the parenteral administration of NCs allows for enhanced bioavailability, reduced toxicity, and site-specific targeting, such as the brain, lungs, liver, kidney, or colon, among other organs. This is made possible by cellular interactions along with nanoparticle uptake [9]. Since they are composed of pure medication particles as well as require low concentrations of surfactants for stability, a scalable production process, and a lower therapeutic dose of the medication at the site of application, NCs are significant in the field of topical delivery. They also do not require the use of co-solvents or an extreme pH range for solubilization [10].

There are several advantages and disadvantages of NCs compared to other drug delivery systems. The relatively high drug to stabilizers proportions of these nano pharmaceuticals, which result in a nearly 100% drug loading capacity, are their greatest advantage. The ability to achieve a high pharmacological concentration in the intended location makes NCs a good choice for drug delivery systems. Topical formulations that have undergone nanonization have the potential to exhibit enhanced drug saturation solubility, dissolving rate, drug release amount, surface adhesion, and cutaneous bioavailability by increasing surface area and decreasing particle size. NCs are thought to be a viable new topical drug delivery technology for cutaneous applications because they can improve the solubility and rate of dissolution of poorly water-soluble medications (particularly BCS class II). In contrast to traditional topical formulations, they could increase skin penetration, improve skin deposition, and increase skin permeability. Drugs with restricted skin penetration, less bioactive medications, and essentially insoluble active medicines can all be delivered using NCs. Some additional noteworthy benefits include improved dosage proportion, ease of process scaling up, a long-term physicochemical stability, aesthetically pleasing topical feeling, cutting down on the requirement for surfactant in nanoformulations, skin bioactivity should be increased, and organic solvents ought to be avoided [11].

The requirement for regular dosage administration is the primary disadvantage. According to a study, dexamethasone encapsulated cellulose NCs, and regular dexamethasone cream did not improve skin penetration as much as dexamethasone NCs. Furthermore, compared with traditional ethyl cellulose nanoformulations, where most of the drug remained in the epidermis, the results showed that in the NC formulation, most of the medication was accumulated in the dermis layer.

Additional drawbacks include the potential for NCs to aggregate and flocculate, which raises the probability of systemic absorption as well as severe adverse reactions, and they are unsuited for use with water-soluble medications [11].

2.2 Properties of Nanocrystals

(a) ***Enhancement of Solubility***: One crucial physicochemical characteristic of pharmaceuticals is their solubility, particularly in water. Drug dissolution is the first stage in it getting into the bloodstream and having a therapeutic impact. A greater concentration gradient between the blood and the stomach lumen and therefore better absorption could be the outcome of the increased saturation solubility. An indication of a small particle's solubility in a bulk solution can be found in the Gibbs–Kelvin equation [12].

$$\ln\left(\frac{xR}{x_\infty}\right) = \frac{2\sigma^\alpha \gamma v_2^\gamma}{k_B RT}$$

where σ^α is the surface tension of the solid particle at its phase α boundary, x_∞ is the relative concentration of solute in phase α, gv^g is the volume per molecule of the solid particle, kB represents the Boltzmann constant, and T is the temperature. xR represents the solubility of tiny solid particles (phase g) in the appropriate bulk phase (phase α) and is related to radius (R). This equation states that the solubility of samples will increase with a decrease in the particle radius. When solubility increased, the relative impact of particle size on the dose's absorption fraction decreased. This suggests that solubility saturation is affected by NCs in addition to the surface area and thickness of the diffusion layer [12].

(b) ***Improvement in Dissolution Velocity***: The Nernst–Brunner equation, which is an improved version of the Noyes–Whitney equation including the integration of Fick's second law, can be used to explain an improvement in the dissolving rate by crystallized materials [13].

$$\frac{dC}{dt} = \frac{DS}{Vh}(Cs - C)$$

where V represents the total volume of the dissolution medium, h represents the thickness of the diffusion section, C_S is the saturation solubility (concentration of medication at dissolving surface), D represents the diffusion coefficient, S represents the effective surface area, or surface area of the particles in contact with the dissolution medium, and C represents the concentration of drug present in the dissolution media at time t. The following equation states that decreasing the size of a particle from micro toward nano enhances its surface area significantly, which raises the dissolving velocity (dC/dt) [14].

(c) ***Increment in Adhesiveness***: One distinct quality of NCs is their adhesiveness, which is attributed to the spectrum of nanosizes in which they are found.

Improved oral absorption is the result of the higher adhesiveness. Analyzing adhesion properties can be done effectively using the kinetics as well as adsorption isotherms. Particle size affects the kinetics of adsorption. The minuscule latexes (230–670 nm) can pass through porous adsorbent and exhibit a distinctive adsorption isotherm of adsorbates. An isotherm plateau denotes the saturation sites, beyond which they are open to further adsorption [9]. The biological mucosa, particularly the mucosa of the gastrointestinal tract, is naturally sticky to NCs. Because of the creation of hydrogen bonds as well as van der Waals bonds connecting the surfaces of the particles and mucus, the NCs in suspension or those produced following the disintegration of a solid dosage form adhere to the gastrointestinal mucosa. Long-term retention of NCs results in a greater concentration gradient throughout the gastrointestinal tract (GIT), which eventually enhances medication absorption and, consequently, bioavailability [13].

2.3 Approaches for the Method of Preparation/Formulation of NCs

(a) *Top-Down Method (Size-Reduction Technique)*

When a suspension is created and driven through an aperture at a very high speed or using shear force to decrease the size of the end crystals, it is referred to as the "top-down" method of mechanical attrition of rough drug powder utilizing a media blender or homogenizer. Utilizing a gentle solvent, such as water, for precipitation rather than toxic organic solvents is one of this technique's main benefits. An uneven size distribution of NCs, uncontrolled crystal development, and the need for sophisticated, costly equipment are a few of the drawbacks. The mechanisms below (Fig. 2.1) demonstrate the breakdown of big aggregates into NCs [15]. In research, the preparation of NCs was done to improve the solubility and bioavailability of midazolam, a sedative with poor water solubility. Using top-down nanotechnology, a stable nanosuspension of midazolam NCs (MDZ/NCs) was created, with an average particle size of 286.6 nm. Pharmacokinetic studies revealed that the bioavailability of MDZ/NCs was 2.72 times higher than the traditional midazolam solution (MDZ/S). Additionally, MDZ/NCs showed more potent anticonvulsant effects and greater neuroprotection in animal models, reducing oxidative stress markers and improving antioxidant levels. These findings highlight the potential of NCs to improve the therapeutic efficacy of poorly soluble drugs like midazolam [14].

(i) *Pearl/Ball Milling (Wet Milling)*

Developed in the 1990s at Sterling Drug Inc., Gary Liversidge and colleagues created wet media-based milling, which gave rise to wet ball milling (WBM). Wet ball milling was first developed, but it has come a long way since then. It became known as NanoCrystal® technology in the pharmaceutical business. For nanosizing pharmaceutical molecules, this strategy has proven to be the most successful. A jar full of clear glass

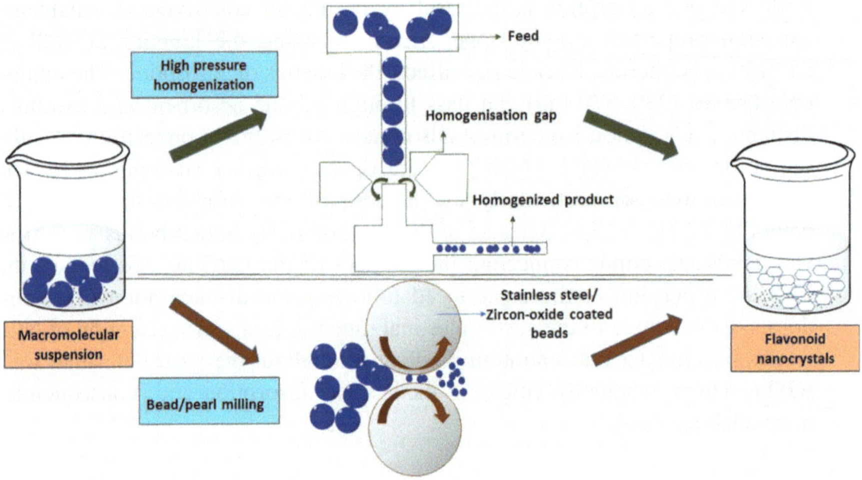

Fig. 2.1 NCs formulated using a top-down approach [15]

beads is used as the grinding medium in wet ball milling. Using this technique, a dispersion medium containing a minimum of one stabilizing agent is added to an equipment containing a rough drug material, preferably micronized. Size reduction through cleavage and abrasion may be the main mechanism of WBM, especially when the process is carried out over an extended length of time, as it is known to produce extremely tiny particles having a narrow size distribution. Note that, irrespective of the use of high-energy (HE-WBM) or low-energy (LE-WBM) techniques, the reported particle sizes for nanosuspensions made by wet ball milling usually lie between 100 and 300 nm [16]. Sharif and Co. had presented a systematic approach to produce ibuprofen (IBF) particles using leucine via wet milling. Utilizing a high shear homogenizer, the study successfully reduced the particle size of IBF from 71.3 to 1.7 μm, with the smallest particles achieved in a 0.05% Tween 80 solution at 17,000 rpm for 15 min [17].

(ii) *High-Pressure Homogenization*

Piston-gap homogenizers are used in the Dissocubes® technology. SkyePharma PLC later purchased the technology, which Müller and associates had created. It is the process of creating room temperature suspensions of nanocrystals in water. Using pressures as high as 4000 bar, usually between 1500 and 2000 bar, a piston is used to push a drug powder that has been disseminated in a solution of aqueous surfactant through the narrow homogenization gap. Depending on the suspension's viscosity and the pressure applied, the homogenization gap's breadth varies between roughly 5 and 20 μm. Bernoulli's law states that the rapid flowing velocity of the suspensions that results raises the dynamic pressure and lowers the static pressure beneath the aqueous phase's vapor pressure.

Because water boils at room temperature, gas bubbles form. As soon as the liquid exits the homogenization gap and returns to the standard atmospheric pressure of one bar, these gas bubbles burst instantly. Shockwaves are produced when gas bubbles develop and implode; this phenomenon is known as cavitation. The massive power of these shockwaves, turbulent flow, and tremendous shear forces causes the drug particles to shrink in size [18].

The *Nanopure*® technology, founded and created by Pharma Sol, headquartered in Berlin, is a further method that makes use of the piston-gap homogenizer. Low-vapor-pressure dispersion medium is used in the technology, and homogenization with low temperatures is optional. There is very little or no cavitation within the homogenization gap. There was enough size reduction even in the absence of cavitation. To produce nanoparticles, the residual shear forces, particle collisions, and turbulences are adequate. Drugs that are sensitive to temperature can be processed while homogenizing at very low temperatures. To prevent medications from hydrolyzing, the entire procedure might be done in nonaqueous fluids. Gelatin or HPMC capsules can be immediately filled with oils, PEG, or hot-melted polyethylene glycols [18]. An investigation focuses on the extraction of cellulose NCs (CNCs) from palm-pressed fiber (PPF), a lignocellulosic biomass from palm oil mills. The study employs a combination of alkaline treatment, bleaching, acid hydrolysis, and high-pressure homogenization (HPH) to enhance the physicochemical properties of CNC. The study concludes that CNC with a gel-like structure has significant potential as a natural pickering emulsifier for stabilizing oil-in-water emulsions in food applications [19].

(b) *Bottom-Up Method (Precipitation Technique)*

The amount of precipitation and evaporation is the fundamental building block of bottom-up technology. The main idea is to take medications from their supersaturated form and cut them into NCs. Then, using the right techniques, you can manage the nanoparticles' size distribution. A tiny and homogeneous NC's creation depends particularly on nucleation. To carefully manage the size of medication particles, the best method is to regulate their growth as crystals. To regulate the formation of crystals, a variety of physical techniques have been employed, including precipitation under high gravity. These approaches offer greater control over particle attributes than top-down approaches do. In conclusion, the low reproducibility of bottom-up technology makes it challenging to scale up, despite its simplicity in theory and operation. Organic solvents may be used in the procedure as well [20].

(i) *Precipitation Method*

For the purpose of effectively precipitating the medications in nanosize particles, this procedure involves dissolving the medication, which is not particularly soluble in water, in a suitable organic solvent first, and then rapidly mixing it with a nonsolvent to form "Hydrosol®." Drug NCs from poorly water-soluble medications are included in hydrosols that are colloidal water-based solutions utilized for intravenous administration. By adding a large amount of water to the medication solution together with "short-term stabilizers" like poloxamer and modified gelatines, the precipitation process is used to create them [21].

The amorphous hydrosol is stable for around 60 minutes following precipitation because of the stabilizers and substantial nonsolvent content. After this, the medication crystallizes. Given that clouding increases with particle size, crystallization as well as particle growth may be identified by a rapid spike in absorbance at an intensity at which the active component does not absorb light. To successfully stabilize the amorphous medicine, the hydrosol is therefore quickly spray-dried using excipients such as lactose or mannitol prior to crystallization occurring. The drug present in hydrosols in particles is of approximately 200 nm size, making them suitable for parenteral delivery. One such drug that can be converted into a hydrosol is cyclosporin [21]. In a research, Zhang H. and Co. examined the impact of various bottom-up methods on the formulation of fenofibrate NCs (FNT-NCs) to enhance solubility and dissolution rates of this poorly water-soluble drug. Three types of FNT-NCs were prepared using antisolvent precipitation and thermal precipitation under different conditions. The study concludes that different preparation methods yield FNT-NCs with varying crystallinity and pharmacokinetic properties, emphasizing the importance of selecting appropriate preparation techniques for optimizing NC formulations [18].

(ii) *Cryo-Vacuum Method*

By quickly cooling water with liquid nitrogen once an active substance has been dissolved in it, the cryo-vacuum process creates ice crystals. The crystals' edges are where the dissolved material nucleates. Then, in a lyophilization unit with constant pressure and temperature of −22 °C and 10–2 mbar, the solvent is eliminated by sublimation. Harmful reagents or surfactants are not required to produce extremely pure NCs with this procedure. The goal of the Nanomorph® technique is to synthesize amorphous nanoparticles from coarsely crystalline medicinal compounds that are poorly soluble in water. The method includes filling a chamber with a solvent-suspended drug suspension, which is then quickly combined with an additional solvent. The drug substance dispersion is transformed into a real molecular solution because of this quick mixing. A polymer-containing aqueous solution is added, which causes the drug material to precipitate. By limiting aggregation or growth, the polymer keeps the drug ingredient molecules in their nanoparticulate condition. This technology is noteworthy because it does away with the necessity of actual milling or grinding operations to produce the required results. Rather, it helps prepare amorphous nanoparticles by transforming coarsely crystalline medicinal compounds into an amorphous state that is spread throughout nanoscales [22]. Research focused on the synthesis and characterization of nanocrystalline calcium sulfate (nCS) for potential use in bone regeneration. The study developed a system where nCS, processed via a cryo-vacuum method and oven drying, was used as a synthetic graft material for treating bone defects. The resulting nCS exhibited a high surface area due to its nanostructure, with acicular crystals measuring 30–80 nm in width and 400–600 nm in length. Additionally, nCS showed promise as a vehicle for the controlled release of growth factors such as bone morphogenetic protein-2 (BMP-2), supporting its potential as a scaffold in bone tissue engineering [20].

(c) *Combination Method*

A new approach for the formulation of NCs with reduced particle sizes—less than 100 nm—is the combination method. In the context of NC technology, the term "combination method" refers to the manufacturing, functionalization, and application of NCs using a variety of techniques or strategies [23].

(i) *NanoEdge Technology*

Baxter's NanoEdge technology was a novel combinative reduction in particle size method created for the manufacturing of drug nanosuspensions. Microprecipitation, a solvent–antisolvent phase in this manufacturing method, is followed by an intense energy process. Initially, the drug is dissolved in an appropriate solvent, often a water-soluble organic solvent. Next, the drug solutions are mixed with another aqueous liquid that has less drug soluble in it. For example, the aqueous liquid is cautiously added to the medication solution using an infuser device. Surfactants might be present to keep it stable. The solubility shift causes the precipitation to happen. One pretreatment technique that yields semi-crystalline or amorphous drug particles is microprecipitation. Next, using a high-energy annealing procedure akin to high-pressure homogenization, the drug particles are shrunk and converted into a more robust crystalline state. By stopping precipitated particles from expanding to micrometer size, the annealing step improves the thermodynamic stability of nanosuspensions. With the use of this technique, medication nanosuspensions with greater drug contents and more adaptable delivery methods—such as oral and injectable—might be produced [23]. An investigation conducted by Alcalá S and Co. explores the generation of recombinant melanin nanoparticles (RMNPs) utilizing preformed recombinant melanin (PRM) as a polymeric raw material. The study employed both bottom-up (nanocrystallization and double emulsion–solvent evaporation) and top-down (high-pressure homogenization—a NanoEdge technology) manufacturing methods to produce the nanoparticles. These findings highlight the potential of RMNPs for various applications, including drug delivery, tissue engineering, diagnostic methods, and sun protection, underscoring their significance in advancing biomedical technologies [21]

(ii) *SmartCrystal Technology*

The second-generation collecting method known as Abbott's SmartCrystal® technology consists of several steps, including combination technology (CT) approaches, H69, H42, and H96 procedures. It incorporates an elevated pressure homogenization (HPH) stage with pretreatment processes such as wet bead grinding, lyophilization, spray drying, as well as precipitation. To aid in precipitation and avoid excessive crystal formation under high pressure, the medication is immersed in a solvent and combined with an aqueous nonsolvent. Particle size is lowered and more stable NCs are formed thanks to the HPH method. Two HPH methods for creating drug NCs are H42 and H96, which combine freeze drying and spray drying. The H42 process produces tiny drug particles and removes organic solvents by using spray drying to perform the precipitation and pretreatment phase. The H96 method requires the manufacture of liquid nitrogen and dissolves pharmaceuticals in chemical solvents with elevated freezing temperatures. The next phase in the HPH process involves homogenizing the freeze-dried product under high pressure while it is diluted in stabilized filtered water. For expensive or thermolabile medications, this method works very well [22]. An investigation was conducted by Hatahet et al. in 2016 on the development of quercetin NCs using smartCrystals® technology to enhance its solubility and bioavailability for dermal applications. Quercetin, a flavonoid with strong antioxidant properties, was prepared using a combination of bead milling and low-pressure high-pressure homogenization (a combination method). The resulting quercetin smartCrystals® were incorporated into nonionic gels and showed a 7.6-fold increase in solubility and dissolution rate compared to crude quercetin. Antioxidant activity was preserved after nanonization, and the quercetin NCs were nontoxic to VERO epithelial cells, even at high concentrations. Also, they provided cellular protection against hydrogen peroxide-induced oxidative damage. This formulation offers a promising, well-tolerated approach for delivering quercetin to the skin [23].

(d) *Other Methods*

(i) *Emulsification-Solvent-Evaporation Technique*

Owing to its simplicity, this approach has gained immense popularity. Its key advantage is that it effectively encapsulates a wide range of lipophilic substances. It takes two processes to evaporate the emulsification solvent. The polymer and medication are dissolved in a volatile chemical solvent in the first step. To reduce residual solvent toxicity issues, ethyl acetate has now mostly replaced dichloromethane and chloroform, which were extensively employed in the past. To emulsify the polymer, there is first an aqueous phase involved. In order to produce nanoparticles, the polymer precipitates on a central core in the second step, when the polymer solvent evaporates [24].

After being collected by ultracentrifugation, the NCs are lyophilized for storage and cleaned with distilled water to get rid of any remaining stabilizer or free medication. The solvent evaporation method was modified recently, and this process is also referred to as high-pressure emulsification. By utilizing high pressure to homogenize the emulsion and then swirling it to remove the organic solvent, this method first prepares the mixture. The size of the particle is greatly influenced by several factors, including temperature, the kind and quantity of the dispersing agent, the pace at which it is stirred, and the viscosity of the organic and aqueous phases. In contrast, the scale-up problem enforces constraints, and this strategy can be used with lipophilic medicines [24]. An experiment conducted by Jaiswal and co. presents the preparation of cyclosporine (CsA)-loaded biodegradable nanoparticles using the high-pressure emulsification-solvent evaporation (HPESE) process to enhance the drug's oral bioavailability and pharmacokinetics. The nanoparticles were formulated with various ratios of poly(DL-lactide-co-glycolide) (DL-PLG) and PEG-modified DL-PLG. The findings suggest that DL-PLG and PEG-DL-PLG nanoparticles could serve as an efficient oral delivery system for CsA, with the method being easy to replicate and scalable for batch production [25].

(ii) *Sonocrystallization*

Sonocrystallization—a term for crystallization induced by ultrasonic waves—was first used in 1927 by Richards and Loomis. Sonocrystallization became more popular in the 1980s due to improvements in ultrasonic equipment, and it is currently employed extensively in the fine chemicals and pharmaceutical industries. Ultrasound power within the 20–100 kHz region is used in sonocrystallization to cause crystallization. In addition to reducing particle growth and agglomeration, sonication increases nucleation rate and enhances micro mixing. A sonocrystallization approach is called "solution atomizing and crystallization by sonication" (SAXS). Utilizing compressed air stress, the solvent gets atomized within the anti-solvent after receiving ultrasonic vibrations from a Sonicator [26]. This technology is very advantageous for heat-sensitive items since its key benefit is that it allows the procedure to be performed at low operating temperature. Also, because the crystalline structure is altered by a shift in solvent composition rather than cooling crystallization, the process demands less energy and control to generate polymorphs [26].

The research conducted by Dhumal and Co. in 2009 explores the use of sonocrystallization to produce fine, elongated crystals of salbutamol sulfate (SS) for improved pulmonary delivery. Sonocrystallization during anti-solvent crystallization resulted in smaller, elongated SS crystals (SCSS) with a narrow particle size distribution, compared to larger, irregular crystals formed without sonication. The study demonstrates that sonocrystallization is an effective method for producing SS crystals tailored for inhalation therapy [26].

(iii) *Supercritical Fluid Process*

By dissolving medications in a supercritical fluid (such as CO2) and atomizing the fluid through a tiny aperture in a nozzle, the supercritical fluid rapidly vaporizes, resulting in the precipitation of NCs through the supercritical fluid (SCF) technique. Supercritical fluids can take on the

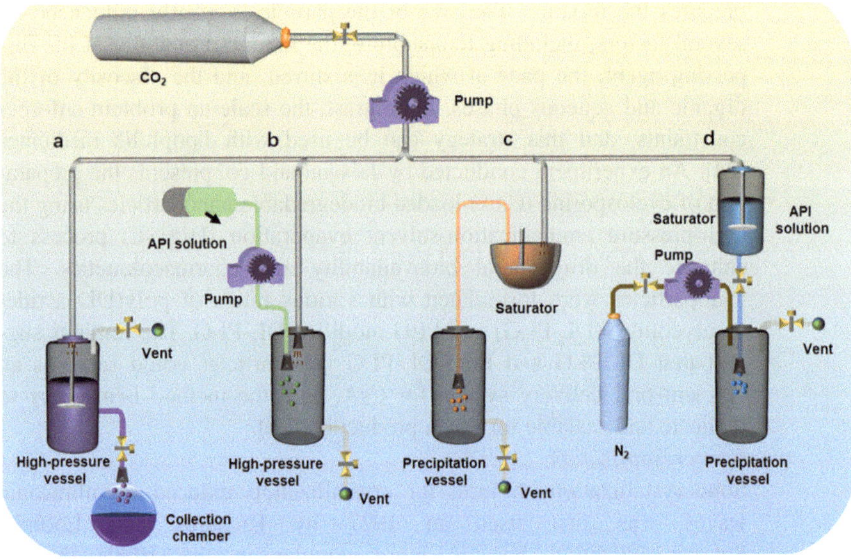

Fig. 2.2 Diagram utilizing SC-CO_2 to illustrate several important particle production processes [27]

characteristics of both a liquid along with a gas because their temperature and pressure are higher than their critical temperatures (Tc) as well as critical pressure (Tp). Because of their considerable compressibility at near-critical temperature, SCFs can significantly modify their solvent power through modest pressure changes that affect the fluid's density and mass transport properties. The medication particles may recrystallize at significantly smaller particle sizes after becoming dissolved in the SCF, which is typically carbon dioxide (Fig. 2.2). Drug particles can be micronized within a limited range of particle sizes, frequently down to submicron levels, because of the flexibility and accuracy provided by SCF techniques.

Sometimes, for processing convenience, little amounts of organic solvents are added to make medications more soluble. These can eventually be eliminated with the use of SCFs. The SCF-based procedures produce smaller-sized parts, leading to higher stability from a colloidal point of view, and entail single-step production, which is extremely challenging by most traditional techniques. These distinctive qualities are attributed to their high solvating strength and quick depressurization [27]. A review conducted by Padrela et al. provides an in-depth analysis of supercritical carbon dioxide (scCO2)-based technologies for producing drug nanoparticles and NCs, addressing the challenge of poor drug solubility and bioavailability. It highlights the advantages of scCO2 processes, which are environmentally friendly, sustainable, and safe for pharmaceutical applications. The article categorizes scCO2-based nanoparticle formation methods into three groups, depending on the role of scCO2: as a solvent, antisolvent/co-antisolvent, or as a "high mobility" additive [28].

(iv) *Drug Dispersion in Carriers*

NCs' drug dispersion is a bottom-up approach based on spray drying that produces solid particles with medication NCs scattered within a small molecule excipient matrix. The nanocrystalline solid dispersion (NSD) has been developed to improve the rate of release and oral bioavailability of medicinal compounds. Drug crystallization is induced by nonpolymeric excipients, which leads to the formation of NCs. By giving a physical barrier to crystal formation, plasticizing, and facilitating primary heterogeneous nucleation, excipients promote the crystallization of drugs. Typically, this technology operates between 40 and 150 °C, is comparatively less expensive, and takes a lot less time to complete. After NSD is dissolved in the dissolving medium, the product made by this technique may show signs of NC aggregation. The right aggregation inhibitors can be used to address this problem. The drug's crystallization characteristics, the kind and amount of an excipient that induces crystallization, as well as process variables, control the formation of NCs. In general, discrete particles measuring between 2 and 50 μm are obtained by spray drying a solution of medicine and additives in a liquid solvent or solvent mixture. Drug NCs ranging in size from 10 to 1000 nm make up each particle [29, 30]. An experiment performed by Peng and Co. in 2020 presents the development of self-dispersing lignin-based drug-loaded materials (SL/CTAB) using sodium lignosulfonate (SL) through self-assembly. These materials were used to create a nanoformulation of avermectin, which self-disperses into nanospheres in water. The nanoformulation demonstrated good physical and chemical stability under both cold and hot storage conditions, with controlled release performance showing cumulative drug release. The formulation also showed increased UV stability, retaining 2.18–2.96 times more avermectin than commercial emulsifiable concentrate (EC) after 50 h of UV exposure. Additionally, the lignin-based nanoformulation exhibited better rainwater scour resistance and higher affinity to plant leaves compared to the EC formulation, indicating its potential for enhanced agricultural applications [31].

(v) *Solvent Evaporation Method*

One of the most popular techniques for creating drug-loaded polymeric systems, or polymeric NCs, for use in pharmaceutical formulations is the solvent evaporation approach. The medication is dissolved into a volatile organic solvent, which is then used to dissolve the polymer, as is customary. Chloroform and dichloromethane are commonly employed, but more frequently in the past. Owing to its toxicity, ethyl acetate has taken their place because it has a superior toxicological profile and is therefore more appropriate for use in biological applications [32]. After achieving high homogeneity, the final solution is introduced to the water phase containing surfactant to create an emulsion. Once a stable emulsion has formed, the chemical solvent is removed either by swirling continuously to cause nanodroplets to disperse or by raising the temperature while applying less pressure. Utilizing an organic solvent to create the core and stabilize it, this technique was also used to create core–shell particles. The resulting aqueous phase was then dispersed using stabilizers and surfactants. The solid NCs can be stored for a long time by freeze-drying them once the solvent had evaporated and then washing and gathering them using centrifugation [33]. An experiment was focused on developing a novel solid lipid nanoparticle (SLN) formulation of docetaxel (DCX) using high-melting-point triglycerides to reduce side effects associated with traditional DCX formulations, particularly those caused by the excipient Tween 80. This new formulation shows potential as an improved, safer alternative for DCX delivery in cancer treatment [34].

(vi) *Melting Solvent/Melt Evaporation/Melt Emulsification Method*
It provides the method for making a solid dispersion by dissolving the medication in an appropriate liquid solvent and adding the mixture straight to the polyethylene glycol melt, which is subsequently evaporated until a transparent, solvent-free film is left. To maintain a steady weight, the film is further dried. Polyethylene glycol 6000 may integrate liquid substances up to a weight percentage of 10% (w/w) without significantly losing its solid qualities. It is possible that the medicine that has been dissolved or the chosen solvent will not melt well with polyethylene glycol. Additionally, the drug's polymorphic form, which precipitates as a solid dispersion, may be impacted using a liquid solvent. This process has the special benefit of both solvent evaporation and fusing. It is restricted to medications with modest therapeutic doses from a practical perspective, for instance, 50 mg [35].

(vii) *Porous Microparticle Technology*
In the 1990s, porous microspheres were discovered to be appropriate materials for medication delivery systems. Typically, active ingredients are either scattered throughout the microparticle matrix or adsorbed on the surface. The pore size of the hollow microspheres affects both the release kinetics and capacity efficiency of medicines. Low density and a large specific area of surface are fundamental characteristics of porous microspheres. Because of these special qualities, they can be used as a vehicle for the delivery of drugs that are gastro- and pulmonary-retentive. The particles have both internal and exterior pores on their surface. Usually, these pores are connected to one another. The following techniques are commonly used to synthesize porous microspheres [36].

Polymers such as methyl methacrylate, or 2-hydroxyethyl methacrylate undergo dissolution in two distinct oil phases using the *polymerization technique*. Membrane emulsification is the method used to prepare homogeneous seed droplets from oil with a higher hydrophobicity. The smaller secondary emulsified droplet forms when another oil phase gets homogenized in water. A homogeneous swell is formed when the hydrophobic seed droplets adsorb the secondary drops of emulsion that diffuse into the water phase during the gentle agitation of these two emulsions. A consistent, porous microsphere is created through the suspension polymerization of swelling droplets [36].

The ***sintering technique*** involves combining an aqueous solution of a hydrophilic polymer, such as chitosan, with a fine power of a solid, such as hydroxyapatite. Oil phase emulsification is applied to this slurry. Glutaraldehyde and other cross-linking chemicals are added while the sphere is being continuously stirred to harden it. To remove any remaining oil, the settled granules are separated by filtering and then cleaned using nonpolar solvents such as petroleum ether and acetone. Heat level as high as 1100 °C is used for sintering after the particles undergo heating to burn off the uncross-linked polymer. Lastly, a thorough water wash and drying are applied to the particles [36].

In the ***phase separation procedure***, a small quantity of solvent, such as dichloromethane, is used to dissolve the medicine to be encapsulated and hydrophobic polymer, such as PLGA. Drop by drop, the resulting solution is added while continuous sonication is applied to a dispersion medium, such as a mineral oil and DCM mixture. To regulate how quickly DCM evaporates, the sonication bath must be regulated at a temperature of not more than 8–10 °C. The system is constantly agitated until all of the DCM has been removed after the polymer phase has been added. At last, to obtain free-flowing power, the microparticles are dried and cleaned using a nonpolar solvent such as petroleum ether [36].

(viii) *Spray Freezing into Liquid (SFL)*

The nozzle in SFL is placed straight into a cold liquid, just below the surface. The liquid can be employed at atmospheric pressure and can be a cryogenic liquid, such as liquid nitrogen, argon, hydrofluoroether, or pentane. An alternative would be to employ a pressurized system, where liquid CO2, propane, or ethane might be utilized. As soon as the droplets appear, they instantly start to freeze. To prevent the particles from clumping together, an impeller can be placed inside the jar and used to agitate the cryogen. After being frozen, the particles are lyophilized to produce powders that are dry and fluid. SFL operates based on liquid interaction between the cryogenic liquid and the pressured feed solution that exits the nozzle. Drug delivery benefits from the application of surface functional lipid (SFL) technology, which yields large surface area nanoparticles with little loss of active ingredients. Very narrow sub-micron domains are produced by SFL, which causes fast nucleation speeds for dissolved compounds [37]. The study conducted by Gol and co. explores the development of NCs-based formulation for risperidone

(RIS) to address its poor solubility. Using lyophilization, the RIS nanosuspension was converted into a stable solid form. The formulation was stabilized with polycaprolactone and Pluronic F-68, and process parameters such as surfactant ratios and stirring speed were optimized. The resulting nanosuspension exhibited a stable particle size of 214 nm and showed a significant improvement in drug release, with 90% released within 10 h, compared to only 11% for the unprocessed drug. The formulation shows promise for enhancing the solubility and therapeutic efficacy of risperidone in treating psychotic disorders [38].

(ix) *Spray Freezing into Vapor*

It entails the atomized droplet coming into contact with the chilly desiccated gas, which causes the droplet to freeze. The intricate process of the spray freezing onto vapor (SFV) approach includes (a) individual creation and movement in relation to other droplets and gas, which is dictated by the spray's hydrodynamics. (b) The transfer of heat from the gas to the droplet, dependent on local parameters including the temperature of the droplet and the gas, the velocity of the droplet's gas slip, and (c) the freezing and crystallization of the droplet. The primary variables that significantly affect the internal structure of the freeze-dried droplet are the rates of nucleation and freezing [37]. Pilcer et al. [39] developed porous nanocrystalline agglomerates of tobramycin embedded in a matrix of amorphous clarithromycin using high-pressure homogenization followed by spray drying. These agglomerates exhibited excellent dispersibility and efficient dissolution in the lungs. The simultaneous local deposition of both drugs could effectively target lung infections, potentially reducing the side effects associated with systemic antibiotic administration. This formulation is promising for localized lung infection therapy [39].

(x) *Spray Freezing into Vapor over Liquid (SFV/L)*

This is the most commonly used method for spray freezing in pharmaceutical applications that involves atomizing a feed solution into a cryogenic vapor above a liquid. The feed solution that is atomized via the nozzle in this process starts to freeze and harden after passing through a vapor gap, and the atomizing spray is placed directly over the boiling cryogenic liquid. Additionally, a modified technique was created in which four nozzles were used to direct spray as well as liquid nitrogen into contact with one another. After being separated, the frozen granules in the solution of nitrogen were then lyophilized using a standard lyophilization procedure to remove the solvent [37]. In research performed by Otake and Co., powders containing hydrophobic amino acids like L-alanine, glycine, L-isoleucine, L-leucine, L-phenylalanine, L-tryptophan, or L-valine obtained by SFV/L were prepared and evaluated for morphology, particle size, crystallinity, and inhalation characteristics. The results demonstrated that these powders' particle diameters ranged between 5 and 10 μm, and they retained their inhalation performance and resisted moisture absorption even in humid conditions, indicating the effectiveness of hydrophobic amino acids in improving the stability of spray freeze-dried (SFD) powders [40].

(xi) *Ultra Rapid Freezing*

Thermal conductivity of 10–20 mK is achieved by using cryogenic solid substrates in the URF process. This method involves applying and then further freezing the drug and carrier solution mixture to the cryogenic substrate. In order to remove any remaining solvent, the frozen crystals are detached from their substrate and then lyophilized once again. Since the crystallized API molecule cannot nucleate due to the extremely quick cooling process, the resulting solid materials are amorphous in character [41]. In this study performed by Overhoff and Co. in 2008, the solid dispersions of tacrolimus (TAC) with various stabilizers were prepared using the ultra-rapid freezing (URF) process to evaluate their ability to enhance supersaturation in aqueous media and improve transport across biological membranes. The stabilizers tested included poly(vinyl alcohol) (PVA), poloxamer 407 (P407), and sodium dodecyl sulfate (SDS). The results showed that URF formulations, particularly those with P407 and SDS, outperformed the commercial product Prograf (PRO) in terms of supersaturation and absorption enhancement. URF-P407 exhibited the highest absorption with a 1.5-fold increase in AUC compared to PRO, and all URF compositions achieved faster absorption rates (shorter Tmax). These findings suggest that nanostructured powders created through the URF process can significantly improve oral absorption of TAC by enhancing supersaturation [42].

2.4 Advanced Characterization of NCs with the Assisted Instruments

Characterization is an essential operation to govern the subsequent use of the NC products when they are obtained. The objective of characterization is to comprehend the characteristics of NCs, including its size, shape, and crystalline form [43]. Advanced characterization techniques for NCs have evolved significantly in recent years, allowing for more precise and comprehensive analysis of their properties, structure, and behavior. These techniques are essential for understanding the performance of NCs in various applications, including drug delivery, catalysis, and energy storage. A wide range of sophisticated tools, each providing a distinct perspective on the physical and chemical properties of NCs, are essential to the advanced characterization of these materials [44]. Below are some of the latest and most advanced characterization methods used for NCs' preparations.

2.4.1 High-Resolution Transmission Electron Microscopy (HR-TEM)

High-resolution transmission electron microscopy (HRTEM) is an imaging technique within transmission electron microscopy (TEM) that enables direct visualization of the atomic structure of biomaterial samples. HR-TEM provides atomic-level imaging by passing electrons through a very thin sample. It offers high spatial resolution, making it ideal for observing the size, shape, crystallinity, and defects of NCs [45]. It achieves this by using both transmitted and scattered beams to form an interference image. A conventional TEM typically offers a resolution better than 0.2 nm, making it highly effective for resolving atomic-scale structures in nanoparticles. Currently, the highest point resolution achieved in HRTEM is approximately 0.5 Ångströms (0.050 nm), allowing for the clear visualization of individual atoms and defects within a crystal [46]. HR-TEM now includes advanced techniques like in situ TEM, which allows real-time observation of NCs growth, phase transitions, and reactions under different environments (e.g., temperature, pressure) [47].

2.4.2 Scanning Transmission Electron Microscopy (STEM) with Energy-Dispersive X-Ray Spectroscopy (EDS)

The principle behind the STEM technique is that it allows high-resolution imaging similar to TEM but focuses the electron beam to a fine point and scans across the sample. Features from both scanning electron microscopy (SEM) and transmission electron microscopy (TEM) are combined in scanning transmission electron microscopy (STEM). Unlike traditional TEM, which captures the full image at once, STEM creates a transmission image by scanning a focused electron beam across the sample. By adding a scanning device that shifts the focused beam across the material to create the image, a TEM can be converted into a STEM. This method provides improved flexibility in the sample examination process and enables high-resolution imaging and elemental analysis [48].

Energy-dispersive X-ray spectroscopy (EDXS), sometimes referred to as EDX or EDS, is an analytical method for determining a sample's elemental composition. When coupled with EDS, it enables elemental mapping and composition analysis at the nanoscale. A concentrated electron beam causes a material to release distinctive X-rays that are related to the elements found in the sample. EDXS offers both quantitative and qualitative information regarding the elemental composition by evaluating these X-rays. It is widely used to evaluate the distribution and concentration of components in a sample at high spatial resolution in materials science, nanotechnology, and biological research. Additionally, STEM-EDS is used to assess the homogeneity and elemental distribution within NCs, which is important for characterizing drug formulations with mixed components [49].

2.4.3 Synchrotron XRD (SXRD)

XRD is a common technique used to determine the crystalline structure, phase purity, and size of NCs by measuring the diffraction pattern of X-rays passing through a sample. Recent advances have developed a technique called synchrotron X-ray diffraction (XRD) that offers high-brilliance, highly collimated X-ray beams, allowing for much more detailed structural analysis. SXRD differs from conventional XRD primarily in its X-ray source and capabilities. Higher resolution and sensitivity are achieved with SXRD than with lower-intensity X-rays from traditional X-ray tubes because it employs strong, collimated, and adjustable synchrotron radiation. Furthermore, SXRD provides tunable wavelengths and quick data collection, enabling sophisticated experimental methods like high-pressure tests and in situ investigations. SXRD is an effective instrument for characterizing NCs. Utilizing the powerful and highly collimated X-ray beams produced by synchrotron radiation facilities, these techniques can offer comprehensive insights into the chemical, physical, and structural characteristics of nanocrystalline materials. Because SXRD has such high sensitivity and resolution, it is perfect for investigating minuscule amounts of NCs [50].

2.4.4 Atomic Force Microscopy (AFM) and Colloidal Probe Microscopy (CPM)

Atomic force microscopy (AFM) enables high-resolution surface topography imaging at the nanometer scale by quantifying the interaction forces between a sharp tip and the sample surface. Colloidal probe microscopy (CPM), an advanced variant of AFM, serves as a powerful method for investigating the surface characteristics of NCs and their interactions with other materials at the nanoscale. In CPM, a colloidal probe—typically a particle or sphere—is affixed to the end of an AFM cantilever, allowing for the measurement of interaction forces between colloidal particles and surfaces. This technique is particularly useful for studying stability and force interactions. CPM facilitates real-time imaging of dynamic processes occurring on NCs' surfaces, while force spectroscopy provides insights into mechanical properties such as stiffness and adhesion forces at the nanoscale. The applications of CPM are especially valuable in pharmaceutical research, where it aids in the evaluation of surface characteristics, mechanical properties, and stability of NCs' formulations. Furthermore, CPM is instrumental in screening nanosuspension stabilizers [51].

2.4.5 Small-Angle X-Ray Scattering (SAXS) and Grazing-Incidence SAXS (GISAXS)

SAXS and GISAXS are powerful techniques for characterizing NCs and nanoscale materials. SAXS operates on the principle that small-angle scattering occurs due to the size and distribution of inhomogeneities within the sample. The scattering intensity at low angles (small q values) is related to the overall shape and size of the scattering entities (e.g., NCs), while higher angles provide information about their internal structure and arrangement. SAXS utilizes a monochromatic X-ray beam, typically from a synchrotron source, to probe the sample at small angles, allowing for the analysis of size, shape, and distribution of nanoparticles based on the intensity of scattered X-rays. A two-dimensional detector captures the intensity of the scattered X-rays. The intensity of the scattered X-rays is analyzed to obtain information about the size, shape, and distribution of the NCs. The resulting scattering profile can be interpreted using various mathematical models to extract structural parameters [52].

In contrast, GISAXS is based on the principle of enhanced surface sensitivity achieved by the grazing incidence of the X-ray beam. This technique allows for the characterization of surface structures, such as layer thickness, interparticle spacing, and ordering of NCs in thin films, as well as their three-dimensional arrangement. By analyzing the scattering patterns in both techniques, researchers can obtain valuable insights into the morphological and structural properties of NCs, making SAXS and GISAXS essential tools in nanotechnology and materials science research [53].

2.4.6 X-Ray Photoelectron Spectroscopy (XPS)

XPS analyzes the surface chemistry of NCs by measuring the energy of electrons emitted from a material when irradiated with X-rays, providing elemental composition and chemical state information. The technique operates on the principle of photoemission, where a monochromatic X-ray beam is directed onto the sample surface, exciting core electrons and causing their ejection. The kinetic energy of the emitted electrons is then measured, allowing for the calculation of their binding energy, which is unique to each element. This enables the identification of elemental composition by analyzing the resulting spectrum, which displays peaks corresponding to different elements present in the sample [54].

In recent years, high-resolution XPS and angle-resolved XPS (ARXPS) were developed that offer more detailed surface chemistry analysis and depth profiling, respectively, allowing for precise analysis of thin films and NCs surfaces. XPS technique provides crucial information for understanding surface modifications, coating stability, and the interaction between NCs and excipients or stabilizers, especially in drug formulations where surface properties influence dissolution rates and bioavailability [55].

2.4.7 Nuclear Magnetic Resonance (NMR) Spectroscopy

A potent analytical method for examining the molecular structure, dynamics, and interactions of NCs is nuclear magnetic resonance (NMR) spectroscopy. The foundation of nuclear magnetic resonance (NMR) lies on the magnetic characteristics of specific atomic nuclei, primarily carbon (^{13}C) and hydrogen (^{1}H), which have a nuclear spin. These nuclei resonate at distinct frequencies related to their respective chemical surroundings when exposed to a strong magnetic field. They emit signals that are detected and converted into an NMR spectrum. The resulting spectrum provides valuable information about the local chemical environment of the nuclei, including information about molecular structure, functional groups, and molecular dynamics [56].

For NCs, NMR can reveal insights into the interactions between the NCs' surface and surrounding molecules, such as solvents or stabilizers, which can significantly influence the properties and behavior of NCs in various applications. Moreover, NMR can be employed to study the motion of atoms within NCs, providing insights into their stability and reactivity. Recent advances present a modified NMR, that is, solid-state NMR and dynamic nuclear polarization (DNP)-enhanced NMR that offers improved resolution and sensitivity, allowing for the study of solid NCs and their interactions with stabilizers or excipients [57].

2.4.8 Cryo-Electron Microscopy (Cryo-EM)

Using transformational imaging, cryo-electron microscopy (cryo-EM) preserves the native structures of NCs in a near-liquid condition while enabling analysis at near-atomic resolution. The quick vitrification of samples—which entails submerging them in liquid ethane or another cryogen to create an amorphous ice layer—is the foundation of the cryo-EM concept. By stopping the production of ice crystals, this procedure preserves the sample's structural integrity and makes it possible to observe inorganic or biological components in their native state. A concentrated electron beam is focused on the sample in cryo-electron microscopy (cryo-EM), and the transmitted electrons are gathered to produce high-resolution pictures. The capacity of cryo-EM to record several projections of an identical NCs from various angles makes it unique [58]. These projections can subsequently be computationally reassembled to provide a three-dimensional (3D) model. The morphology, internal structures, and surfaces of NCs can be visualized in depth as a result, offering insights into their size, shape, and crystallinity. This technique is also very useful for understanding the surface defects, which is crucial for controlling drug release and bioavailability. Recent advances in this technique are single-particle cryo-EM and cryo-electron tomography, which has enabled atomic-resolution imaging of complex nanostructures without requiring crystallization, a significant advancement for soft and biological NCs [59].

2.4.9 Fourier Transform Infrared Spectroscopy (FTIR) and Raman Spectroscopy

Formulation of NCs is often accepted by virtue of the compatibility of the chemical properties of the different materials used in the formulation. FTIR spectrometry being a real-time, nondestructive analytical technique is used for the measurement of the chemical nature of the sample [60]. The utilization of FTIR is crucial in the characterization and formulation of NCs as it offers valuable insights into their molecular structure, chemical composition, and internal interactions. It assists in making it possible to identify unknown components from samples of solid, liquid, or gas, as well as determine their concentration from samples of organic and inorganic chemicals (qualitative determination) [61]. For instance, the absence of drug-excipient interactions was confirmed for fenofibrate NCs by FTIR analysis, which showed no change in the functional groups between pure fenofibrate and the NC formulation. To sum up, FTIR is a flexible instrument that can be used to investigate many facets of NC formulation. It offers vital information for refining and comprehending the characteristics of NCs for a range of purposes [62].

Surface-enhanced Raman spectroscopy (SERS) is a highly sensitive analytical technique used to study the vibrational modes of molecules adsorbed on or near metallic nanostructures, particularly NCs [63]. The mechanism behind SERS involves two primary phenomena: electromagnetic enhancement and chemical enhancement. Electromagnetic enhancement occurs due to localized surface plasmon resonance (LSPR), where incident light induces oscillations of free electrons on metallic nanoparticles, significantly increasing the electromagnetic field near their surfaces. This results in a dramatic enhancement of the Raman signal from nearby molecules. SERS allows researchers to investigate surface chemistry, functionalization, and molecular interactions of NCs with high sensitivity, making it valuable for applications in biosensing, environmental monitoring, and pharmaceutical analysis [64].

2.4.10 Solubility and Dissolution Studies of NCs

Solubility issue with nanoformulations has been a headache to the researchers and scientists from the beginning of this technology. Solubility studies of NCs help in understanding and optimizing their efficiency performance, especially in pharmaceutical and material science applications. The determination of solubility for NC is DLS, which is also used for the measurement of particle sizes. Alongside this, other instruments used to determine the solubility include UV-Vis spectrophotometry, equilibrium dialysis, and flask method. Using the UV-Vis spectrophotometer, measure the absorbance of known concentrations of the NC medication that have been dissolved in the same solvent to create a calibration curve. To build the curve, plot absorbance against concentration.

Based on the sample solution's absorbance, use the calibration curve to determine the concentration of dissolved NCs in it. However, due to the fact that it is a relatively novel delivery mechanism, and the abovementioned approaches have not been validated for use in solubility studies of NCs, their application is profoundly limited [65]. Choi and co. have illustrated how well particle size reduction works by experimenting on the cilostazol NCs, which are made without surfactants. It showed a 10.7-fold increase in solubility over a range of pH 1.2 and pH 6.8 buffers, respectively, for 24 h compared to the pure medication [66, 67].

The pharmaceutical industry has undergone a revolution because of advancements in drug NC technologies, which have improved the bioavailability and dissolution rates of drugs classified as class II and IV in the Biopharmaceutics Classification System (BCS) and solved the problem of poor aqueous solubility in drug development [68, 69]. Regarding dissolution testing, the US Pharmacopeia has recognized seven different dissolution equipment (1–7): rotating basket, paddle assembly, reciprocating cylinder, flow-through cell, paddle over disk, cylinder, and reciprocating holder, in that order. The device is utilized in accordance with several types of drugs [70].

The mechanism associated with the enhancement of the dissolution of NC formulation is discussed in detail in the following context. The basic mechanism of dissolution study is the reduction of the particle size, thereby increasing the surface area. Hence, the insight mechanism is discussed in three points: saturated solubility, dissolution pressure, and dissolution rate. First is the saturated solubility. The mass at which a given solid substance can dissolve completely in each solvent at a given temperature is known as saturation solubility. According to the Ostwald–Freundlich equation, $\text{Log}\ (\text{Cs}/C) = 2\sigma V/(2.303RT\rho r)$, where V is the molar volume of the particle material, R is the gas constant, T is the absolute temperature, ρ is the density of the solid, r is the radius, Cs is the saturation solubility, C is the solubility of the solid made up of large particles, and σ is the substance's interfacial tension; the decrease in the particle size causes increase in the saturated solubility of a drug [71]. Second is the dissolution pressure that gets increased on effective increase in surface area due to the reduction in particle size of NC. Initially, the vapor pressure on the surface of droplets in the gas was explained by Kelvin's equation. According to this equation, the vapor pressure is equal to the dissolving pressure and surface curvature, which are both increased by the decrease in droplet size [72]. Finally, the third point is the dissolution rate of NCs was increased according to the Noyes–Whitney law: $dC/dT = kS\ (\text{Cs} - \text{Ct})$, where k is the dissolution rate constant, S is the surface area, Cs is the saturation solubility, Ct is the concentration in the bulk fluid, and (dC/dT) is the change in concentration over time. Particle dissolution rate rises when particle size decreases because more surface area is available for interaction with the dissolving solvent. As demonstrated by Mauludin and co. in 2009, compared to commercially available tablets and tablets filled with rutin microcrystals, the rutin NC-laden tablets dissolved more quickly. Furthermore, the enhanced solubility of rutin NC-loaded tablets could end up in increased rutin's bioavailability in vivo [73]. The solubility behavior of NCs differs from that of bulk materials due to their small size and huge surface area. NCs increase the dissolution rates and surface area simultaneously, thereby significantly improving the formulation's solubility [74].

2.4.11 Stability Study of NCs

A new era of nanomedicine and pharmaceutical products fortune has been brought about by nanotechnology in all sectors, including physics, chemistry, and biological sciences. NCs being one of the renowned parts of nanotechnology-based delivery

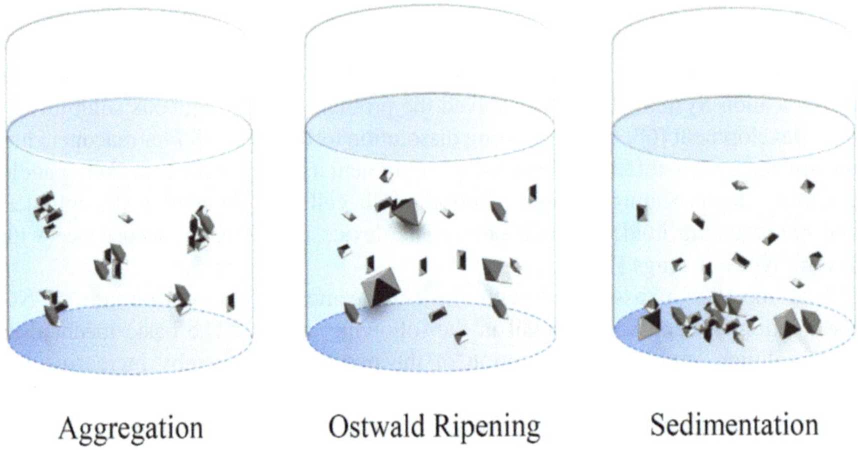

Fig. 2.3 Types of instability caused in the NC preparation

system suffers a great setback due to the instability under different conditions. The causes of instability are many, with each having a different mechanism and significance. The causes of instability include small nanoscale particle size, large surface area, and greater interaction with the environment. Due to the small particle size, the surface energy becomes high. In order to reduce the surface energy, the particle's size starts increasing by attracting nearby particles, leading to the thermodynamic instability. The Ostwald ripening, aggregation, and sedimentation are the result of thermodynamic instability (Fig. 2.3) [75, 76].

Ostwald ripening is the phenomenon of instability of NC due to the rise in the particle size because of the differences in the solubility of different types of substances in the formulation. Small-sized particles have a high surface free energy and thereby higher saturation solubility compared to the larger crystal. Therefore, the diffused mass exchange takes place as the large and small size crystals interact, leading to the growth of large crystals and shrinkage of the small crystals [77].

Aggregation is another form of instability of the NCs. The principal mechanism by which the aggregation occurs is the reduced particle size, causing the high surface-to-volume ratio. Due to the increased surface-to-volume ratio, the Gibbs free energy is increased, facilitating the collision and adhesion between the NCs, thereby causing the enhanced Brownian movement between individual NCs. It is interesting to know that the enhanced particle adhesiveness causes the agglomeration of the NCs, thereby reducing the redundant Gibbs free energy. Therefore, this process severely affects the stability and efficacy of the NCs [78].

Sedimentation of the nanoformulation is the well-known yet problematic cause of the instability of the NC formulation having suspension type of dosage form. It is divided into two basic categories, that is, flocculated and deflocculated nanosized suspension. The flocculated type has a fast, rapid sedimentation rate but loses cake formation that can be redispersed easily, whereas the deflocculated type follows slow sedimentation rate and forms a hard cake, which is difficult to redisperse. The rate of setting velocity can be obtained using Stoke's law. Therefore, it is very important to use the stabilizers to inhibit the deposition of NCs and improve the stability, quality, and efficacy of the formulation [79].

Commonly incorporated with the active ingredients, stabilizers with a predominance of polymers and surfactants help increase the stability of the formulation through steric hindrance or physical repulsion. To identify the best-fit category and dose of stabilizers for a desired NC simulation, numerous cycles of phenomenon-based experiments must be conducted because the interaction between the stabilizers chosen is unpredictable and no uniform standards for selecting appropriate stabilizers have been established thus far. To put it briefly, the ideal molecular weight, amphiphilicity, and an adequate hydrophobic abutting surface are crucial requirements for the first screening of stabilizers [80]. Table 2.1 shows the types of stabilizers used in NCs with their special properties.

2.4.12 Estimation of Drug Content in NC Formulation Using UV-Vis Spectrophotometer

The amount of active pharmaceutical ingredient (API) in the NC formulation is referred to as the drug content of the NCs. For medical products to be consistent, safe, and effective, the medication content must be determined accurately. The drug content is checked using the UV-Vis spectrophotometer, which helps determine the purity and efficacy of the drug upon administration [93]. To determine the drug content of NC formulation, it is dissolved in a suitable solvent (mostly distilled water at 2.5 mg/mL concentration) and filtered through 0.8 μm filter. The filtrate is then dissolved with 4% SLS solution and thereafter subjected to the UV-Vis spectrophotometry to determine the drug content of the NC formulation [94]. To achieve the intended therapeutic effects and optimize the formulation of NCs, it is imperative to comprehend their pharmacological content.

2.4.13 In Vitro Drug Release of NC

In vitro dissolution testing has drawn more attention as a stand-in test for product performance during the drug product development process. It is an effort to potentially reduce manufacturing costs, labor costs, and the need for in vivo assessments of drug release kinetics [95]. The most widely used techniques for in vitro drug release study in case of NCs are dialysis bag diffusion/dialysis sac technique, Franz diffusion cell, or USP type II (paddle) and IV (flow through cell) dissolution apparatus. The sample (5 mL) to be analyzed should be kept in the dialysis bag and sealed and kept in the release media (250 mL of 0.1 N HCl). The temperature to be maintained is the room temperature. Then, it is subjected to the magnetic stirrer at 500 rpm [96]. The sampling can be done at predetermined time interval and subject to HPLC quantitative analysis for the development of the dissolution curve. To

Table 2.1 Compilation of various frequently used categories of stabilizers with their special properties

Category	Types	Examples	Mechanism	References
Surfactants	Ionic	Sodium deoxycholate	Electrostatic repulsion	[81]
		SDS	Electrostatic repulsion	[82]
		SLS	Electrostatic repulsion	[83]
	Nonionic	Poloxamer 188/407/2000	Steric hindrance	[84]
		PNS	Steric hindrance	[85]
		TPGS	Steric hindrance	[86]
		TW80	Steric hindrance	
	Amphoteric	Lecithin	Steric hindrance	[87]
Polymers	Charged	Carbopol	Electrostatic repulsion and steric hindrance	[83]
		PAH	Electrostatic repulsion and steric hindrance	[86]
		PEI	Electrostatic repulsion and steric hindrance	[88]
	Uncharged	HPMC	Steric hindrance	[80]
		Tragacanth	Steric hindrance	[18]
		PVP	Steric hindrance	[89]
	Co-polymer	POVACOAT Type F	Electrostatic repulsion and steric hindrance	[89]
		Pl-g-CH	Electrostatic repulsion and steric hindrance	[90]
Others	Proteins	Transferrin	Steric hindrance	[91]
		SPI, WPI	Electrostatic repulsion and steric hindrance	[92]

SDS sodium dodecyl sulfate, *SLS* sodium lauryl sulfate, *PNS* panax notoginseng saponins, *TPGS* D-α-tocopheryl polyethylene glycol 1000, *TW80* Tween 80, *PAH* poly (allylamine hydrochloride), *PEI* polyethylenimine, *HPMC* hydroxypropyl methylcellulose, *PVP* polyvinyl pyrrolidone, *POVACOAT Type F* polyvinyl alcohol/acrylic acid/methyl methacrylate copolymer, *Pl-g-CH* pluronic-grafted chitosan, *SPI* soybean protein isolate, *WPI* whey protein isolate

comprehend the release mechanism, release data are fitted to a variety of kinetic models (such as zero-order, first-order, Higuchi, or Korsmeyer–Peppas models). When comparing NC medication release to conventional formulations, one can assess whether bioavailability or controlled release characteristics are improved or remain unaltered [97].

The advanced characterization techniques discussed in this chapter provide a comprehensive understanding of NCs, encompassing their size, shape, crystallinity, surface chemistry, and interactions with stabilizers or excipients. Such insights are vital for optimizing NC formulations, particularly in pharmaceutical applications where stability, dissolution rate, and bioavailability are critical factors. The integration of traditional methods with more sophisticated approaches has enabled precise control and engineering of NCs for a diverse array of applications.

2.5 Conclusion and Future Perspectives of NCs in Human Health

In conclusion, this chapter has explored the formulation-based composites in nanocrystal preparation and the advanced characterization techniques that are pivotal for understanding and optimizing nanocrystals for various applications. We have discussed the distinct properties of nanocrystals, diverse preparation methods—including top-down, bottom-up, and combination techniques—and advanced characterization instruments that facilitate a deeper understanding of nanocrystals at the molecular level.

With advancements in precision medicine and genomics, NCs hold significant promise in developing personalized medication compositions tailored to an individual's genetic makeup. Personalized drug delivery based on specific patient biomarkers could revolutionize healthcare by offering more effective therapies. The combination of NCs with other nanotechnologies, such as micelles, polymeric nanoparticles, or liposomes, can create hybrid drug delivery systems that leverage the strengths of multiple technologies. In light of the discussions presented, it is evident that the use of NCs in drug delivery is poised for remarkable growth in the near future. Achieving broad access to fully leverage NCs in healthcare will require further research, enhanced regulations, and collaborative efforts within the industry. The collaborative efforts among researchers, industry stakeholders, and regulatory bodies will be essential for unlocking the full potential of nanocrystals in therapeutic applications, ultimately leading to enhanced patient outcomes and improved healthcare solutions.

References

1. Malik S, Muhammad K, Waheed Y (2023) Nanotechnology: a revolution in modern industry. Molecules 28(2):661. https://doi.org/10.3390/molecules28020661
2. Bayda S, Adeel M, Tuccinardi T, Cordani M, Rizzolio F (2019) The history of nanoscience and nanotechnology: from chemical-physical applications to nanomedicine. Molecules 25(1):112. https://doi.org/10.3390/molecules25010112
3. Montanarella F, Kovalenko MV (2022) Three millennia of nanocrystals. ACS Nano 16(4): 5085–5102. https://doi.org/10.1021/acsnano.1c11159
4. Boles MA, Ling D, Hyeon T, Talapin DV (2016) The surface science of nanocrystals. Nat Mater 15(2):141–153. https://doi.org/10.1038/nmat4526
5. Patel A, Patel K, Patel V, Rajput MS, Patel R, Rajput A (2024) Nanocrystals: an emerging paradigm for cancer therapeutics. Futur J Pharm Sci 10(1):4. https://doi.org/10.1186/s43094-024-00579-4
6. Junyaprasert VB, Morakul B (2015) Nanocrystals for enhancement of oral bioavailability of poorly water-soluble drugs. Asian J Pharm Sci 10(1):13–23. https://doi.org/10.1016/j.ajps.2014.08.005
7. Junghanns JUAH, Müller RH (2008) Nanocrystal technology, drug delivery and clinical applications. Int J Nanomedicine 3(3):295–310
8. Jarvis M, Krishnan V, Mitragotri S (2019) Nanocrystals: a perspective on translational research and clinical studies. Bioeng Transl Med 4(1):5–16. https://doi.org/10.1002/btm2.10122

9. Mohammad IS, Hu H, Yin L, He W (2019) Drug nanocrystals: fabrication methods and promising therapeutic applications. Int J Pharm 562:187–202. https://doi.org/10.1016/j.ijpharm.2019.02.045
10. Parmar PK, Wadhawan J, Bansal AK (2021) Pharmaceutical nanocrystals: a promising approach for improved topical drug delivery. Drug Discov Today 26(10):2329–2349. https://doi.org/10.1016/j.drudis.2021.07.010
11. Ghasemiyeh P, Mohammadi-Samani S (2020) Potential of nanoparticles as permeation enhancers and targeted delivery options for skin: advantages and disadvantages. Drug Des Devel Ther 14:3271–3289. https://doi.org/10.2147/DDDT.S264648
12. Zhou Y, Du J, Lulu W, Yancai W (2017) Nanocrystals technology for improving bioavailability of poorly soluble drugs: a mini-review. J Nanosci Nanotechnol 17(1):18–28. https://doi.org/10.1166/jnn.2017.13108
13. Dhaval M, Makwana J, Sakariya E, Dudhat K (2020) Drug nanocrystals: a comprehensive review with current regulatory guidelines. Curr Drug Deliv 17(6):470–482. https://doi.org/10.2174/1567201817666200512104833
14. Zhang X et al (2020a) Preparation of nanocrystals for insoluble drugs by top-down nanotechnology with improved solubility and bioavailability. Molecules 25(5):1080. https://doi.org/10.3390/molecules25051080
15. Gujar K, Wairkar S (2020) Nanocrystal technology for improving therapeutic efficacy of flavonoids. Phytomedicine 71:153240. https://doi.org/10.1016/j.phymed.2020.153240
16. Chary PS, Shaikh S, Bhavana V, Rajana N, Vasave R, Mehra NK (2024) Emerging role of nanocrystals in pharmaceutical applications: a review of regulatory aspects and drug development process. Appl Mater Today 40:102334. https://doi.org/10.1016/j.apmt.2024.102334
17. Sharif S, Muneer S, Wang T, Izake EL, Islam N (2023) Robust wet milling technique for producing micronized ibuprofen particles with improved solubility and dissolution. AAPS PharmSciTech 24(1):16. https://doi.org/10.1208/s12249-022-02480-w
18. Zhang H, Meng Y, Wang X, Dai W, Wang X, Zhang Q (2014) Pharmaceutical and pharmacokinetic characteristics of different types of fenofibrate nanocrystals prepared by different bottom-up approaches. Drug Deliv 21(8):588–594. https://doi.org/10.3109/10717544.2013.865815
19. Ng SW et al (2022) Pickering emulsion stabilized by palm-pressed fiber cellulose nanocrystal extracted by acid hydrolysis-assisted high pressure homogenization. PLoS One 17(8):e0271512. https://doi.org/10.1371/journal.pone.0271512
20. Park YB et al (2011) Synthesis and characterization of nanocrystalline calcium sulfate for use in osseous regeneration. Biomed Mater 6(5):055007. https://doi.org/10.1088/1748-6041/6/5/055007
21. Alcalá-Alcalá S et al (2023) Melanin nanoparticles obtained from preformed recombinant melanin by bottom-up and top-down approaches. Polymers (Basel) 15(10):2381. https://doi.org/10.3390/polym15102381
22. Ding Y et al (2024) Recent developments in the use of nanocrystals to improve bioavailability of APIs. Wiley Interdiscip Rev Nanomed Nanobiotechnol 16:e1958. https://doi.org/10.1002/wnan.1958
23. Hatahet T, Morille M, Hommoss A, Dorandeu C, Müller RH, Bégu S (2016) Dermal quercetin smartCrystals®: formulation development, antioxidant activity and cellular safety. Eur J Pharm Biopharm 102:51–63. https://doi.org/10.1016/j.ejpb.2016.03.004
24. Chaudhary SA, Patel DM, Patel JK, Patel DH (2021) Solvent emulsification evaporation and solvent emulsification diffusion techniques for nanoparticles. In: Emerging technologies for nanoparticle manufacturing. Springer International Publishing, Cham, pp 287–300. https://doi.org/10.1007/978-3-030-50703-9_12
25. Jaiswal J, Gupta SK, Kreuter J (2004) Preparation of biodegradable cyclosporine nanoparticles by high-pressure emulsification-solvent evaporation process. J Control Release 96(1):169–178. https://doi.org/10.1016/j.jconrel.2004.01.017

26. Dhumal RS, Biradar SV, Paradkar AR, York P (2009) Particle engineering using sonocrystallization: salbutamol sulphate for pulmonary delivery. Int J Pharm 368(1–2): 129–137. https://doi.org/10.1016/j.ijpharm.2008.10.006
27. Kankala RK, Xu PY, Chen BQ, Bin Wang S, Chen AZ (2021) Supercritical fluid (SCF)-assisted fabrication of carrier-free drugs: an eco-friendly welcome to active pharmaceutical ingredients (APIs). Adv Drug Deliv Rev 176:113846. https://doi.org/10.1016/j.addr.2021.113846
28. Padrela L, Rodrigues MA, Duarte A, Dias AMA, Braga MEM, de Sousa HC (2018) Supercritical carbon dioxide-based technologies for the production of drug nanoparticles/nanocrystals – a comprehensive review. Adv Drug Deliv Rev 131:22–78. https://doi.org/10.1016/j.addr.2018.07.010
29. Malamatari M, Charisi A, Malamataris S, Kachrimanis K, Nikolakakis I (2020) Spray drying for the preparation of nanoparticle-based drug formulations as dry powders for inhalation. Processes 8(7):788. https://doi.org/10.3390/pr8070788
30. Shete G, Bansal AK (2016) NanoCrySP technology for generation of drug nanocrystals: translational aspects and business potential. Drug Deliv Transl Res 6(4):392–398. https://doi.org/10.1007/s13346-016-0286-y
31. Peng R, Yang D, Qiu X, Qin Y, Zhou M (2020) Preparation of self-dispersed lignin-based drug-loaded material and its application in avermectin nano-formulation. Int J Biol Macromol 151: 421–427. https://doi.org/10.1016/j.ijbiomac.2020.02.114
32. Zielinska A et al (2020) Polymeric nanoparticles: production, characterization, toxicology and ecotoxicology. Molecules 25(16):3731. https://doi.org/10.3390/molecules25163731
33. Cheaburu-Yilmaz CN, Karasulu HY, Yilmaz O (2018) Nanoscaled dispersed systems used in drug-delivery applications. In: Polymeric nanomaterials in nanotherapeutics, pp 437–468. https://doi.org/10.1016/B978-0-12-813932-5.00013-3
34. Naguib YW, Rodriguez BL, Li X, Hursting SD, Williams RO, Cui Z (2014) Solid lipid nanoparticle formulations of docetaxel prepared with high melting point triglycerides: in vitro and in vivo evaluation. Mol Pharm 11(4):1239–1249. https://doi.org/10.1021/mp4006968
35. Patel K, Shah S, Patel J (2022) Solid dispersion technology as a formulation strategy for the fabrication of modified release dosage forms: a comprehensive review. DARU J Pharm Sci 30(1):165–189. https://doi.org/10.1007/s40199-022-00440-0
36. Ghosh Dastidar D, Saha S, Chowdhury M (2018) Porous microspheres: synthesis, characterisation and applications in pharmaceutical & medical fields. Int J Pharm 548(1): 34–48. https://doi.org/10.1016/j.ijpharm.2018.06.015
37. Pardeshi S et al (2021) A meticulous overview on drying-based (spray-, freeze-, and spray-freeze) particle engineering approaches for pharmaceutical technologies. Dry Technol 39(11): 1447–1491. https://doi.org/10.1080/07373937.2021.1893330
38. Gol D, Thakkar S, Misra M (2018) Nanocrystal-based drug delivery system of risperidone: lyophilization and characterization. Drug Dev Ind Pharm 44(9):1458–1466. https://doi.org/10.1080/03639045.2018.1460377
39. Pilcer G, Rosière R, Traina K, Sebti T, Vanderbist F, Amighi K (2013) New co-spray-dried tobramycin nanoparticles-clarithromycin inhaled powder systems for lung infection therapy in cystic fibrosis patients. J Pharm Sci 102(6):1836–1846. https://doi.org/10.1002/jps.23525
40. Otake H, Okuda T, Okamoto H (2016) Development of spray-freeze-dried powders for inhalation with high inhalation performance and antihygroscopic property. Chem Pharm Bull (Tokyo) 64(3):239–245. https://doi.org/10.1248/cpb.c15-00824
41. Nair AR, Lakshman YD, Anand VSK, Sree KSN, Bhat K, Dengale SJ (2020) Overview of extensively employed polymeric carriers in solid dispersion technology. AAPS PharmSciTech 21:1–20. https://doi.org/10.1208/s12249-020-01849-z
42. Overhoff KA, McConville JT, Yang W, Johnston KP, Peters JI, Williams RO (2008) Effect of stabilizer on the maximum degree and extent of supersaturation and oral absorption of tacrolimus made by ultra-rapid freezing. Pharm Res 25(1):167–175. https://doi.org/10.1007/s11095-007-9417-y

43. Ahmad W, Chandra Bhatt S, Verma M, Kumar V, Kim H (2022) A review on current trends in the green synthesis of nickel oxide nanoparticles, characterizations, and their applications. Environ Nanotechnol Monit Manag 18:100674. https://doi.org/10.1016/j.enmm.2022.100674
44. Dhibar M, Chakraborty S, Kundu A, Laha P (2023) Chemistry characterization and application of nanocrystals-based drug delivery system: present to future perspective. Pharm Nanotechnol 11(3):265–275. https://doi.org/10.2174/2211738511666230109151410
45. Mitić Ž et al (2017) Instrumental methods and techniques for structural and physicochemical characterization of biomaterials and bone tissue: a review. Mater Sci Eng C 79:930–949. https://doi.org/10.1016/j.msec.2017.05.127
46. Kisielowski C et al (2008) Detection of single atoms and buried defects in three dimensions by aberration-corrected electron microscope with 0.5-Å information limit. Microsc Microanal 14(5):469–477. https://doi.org/10.1017/S1431927608080902
47. Zhao H et al (2023) Atomic-scale structure dynamics of nanocrystals revealed by in situ and environmental transmission electron microscopy. Adv Mater 35(50):2206911. https://doi.org/10.1002/adma.202206911
48. Zhou X, Thompson GE (2017) Electron and photon based spatially resolved techniques. In: Reference module in materials science and materials engineering, pp 1–30. https://doi.org/10.1016/b978-0-12-803581-8.10140-7
49. Herzing AA et al (2008) Energy dispersive X-ray spectroscopy of bimetallic nanoparticles in an aberration corrected scanning transmission electron microscope. Faraday Discuss 138:337–351. https://doi.org/10.1039/b706293c
50. Peng S et al (2012) Real-time probing of the synthesis of colloidal silver nanocubes with time-resolved high-energy synchrotron X-ray diffraction. J Phys Chem C 116(21):11842–11847. https://doi.org/10.1021/jp304557p
51. Zou Y, Jayasuriya S, Manke CW, Mao G (2015) Influence of nanoscale surface roughness on colloidal force measurements. Langmuir 31(38):10341–10350. https://doi.org/10.1021/acs.langmuir.5b02672
52. Giannini C, Ladisa M, Altamura D, Siliqi D, Sibillano T, De Caro L (2016) X-ray diffraction: a powerful technique for the multiple-length-scale structural analysis of nanomaterials. Crystals 6(8):87. https://doi.org/10.3390/cryst6080087
53. Juraić K et al (2010) GISAXS and GIWAXS analysis of amorphous-nanocrystalline silicon thin films. Nucl Instrum Methods Phys Res B 268(3–4):259–262. https://doi.org/10.1016/j.nimb.2009.09.046
54. Katari JEB, Colvin VL, Alivisatos AP (1994) X-ray photoelectron spectroscopy of CdSe nanocrystals with applications to studies of the nanocrystal surface. J Phys Chem 98(15):4109–4117. https://doi.org/10.1021/j100066a034
55. Huang K, Demadrille R, Silly MG, Sirotti F, Reiss P, Renault O (2010) Internal structure of InP/ZnS nanocrystals unraveled by high-resolution soft X-ray photoelectron spectroscopy. ACS Nano 4(8):4799–4805. https://doi.org/10.1021/nn100581t
56. Hens Z (2023) Ligands on nanocrystal surfaces, the 1H nuclear magnetic resonance fingerprint. Acc Chem Res 56(12):1623–1633. https://doi.org/10.1021/acs.accounts.3c00170
57. Hens Z, Martins JC (2013) A solution NMR toolbox for characterizing the surface chemistry of colloidal nanocrystals. Chem Mater 25(8):1211–1221. https://doi.org/10.1021/cm303361s
58. Wu Y et al (2016) Core size determination and structural characterization of intravenous iron complexes by cryogenic transmission electron microscopy. Int J Pharm 505(1–2):167–174. https://doi.org/10.1016/j.ijpharm.2016.03.029
59. Li Y, Huang W, Li Y, Chiu W, Cui Y (2020) Opportunities for cryogenic electron microscopy in materials science and nanoscience. ACS Nano 14(8):9263–9276. https://doi.org/10.1021/acsnano.0c05020
60. Hu L, Kong D, Hu Q, Gao N, Pang S (2015) Evaluation of high-performance curcumin nanocrystals for pulmonary drug delivery both in vitro and in vivo. Nanoscale Res Lett 10(1):1–9. https://doi.org/10.1186/s11671-015-1085-y

61. Ţucureanu V, Matei A, Avram AM (2016) FTIR spectroscopy for carbon family study. Crit Rev Anal Chem 46(6):502–520. https://doi.org/10.1080/10408347.2016.1157013
62. As'adi MHA, Kuncahyo I, Sulaiman TNS (2023) Formulation of fenofibrate nanocrystals with wet milling method. Riset Inf Kesehat 12(1):50–60. https://doi.org/10.30644/rik.v12i1.761
63. Lin S, Lin X, Shang Y, Han S, Hasi W, Wang L (2019) Self-assembly of faceted gold nanocrystals for surface-enhanced Raman scattering application. J Phys Chem C 123(40): 24714–24722. https://doi.org/10.1021/acs.jpcc.9b06686
64. Tao AR (2009) Nanocrystal assembly for bottom-up plasmonic materials and surface-enhanced Raman spectroscopy (SERS) sensing. Pure Appl Chem 81(1):61–71. https://doi.org/10.1351/PAC-CON-08-08-38
65. Hartmann NB et al (2015) Techniques and protocols for dispersing nanoparticle powders in aqueous media – is there a rationale for harmonization? J Toxicol Environ Health B Crit Rev 18(6):299–326. https://doi.org/10.1080/10937404.2015.1074969
66. Choi JS (2020) Design of cilostazol nanocrystals for improved solubility. J Pharm Innov 15(3). https://doi.org/10.1007/s12247-019-09391-7
67. Patel V, Agrawal Y (2011) Nanosuspension: an approach to enhance solubility of drugs. J Adv Pharm Technol Res 2(2):81–87. https://doi.org/10.4103/2231-4040.82950
68. Alykova AF, Zavestovskaya IN, Yakunin VG, Timoshenko VY (2018) Raman diagnostics of silicon nanocrystals dissolution in aqueous medium. J Phys Conf Ser 945(1):012002. https://doi.org/10.1088/1742-6596/945/1/012002. IOP Publishing
69. Pardhi E, Vasave R, Srivastava V, Yadav R, Mehra NK (2024) Nanocrystal technologies in biomedical science: from the bench to the clinic. Drug Discov Today 29(3):103913. https://doi.org/10.1016/j.drudis.2024.103913
70. Azarmi S, Roa W, Löbenberg R (2007) Current perspectives in dissolution testing of conventional and novel dosage forms. Int J Pharm 328(1):12–21. https://doi.org/10.1016/j.ijpharm.2006.10.001
71. Mertins O, Mathews PD, Angelova A (2020) Advances in the design of ph-sensitive cubosome liquid crystalline nanocarriers for drug delivery applications. Nanomaterials 10(5):963. https://doi.org/10.3390/nano10050963
72. Mudunkotuwa IA, Rupasinghe T, Wu CM, Grassian VH (2012) Dissolution of ZnO nanoparticles at circumneutral pH: a study of size effects in the presence and absence of citric acid. Langmuir 28(1):396–403. https://doi.org/10.1021/la203542x
73. Mauludin R, Müller RH, Keck CM (2009) Development of an oral rutin nanocrystal formulation. Int J Pharm 370(1–2):202–209. https://doi.org/10.1016/j.ijpharm.2008.11.029
74. Tian H et al (2021) Consideration of the dissolution media for drug nanocrystal evaluation. Powder Technol 392:179–190. https://doi.org/10.1016/j.powtec.2021.07.016
75. Cai H et al (2020) A nanostrategy for efficient imaging-guided antitumor therapy through a stimuli-responsive branched polymeric prodrug. Adv Sci 7(6):1903243. https://doi.org/10.1002/advs.201903243
76. Zhang X et al (2020b) Glycodendron/pyropheophorbide-a (Ppa)-functionalized hyaluronic acid as a nanosystem for tumor photodynamic therapy. Carbohydr Polym 247:116749. https://doi.org/10.1016/j.carbpol.2020.116749
77. Singh S, Vaidya Y, Gulati M, Bhattacharya S, Garg V, Pandey N (2016) Nanosuspension: principles, perspectives and practices. Curr Drug Deliv 13(8):1222–1246. https://doi.org/10.2174/1567201813666160101120452
78. Le Berre F, Chauveteau G, Pefferkorn E (1998) Perikinetic and orthokinetic aggregation of hydrated colloids. J Colloid Interface Sci 199(1):1–12. https://doi.org/10.1006/jcis.1997.5307
79. Li J et al (2021) Preparation of loratadine nanocrystal tablets to improve the solubility and dissolution for enhanced oral bioavailability. J Pharm Pharmacol 73(7):937–946. https://doi.org/10.1093/jpp/rgab043
80. Yang H et al (2018) Pharmaceutical strategies for stabilizing drug nanocrystals. Curr Pharm Des 24(21):2362–2374. https://doi.org/10.2174/1381612824666180515125247

81. Mu S et al (2016) Spironolactone nanocrystals for oral administration: different pharmacokinetic performances induced by stabilizers. Colloids Surf B Biointerfaces 147:73–80. https://doi.org/10.1016/j.colsurfb.2016.07.051
82. Qu C et al (2018) Preparation and evaluation of wet-milled usnic acid nanocrystal suspension for better bioaffinity. Drug Dev Ind Pharm 44(5):707–712. https://doi.org/10.1080/03639045.2017.1409756
83. Rao Y, Kumar M, Apte S (2008) Formulation of nanosuspensions of albendazole for oral administration. Curr Nanosci 4(1):53–58. https://doi.org/10.2174/157341308783591807
84. Park JJ, Meghani N, Choi JS, Lee BJ (2016) Development and evaluation of decorated aceclofenac nanocrystals. Colloids Surf B Biointerfaces 143:206–212. https://doi.org/10.1016/j.colsurfb.2016.03.022
85. Xie Y et al (2016) Panax notoginseng saponins as a novel nature stabilizer for poorly soluble drug nanocrystals: a case study with baicalein. Molecules 21(9):1149. https://doi.org/10.3390/molecules21091149
86. Fan M et al (2018) Nanocrystal technology as a strategy to improve drug bioavailability and antitumor efficacy for the cancer treatment. Curr Pharm Des 24(21):2416–2424. https://doi.org/10.2174/1381612824666180515154109
87. Singh SK et al (2017) Preparation and in-vitro/in-vivo characterization of trans-resveratrol nanocrystals for oral administration. Drug Deliv Transl Res 7(3):395–407. https://doi.org/10.1007/s13346-017-0362-y
88. Dhapte V, Pokharkar V (2011) Polyelectrolyte stabilized antimalarial nanosuspension using factorial design approach. J Biomed Nanotechnol 7(1):139–141. https://doi.org/10.1166/jbn.2011.1239
89. Ochi M et al (2014) Development of nanocrystal formulation of meloxicam with improved dissolution and pharmacokinetic behaviors. Int J Pharm 474(1–2):151–156. https://doi.org/10.1016/j.ijpharm.2014.08.022
90. Sharma S, Verma A, Pandey G, Mittapelly N, Mishra PR (2015) Investigating the role of Pluronic-g-Cationic polyelectrolyte as functional stabilizer for nanocrystals: impact on paclitaxel oral bioavailability and tumor growth. Acta Biomater 26:169–183. https://doi.org/10.1016/j.actbio.2015.08.005
91. Choi JS, Park JS (2017) Development of docetaxel nanocrystals surface modified with transferrin for tumor targeting. Drug Des Devel Ther 11:17–26. https://doi.org/10.2147/DDDT.S122984
92. He W, Lu Y, Qi J, Chen L, Hu F, Wu W (2013) Food proteins as novel nanosuspension stabilizers for poorly water-soluble drugs. Int J Pharm 441(1–2):269–278. https://doi.org/10.1016/j.ijpharm.2012.11.033
93. Chen Z, Wu W, Lu Y (2020) What is the future for nanocrystal-based drug-delivery systems? Ther Deliv 11(4):225–229. https://doi.org/10.4155/tde-2020-0016
94. Rossier B, Jordan O, Allémann E, Rodríguez-Nogales C (2024) Nanocrystals and nanosuspensions: an exploration from classic formulations to advanced drug delivery systems. Drug Deliv Transl Res 14(12):3438–3451. https://doi.org/10.1007/s13346-024-01559-0
95. Rudd ND et al (2020) Interpreting in vitro release performance from long-acting parenteral nanosuspensions using USP-4 dissolution and spectroscopic techniques. Mol Pharm 17(5):1734–1747. https://doi.org/10.1021/acs.molpharmaceut.0c00208
96. Sievens-Figueroa L et al (2012) Using USP I and USP IV for discriminating dissolution rates of nano- and microparticle-loaded pharmaceutical strip-films. AAPS PharmSciTech 13(4):1473–1482. https://doi.org/10.1208/s12249-012-9875-3
97. Gao L et al (2013) Application of drug nanocrystal technologies on oral drug delivery of poorly soluble drugs. Pharm Res 30:307–324. https://doi.org/10.1007/s11095-012-0889-z

Stabilizer Selection in Nanocrystal Preparation: Traditional and Advanced Methods

3

Rishi Paliwal, Akanksha Malaiya, Puja Nayak, and Shivani Rai Paliwal

Abstract

Nanocrystals, which are pure drug particles sized between 10 and 1000 nm, offer a promising approach for improving the solubility, dissolution rate, and bioavailability of poorly water-soluble drugs. Due to their high surface energy, nanocrystals tend to be unstable and prone to agglomeration, making stabilizers crucial for maintaining their physical stability and controlling particle growth. This chapter thoroughly reviews stabilizer selection, a vital aspect of nanocrystal formulation. Stabilizers are broadly categorized into polymeric, ionic, and non-ionic types, with their functional performance significantly influenced by physicochemical parameters such as concentration, molecular weight, hydrophilic–lipophilic balance (HLB), drug solubility in the dispersion medium, surface free energy compatibility, and pH. Specific stabilizers like poloxamers, vitamin E TPGS, hydroxypropyl methylcellulose (HPMC), Soluplus®, and cyclodextrins have been extensively investigated for their unique roles in improving drug delivery by enhancing solubility, stability, and bioavailability of poorly water-soluble drugs. These stabilizers not only prevent particle aggregation and Ostwald ripening but also modulate interactions with biological membranes,

R. Paliwal (✉)
Nanomedicine and Bioengineering Research Laboratory, Department of Pharmacy, Indira Gandhi National Tribal University, Amarkantak, Madhya Pradesh, India

A. Malaiya
Nanomedicine and Bioengineering Research Laboratory, Department of Pharmacy, Indira Gandhi National Tribal University, Amarkantak, Madhya Pradesh, India

Bhagyoday Tirth Pharmacy College, Sagar, Madhya Pradesh, India

P. Nayak · S. R. Paliwal
Department of Pharmacy, Guru Ghasidas Vishwavidyalaya (A Central University), Bilaspur, Chhattisgarh, India

N. K. Mehra, K. K. Bansal (eds.), *Nanocrystal Technologies in Biomedical Sciences*,
https://doi.org/10.1007/978-981-95-2968-1_3

potentially enhancing permeability and cellular uptake. Understanding the interplay between stabilizer properties and formulation parameters is critical for optimizing nanocrystal performance in targeted and sustained drug delivery systems.

Keywords

Nanocrystal · Stabilizer · Surfactants

3.1 Introduction

Drug nanocrystals are pure drug particles (10–1000 nm) stabilized by surfactants or polymeric stabilizers to enhance solubility and bioavailability. The size reduction approach, supported by theories like Ostwald–Freundlich, Kelvin, and Noyes–Whitney, improves drug dissolution and absorption. Nanocrystals have been successfully applied in oral and parenteral drug delivery, with marketed formulations such as Triglide® (fenofibrate), Rapamune® (sirolimus), and Invega Sustenna™ (paliperidone palmitate). Many nanocrystal-based drugs are also in clinical trials for various administration routes [1]. Depending on the method of preparation, nanocrystals can exist in either crystalline or amorphous forms, with the latter offering higher solubility but posing a risk of crystallization over time, which can affect drug release. Their increased surface area enhances dissolution, interaction with biological membranes, and muco-adhesion while reducing food-related absorption variability and drug dosage requirements [2]. Post-production processes like freeze-drying and spray-drying enable formulation into solid, liquid, semisolid, and aerosol dosage forms [3]. Various formulation strategies, such as solid dispersions, co-amorphous systems, emulsions, self-nanoemulsifying drug delivery systems, liposomes, complexation, salt formation, and cyclodextrin inclusion complexes, have been employed to enhance the solubility and dissolution rate of poorly water-soluble drugs [4, 5]. However, nanocrystals have become more desirable than other techniques due to high drug loading, reduced toxicity, and positive effects on pharmacokinetics, drug absorption, and biodistribution [6]. The selection and design of stabilizers for achieving stable nanocrystals have significantly advanced with the evolution of computational technologies and the increasing availability of high-performance computing resources.

3.1.1 Techniques for the Production of Nanocrystals

Currently, two main approaches are used for formulating nanocrystals: the *top-down* approach and the *bottom-up* approach. The *top-down* method involves size reduction of bulk drugs into nanosized particles through various homogenization techniques, resulting in a liquid suspension. In contrast, the *bottom-up* approach builds nanosized particles molecule by molecule through precipitation [7]. Most commercial nanocrystal products are developed using the *top-down* approach due to its high reproducibility and ease of execution. However, this method requires significant energy input compared to bottom-up techniques. Additionally, literature on process yield is limited [8].

For instance, in milling-based batch processes, initial yields may be high, but material losses can occur due to adhesion to container surfaces or atomization chamber walls in liquid atomization-based processes [6].

3.2 Stability of Nanocrystals

The meticulous selection of an optimal nanocrystal formulation remains a critical and foundational step in the development process, despite being inherently time-consuming and labor-intensive [9]. Numerous studies have underscored the risks associated with the inappropriate choice of stabilizers, which can result in undesirable phenomena such as particle aggregation, precipitation, or Ostwald ripening during milling and storage [10]. To address these challenges, a structured screening approach grounded in physicochemical principles has been proposed as a rapid and efficient strategy for evaluating a broad range of stabilizers [11, 12].

This methodology is typically divided into two main phases: a qualitative screening step aimed at identifying promising stabilizer candidates for further investigation, followed by a quantitative phase focused on optimizing the ratios of wetting and dispersing agents based on parameters such as wettability, surface charge, adsorption behavior, processability, and storage stability [13]. Stabilizers serve a pivotal function in NC formulations by adsorbing onto the drug particle surface and preventing aggregation within the dispersion medium, thereby ensuring physical stability. The type and concentration of the selected stabilizer significantly influence the final particle size and the degree of aggregation prevention, making this decision a crucial determinant of formulation success [14]. Key considerations during stabilizer selection include polymer chain length and molecular weight, which affect the kinetics and thermodynamics of adsorption onto particle surfaces. Higher molecular weight polymers typically exhibit slower adsorption rates, and their excessive use, especially in the case of long-chain polymers, can impair dissolution rates, thus diminishing the advantages conferred by nanomilling, particularly for poorly water-soluble drugs. Furthermore, elevated concentrations of certain surfactant-based stabilizers, such as sodium lauryl sulfate and Pluronic®, may pose formulation challenges in developing patient-friendly dosage forms due to the risk of gastrointestinal irritation, especially in pediatric populations.

The conversion of drug particles to nano-size creates new interfaces, leading to an increase in Gibbs free energy, making the system thermodynamically unstable. To regain stability, nanocrystals tend to agglomerate, resulting in crystal growth. Even subtle changes in particle size can significantly alter the properties of nanocrystals [15]. To prevent this instability, stabilizers are used to lower the system's free energy and maintain stability over time. These stabilizers adsorb onto the surface of nanocrystals, forming either a rough or charged surface, depending on their properties. This prevents particle coalescence, thereby inhibiting crystal growth and an increase in particle size [16, 17]

Stabilizers are categorized into three types:

1. *Polymeric stabilizers*—such as polyvinyl alcohol (PVA), polyvinylpyrrolidone (PVP), hydroxypropyl cellulose (HPC), and hydroxypropylmethyl cellulose (HPMC).
2. *Ionic surfactants*—including sodium lauryl sulfate (SLS), sodium dodecyl sulfate (SDS), polyethylene imine (PEI), cetyltrimethylammonium bromide (CTAB), tetraoctylammonium bromide (TOAB), and cetyltrimethylammonium chloride (CTAC)
3. *Nonionic surfactants*—such as poly (ethylene oxide)-poly(propylene oxide) (PEO-PPO-PEO) (Pluronics®), Tween, Soluplus®, and D-α-tocopherol polyethylene glycol succinate (TPGS).

These stabilizers enhance stability through steric repulsion (polymeric and nonionic stabilizers) or electrostatic repulsion (ionic stabilizers). The interaction between the drug molecules and the stabilizer, and consequently the stability of the system, is influenced by factors such as the drug's properties (logP, enthalpy, presence of hydrophilic, hydrophobic, or charged moieties) and the stabilizer's characteristics (molecular weight, chain length, and charge) [15].

For example, HPC and PVP are recommended for drug molecules with hydrophobic surfaces that lack polar functional groups. Strong and rapid interactions between the drug and stabilizer are essential for maintaining a stable system. However, selecting the most suitable stabilizer or combination of stabilizers requires systematic formulation of nanocrystal batches, followed by comparison of their properties, such as particle size and distribution, zeta potential, crystallinity, and overall stability [18]. Figure 3.1 represents the risk of agglomeration of nanocrystals with various types of stabilizers chosen in the gastric medium [19].

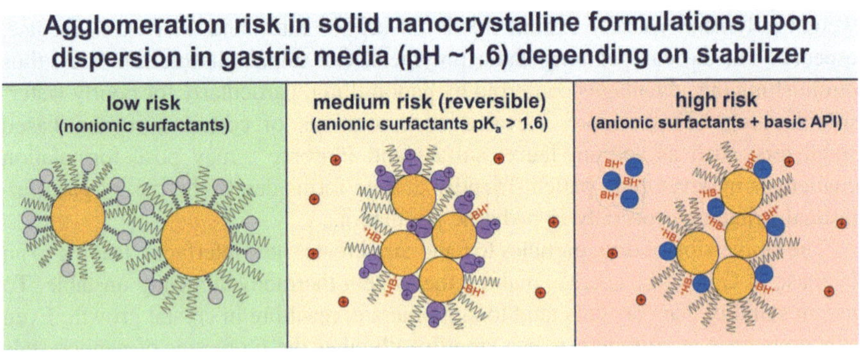

Fig. 3.1 Pictorial representation of agglomeration risk behavior associated with different types of stabilizers used in nanocrystal development under gastric pH conditions. (Reproduced with copyright permission from Ref. [19])

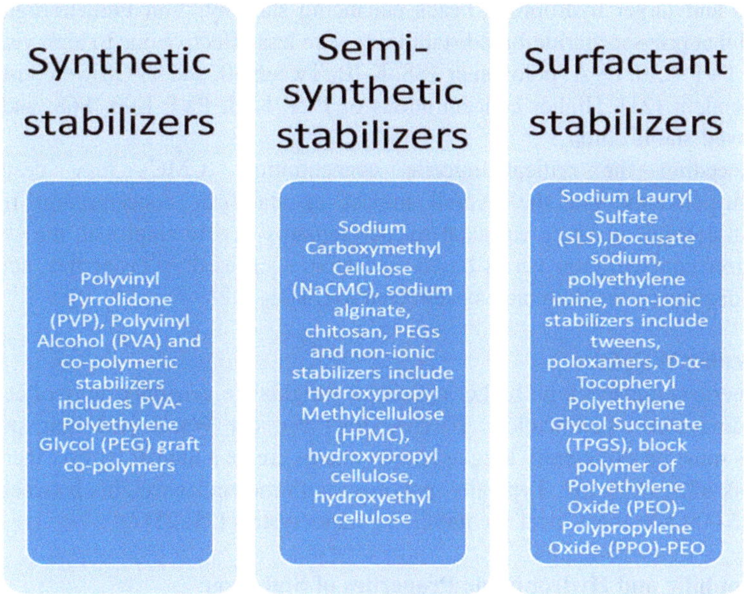

Fig. 3.2 Representative types of stabilizers commonly used in nanocrystal development, including synthetic, semi-synthetic, and surfactant-based agents. (Adapted and modified from Ref. [20])

Stabilizers play a crucial role in the formulation and stability of nanocrystals by preventing particle aggregation and controlling crystal growth. An overview of the three primary types of stabilizers commonly employed in nanocrystal development: synthetic, semi-synthetic, and surfactant-based is summarized in Fig. 3.2. Synthetic stabilizers, such as polyvinyl alcohol (PVA) and polyvinylpyrrolidone (PVP), offer consistent molecular properties and high stability. Semi-synthetic polymers like hydroxypropyl methylcellulose (HPMC) and methylcellulose (MC) provide biocompatibility and enhanced solubility. Surfactant-based stabilizers, including both ionic (e.g., sodium dodecyl sulfate) and nonionic types (e.g., polysorbates), are widely used to reduce surface tension and improve wettability. The appropriate selection and combination of these stabilizers are essential to achieve the desired particle size, stability, and bioavailability in nanocrystal formulations [20].

3.3 Factors Affecting Stabilizer Selection

3.3.1 Concentration of Stabilizer

The stability of colloidal systems depends on the optimal concentration of stabilizers, as insufficient amounts reduce steric repulsion, leading to aggregation. Surfactant structure influences the required concentration, with longer hydrophobic chains and larger hydrophilic heads enhancing stability. Van Eerdenbrugh et al. found that polysaccharide-based stabilizers were less effective due to high viscosity, while PVP K30, PVA, poloxamer 188, K-IR, Tween 80, and TPGS were viscosity-independent [21]. Higher concentrations of PVP K30, PVP K90, F68, and K-IR improved stabilization.

Exceeding the critical micelle concentration (CMC) may destabilize nanosuspensions, but the overall impact of stabilizer concentration remains system-dependent. Since colloidal science remains largely empirical, the effect of stabilizer concentration varies based on factors such as drug properties, stabilizer type, dispersion medium, and system characteristics [22].

Molecular Weight

Polymeric stabilizers with higher molecular weights are generally more effective in providing steric stabilization. To counteract van der Waals attraction, polymer chains must be sufficiently long as shorter chains create a thinner barrier, increasing the risk of aggregation. Typically, polymers with molecular weights between 5000 and 25,000 g/mol are ideal for stabilizing nanocrystals [21, 23].

Hydrophilic and Hydrophobic Properties of Stabilizer

The hydrophilic–lipophilic balance (HLB) is an empirical measure of a surfactant's affinity for water or oil. A lower HLB indicates a more hydrophobic surfactant, while a higher HLB signifies greater hydrophilicity. HLB values provide a quick and practical way to estimate a surfactant's effectiveness in stabilizing drug nanocrystals. Lee et al. (2005) studied the use of amphiphilic amino acid copolymers as stabilizers for naproxen nanocrystal dispersions [24].

3.4 Drug Solubility in Dispersion Media

The solubility of a drug in aqueous stabilizer solutions can influence nanocrystal stabilization. In microfluidization, increased ibuprofen solubility in the presence of a stabilizer promoted Ostwald ripening, leading to larger particle sizes. Ostwald ripening occurs when small crystals dissolve and redeposit onto larger ones, a process linked to the drug concentration in the dispersed phase. Aqueous surfactant solutions enhance the apparent solubility of low molecular weight compounds by forming micelle-like structures, which can accelerate Ostwald ripening and potentially compromise storage stability [25].

3.4.1 Surface Free Energy

The key challenge in processing drug nanocrystals is stabilizing the high free energy of newly formed surfaces. Effective stabilization requires rapid adsorption of stabilizers. Surface free energy measurements help in selecting suitable stabilizers, as Van Eerdenbrugh et al. demonstrated for various drugs. For polymeric stabilizers, optimal stabilization occurs when the drug and stabilizer have similar surface free energies. When the energy difference is significant, adding a surfactant improves interactions, enhancing stability.

3.4.2 pH of the Dispersion Media

The pH of the dispersion medium plays a key role in stabilizing nanocrystals, especially when using ionizable polymers. It affects how well stabilizers work. He et al. (2013) studied food proteins like whey protein, soybean protein isolate, and β-lactoglobulin to stabilize indomethacin nanosuspensions. These proteins have isoelectric points (pI) between 4 and 6, where they have no net charge, making them less effective and causing particle clumping. When the pH is higher or lower than the pI, the proteins act as better stabilizers due to their amphiphilic nature. However, extremely high or low pH levels break the proteins into smaller parts, reducing their effectiveness [26].

3.5 Common Stabilizers Used in Nanocrystal Formulations

3.5.1 Poloxamer

Poloxamers are synthetic block copolymers of ethylene oxide and propylene oxide, widely used in pharmaceuticals as polymeric stabilizers, especially for drug nanocrystals. They are amphiphilic, consisting of hydrophilic polyethylene oxide (PEO) and hydrophobic polypropylene oxide (PPO) segments. Common variants include poloxamer 407 (F127) and poloxamer 188 (F68), which prevent particle aggregation through steric hindrance. Their stabilizing efficiency depends on factors such as molecular weight, hydrophobicity, and polymer morphology. While effective for some drugs like indomethacin and itraconazole, they fail with others like brinzolamide due to weak surface adsorption [27].

Danhier et al. (2014) conducted a comparative study on poloxamer-stabilized nanocrystalline formulations of the anti-cancer multi-targeted kinase inhibitor (MTKi-327) against other nanocarrier systems, including PEGylated PLGA nanoparticles, polymeric micelles, and Captisol solution. All nanoformulations had particle sizes below 200 nm and were well-tolerated in mice, even at the highest administered doses. In contrast, the maximum tolerated dose of the Captisol solution was significantly lower, being 20 times less than that of the nanoformulations. Pharmacokinetic analysis revealed that, following both intravenous and oral administration, the area under the curve (AUC) value for the nanocrystalline formulation was 2.4 times greater than that observed with Captisol solution, indicating superior bioavailability and systemic drug exposure [28].

3.5.2 Vitamin E TPGS

Vitamin E TPGS is a versatile excipient used in pharmaceuticals, nutraceuticals, and cosmetics. It is synthesized by esterifying D-α-tocopheryl succinate with polyethylene glycol 1000 and serves as an absorption enhancer, emulsifier, and bioavailability booster for poorly soluble drugs. Its amphiphilic nature allows applications in various drug delivery systems, including micelles, liposomes, and nanoparticles, while also acting as a P-glycoprotein (P-gp) inhibitor to overcome multidrug resistance (MDR) in cancer treatment [29]. Vitamin E TPGS is stable against heat, light, and oxidation but degrades in alkaline conditions. Its long shelf life and ability to enhance cellular drug uptake make it valuable for controlled and targeted drug delivery [18]. Studies have demonstrated its role in improving the bioavailability of anticancer drugs like paclitaxel, docetaxel, and SN-38 by promoting clathrin-mediated endocytosis and modulating P-gp activity [30].

In nanocrystal formulations, vitamin E TPGS has shown remarkable effects on drug solubility and bioavailability. For example, ursolic acid nanocrystals stabilized with TPGS1000 exhibited a 27.5-fold increase in bioavailability and a nine-fold higher Cmax compared to bulk drug dispersion, primarily due to P-gp inhibition in intestinal cells. These properties make vitamin E TPGS a highly effective excipient for advanced drug delivery systems [31].

3.5.3 Cellulose

Hydroxypropyl methylcellulose (HPMC) is a semisynthetic, nonionic cellulose polymer widely used in pharmaceutical applications due to its nontoxicity, hydrophilicity, and swelling properties. It is produced by treating alkali cellulose with chloromethane and propylene oxide, with different molecular weights achieved through depolymerization. HPMC is commonly used in oral controlled drug delivery and ophthalmic formulations as a stabilizer, lubricant, and wound-healing agent. In nanocrystalline applications, low-molecular-weight HPMC stabilizes drug formulations, such as fenofibrate nanocrystals, thereby enhancing dissolution and bioavailability. It has also been used in brinzolamide nanocrystal formulations for ocular drug delivery, effectively reducing intraocular pressure, particularly at pH 4.5, due to its hydrogen bonding and steric stabilization properties [32].

3.5.4 Soluplus®

Soluplus® (BASF) is a graft copolymer (PVAc-PVCap-PEG) designed to enhance the solubility of poorly soluble drugs [33]. It is used in various drug formulation techniques, including hot melt extrusion, spray drying, and solvent evaporation. Soluplus® offers advantages like high flowability, extrudability, and the ability to act as a binder. Studies have shown its superiority over HPMC in stabilizing fenofibrate nanosuspensions, producing smaller particles and reducing Ostwald ripening. Additionally, Soluplus® enhances intestinal drug permeability, as demonstrated in studies with danazol, fenofibrate, and itraconazole, leading to increased bioavailability. It also significantly improves the solubility and absorption of repaglinide nanocrystals, increasing saturation solubility up to 25.67 times compared to the pure drug [34].

3.5.5 Cyclodextrin

Cyclodextrins are commonly used in pharmaceutical formulations to enhance the solubility of poorly water-soluble drugs through complex formation [35]. In a study on indomethacin, cyclodextrins served as stabilizers for nanocrystals produced via the emulsion solvent diffusion method. The resulting indomethacin nanocrystals measured between 300 and 500 nm, with stabilization achieved through intermolecular interactions within the cyclodextrin network. The nanocrystals exhibited the α-form of indomethacin, in contrast to the γ-form found in the bulk drug, leading to improved dissolution [36].

3.6 Conclusion

Nanocrystals represent a promising drug delivery system owing to their facile production and substantial drug loading capacity. A pivotal aspect of their formulation involves the selection of stabilizers, which influence not only physical stability but also cellular interactions, permeability, and bioavailability. Certain stabilizers can modulate drug transport by inhibiting efflux pumps or enhancing mucoadhesion. Although nanocrystals facilitate a supersaturated state that improves solubility, rapid dissolution does not invariably ensure higher bioavailability due to possible precipitation or modifications in transit time. Additionally, stabilizers may fulfill supplementary roles, such as cryoprotectants or disintegrants. Future investigations should prioritize the integration of nanocrystal formation with stabilizer selection to develop optimized drug delivery systems.

References

1. Tuomela A, Hirvonen J, Peltonen L (2016) Stabilizing agents for drug nanocrystals: effect on bioavailability. Pharmaceutics 8:16. https://doi.org/10.3390/pharmaceutics8020016
2. Huang S, Zhang T, Shi C, Zi Y, Xu J, Su J, Wei D, Wang X, Zhong J (2025) Nanocrystal technology. In: Drug delivery to tumors. Springer Nature Singapore, Singapore, pp 229–258. https://doi.org/10.1007/978-981-19-8930-8_9
3. Miao Y, Xie R, Kan Q, Yu Y, Dong S, Wang S, Mao L (2025) Silica-stabilized lead halide perovskite nanocrystals: advantages, progress, and future directions. Adv Opt Mater 13: 2402320. https://doi.org/10.1002/adom.202402320

4. Khemchandani R, Pardhi E, Goud S, Tomar DS, Medtiya P, Khushwaha BS, Dhiman V, Golla VM, Shaikh N, Godugu C, Mehra NK, Samanthula G (2025) Multi-method coamorphous systems of lumefantrine with alpha ketoglutaric acid: comprehensive characterization, biological evaluation and stability analysis. Int J Pharm 669:125056. https://doi.org/10.1016/j.ijpharm.2024.125056
5. Ding Y, Zhao T, Fang J, Song J, Dong H, Liu J, Li S, Zhao M (2024) Recent developments in the use of nanocrystals to improve bioavailability of APIs. WIREs Nanomed Nanobiotechnol 16:e1958. https://doi.org/10.1002/wnan.1958
6. Müller RH, Shegokar R, Gohla S, Keck CM (2011) Nanocrystals: production, cellular drug delivery, current and future products. In: Intracellular delivery. Fundamental biomedical technologies, vol 5. Springer, Dordrecht, pp 411–432. https://doi.org/10.1007/978-94-007-1248-5_15
7. Chary PS, Shaikh S, Bhavana V, Rajana N, Vasave R, Mehra NK (2024) Emerging role of nanocrystals in pharmaceutical applications: a review of regulatory aspects and drug development process. Appl Mater Today 40:102334. https://doi.org/10.1016/j.apmt.2024.102334
8. Patel A, Patel K, Patel V, Rajput MS, Patel R, Rajput A (2024) Nanocrystals: an emerging paradigm for cancer therapeutics. Futur J Pharm Sci 10:4. https://doi.org/10.1186/s43094-024-00579-4
9. Capretto L, Carugo D, Mazzitelli S, Nastruzzi C, Zhang X (2013) Microfluidic and lab-on-a-chip preparation routes for organic nanoparticles and vesicular systems for nanomedicine applications. Adv Drug Deliv Rev 65:1496–1532. https://doi.org/10.1016/j.addr.2013.08.002
10. Wang Y, Zheng Y, Zhang L, Wang Q, Zhang D (2013) Stability of nanosuspensions in drug delivery. J Control Release 172:1126–1141. https://doi.org/10.1016/j.jconrel.2013.08.006
11. Guan W, Ma Y, Ding S, Liu Y, Song Z, Liu X, Tang L, Wang Y (2022) The technology for improving stability of nanosuspensions in drug delivery. J Nanopart Res 24:14. https://doi.org/10.1007/s11051-022-05403-9
12. Pardhi E, Vasave R, Srivastava V, Yadav R, Mehra NK (2024) Nanocrystal technologies in biomedical science: from the bench to the clinic. Drug Discov Today 29:103913. https://doi.org/10.1016/j.drudis.2024.103913
13. Ghosh I, Schenck D, Bose S, Ruegger C (2012) Optimization of formulation and process parameters for the production of nanosuspension by wet media milling technique: effect of vitamin E TPGS and nanocrystal particle size on oral absorption. Eur J Pharm Sci 47:718–728. https://doi.org/10.1016/j.ejps.2012.08.011
14. Ahmadi Tehrani A, Omranpoor MM, Vatanara A, Seyedabadi M, Ramezani V (2019) Formation of nanosuspensions in bottom-up approach: theories and optimization, DARU. J Pharm Sci 27:451–473. https://doi.org/10.1007/s40199-018-00235-2
15. Li J, Wang Z, Zhang H, Gao J, Zheng A (2021) Progress in the development of stabilization strategies for nanocrystal preparations. Drug Deliv 28:19–36. https://doi.org/10.1080/10717544.2020.1856224
16. Gryaznov VG, Polonsky IA, Romanov AE, Trusov LI (1991) Size effects of dislocation stability in nanocrystals. Phys Rev B 44:42–46. https://doi.org/10.1103/PhysRevB.44.42
17. Purohit D, Sharma S, Lamba AK, Saini S, Minocha N, Vashist N, Kaushik D, Pandey P (2023) Nanocrystals: a deep insight into formulation aspects, stabilization strategies, and biomedical applications. Recent Pat Nanotechnol 17:307–326. https://doi.org/10.2174/1872210516666220523120313
18. Yang C, Wu T, Qi Y, Zhang Z (2018) Recent advances in the application of vitamin E TPGS for drug delivery. Theranostics 8:464–485. https://doi.org/10.7150/thno.22711
19. Böck NC, Sundermann J, Koziolek M, Keller B-L, Mäder K (2025) Impact of stabilizers on particle size and dispersion behavior in biorelevant media in solid nanocrystal formulations. Eur J Pharm Biopharm 208:114651. https://doi.org/10.1016/j.ejpb.2025.114651
20. Pujitha R, Chellakumari SD, Damayanthi RD, Aakash NS, Aswin Kumar AM (2024) Engineered nanocrystals for poorly soluble drug delivery: a review. Indian J Pharm Sci 86:742–754. https://doi.org/10.36468/pharmaceutical-sciences.1332

21. Van Eerdenbrugh B, Vermant J, Martens JA, Froyen L, Van Humbeeck J, Augustijns P, Van den Mooter G (2009) A screening study of surface stabilization during the production of drug nanocrystals. J Pharm Sci 98:2091–2103. https://doi.org/10.1002/jps.21563
22. Wu L, Zhang J, Watanabe W (2011) Physical and chemical stability of drug nanoparticles. Adv Drug Deliv Rev 63:456–469. https://doi.org/10.1016/j.addr.2011.02.001
23. Baines FL, Dionisio S, Billingham NC, Armes SP (1996) Use of block copolymer stabilizers for the dispersion polymerization of styrene in alcoholic media. Macromolecules 29:3096–3102. https://doi.org/10.1021/ma951767s
24. Lee J, Lee S-J, Choi J-Y, Yoo JY, Ahn C-H (2005) Amphiphilic amino acid copolymers as stabilizers for the preparation of nanocrystal dispersion. Eur J Pharm Sci 24:441–449. https://doi.org/10.1016/j.ejps.2004.12.010
25. Ouranidis A, Gkampelis N, Vardaka E, Karagianni A, Tsiptsios D, Nikolakakis I, Kachrimanis K (2020) Overcoming the solubility barrier of ibuprofen by the rational process design of a nanocrystal formulation. Pharmaceutics 12:969. https://doi.org/10.3390/pharmaceutics12100969
26. He W, Lu Y, Qi J, Chen L, Hu F, Wu W (2013) Food proteins as novel nanosuspension stabilizers for poorly water-soluble drugs. Int J Pharm 441:269–278. https://doi.org/10.1016/j.ijpharm.2012.11.033
27. Gigliobianco MR, Casadidio C, Censi R, Di Martino P (2018) Nanocrystals of poorly soluble drugs: drug bioavailability and physicochemical stability. Pharmaceutics 10:134. https://doi.org/10.3390/pharmaceutics10030134
28. Danhier F, Ucakar B, Vanderhaegen M-L, Brewster ME, Arien T, Préat V (2014) Nanosuspension for the delivery of a poorly soluble anti-cancer kinase inhibitor. Eur J Pharm Biopharm 88:252–260. https://doi.org/10.1016/j.ejpb.2014.05.014
29. Guo Y, Luo J, Tan S, Otieno BO, Zhang Z (2013) The applications of vitamin E TPGS in drug delivery. Eur J Pharm Sci 49:175–186. https://doi.org/10.1016/j.ejps.2013.02.006
30. Wang Y, Guo M, Lu Y, Ding L-Y, Ron W-T, Liu Y-Q, Song F-F, Yu S-Q (2012) Alpha-tocopheryl polyethylene glycol succinate-emulsified poly(lactic- *c o* -glycolic acid) nanoparticles for reversal of multidrug resistance *in vitro*. Nanotechnology 23:495103. https://doi.org/10.1088/0957-4484/23/49/495103
31. Ge Z-Q, Du X-Y, Huang X-N, Qiao B (2015) Enhanced oral bioavailability of ursolic acid nanoparticles via antisolvent precipitation with TPGS1000 as a stabilizer. J Drug Deliv Sci Technol 29:210–217. https://doi.org/10.1016/j.jddst.2015.08.001
32. El-Sousi S, Nácher A, Mura C, Catalán-Latorre A, Merino V, Merino-Sanjuán M, Díez-Sales O (2013) Hydroxypropylmethylcellulose films for the ophthalmic delivery of diclofenac sodium. J Pharm Pharmacol 65:193–200. https://doi.org/10.1111/j.2042-7158.2012.01587.x
33. Zhang K, Yu H, Luo Q, Yang S, Lin X, Zhang Y, Tian B, Tang X (2013) Increased dissolution and oral absorption of itraconazole/Soluplus extrudate compared with itraconazole nanosuspension. Eur J Pharm Biopharm 85:1285–1292. https://doi.org/10.1016/j.ejpb.2013.03.002
34. Gadadare R, Mandpe L, Pokharkar V (2015) Ultra rapidly dissolving repaglinide nanosized crystals prepared via bottom-up and top-down approach: influence of food on pharmacokinetics behavior. AAPS PharmSciTech 16:787–799. https://doi.org/10.1208/s12249-014-0267-8
35. Becket G, Schep LJ, Tan MY (1999) Improvement of the in vitro dissolution of praziquantel by complexation with α-, β- and γ-cyclodextrins. Int J Pharm 179:65–71. https://doi.org/10.1016/S0378-5173(98)00382-2
36. Makhlof A, Miyazaki Y, Tozuka Y, Takeuchi H (2008) Cyclodextrins as stabilizers for the preparation of drug nanocrystals by the emulsion solvent diffusion method. Int J Pharm 357: 280–285. https://doi.org/10.1016/j.ijpharm.2008.01.025

Nanocrystals: Optimization Techniques, Solubility Considerations, and Stability Issues

4

Atte Junnila, Erika Espo, Maha F. Emam, Irem Namli, Eetu Myöhänen, Anssi-Pekka Karttunen, Tapani Viitala, and Leena Peltonen

Abstract

Nanocrystals represent an advanced approach to improve the solubility and bioavailability of drugs with poor aqueous solubility. Their formulation requires a well-structured optimization process that leverages techniques such as high-pressure homogenization, wet milling, and precipitation to achieve optimal particle size and stability. Enhancing drug solubility primarily depends on reducing particle size, modifying surface properties, and selecting appropriate stabilizers to facilitate dissolution and absorption. However, maintaining stability remains a challenge as drug nanocrystals are susceptible to aggregation, Ostwald ripening, and structural transformations over time. To counteract these issues, stabilization strategies such as polymeric coatings, lyophilization, and surfactant incorporation are employed to preserve their physicochemical properties and extend shelf life. This chapter emphasizes key optimization strategies, and solubility and stability considerations for drug nanocrystal formulations, which are essential for maximizing their therapeutic potential.

Keywords

Dissolution · Nanocrystals · Optimization · Solubility · Stability

A. Junnila · E. Espo · M. F. Emam · I. Namli · E. Myöhänen · A.-P. Karttunen · L. Peltonen (✉)
Division of Pharmaceutical Chemistry and Technology, University of Helsinki, Helsinki, Finland
e-mail: leena.peltonen@helsinki.fi

T. Viitala
Division of Pharmaceutical Chemistry and Technology, University of Helsinki, Helsinki, Finland

Pharmaceutical Sciences Laboratory, Åbo Akademi University, Turku, Finland

N. K. Mehra, K. K. Bansal (eds.), *Nanocrystal Technologies in Biomedical Sciences*,
https://doi.org/10.1007/978-981-95-2968-1_4

Abbreviations

A	Hamaker constant
$A/\Delta A$	Surface area/surface area change
API	Active Pharmaceutical Ingredient
BBD	Box–Behnken design
BCS	Biopharmaceutics Classification System
CAP	Cellulose Acetate Phthalate
C_b	Dissolved concentration
CCD	Central composite design
CMC-Na	Croscarmellose sodium
CPP	Critical process parameter
CQA	Critical quality attribute
C_s	Saturated solubility
D	Diffusion coefficient
d	Interparticulate distance or particle diameter
DLVO	Derjaguin–Landau–Verwey–Overbeek
dm	Dissolved mass
DoE	Design of experiment
dt	Time
EDL	Electrical double layer
FBS	Fetal bovine serum
g	Gravitational acceleration
γ	Surface tension
h	Diffusional layer thickness
HPC	Hydroxypropyl cellulose
HPH	High-pressure homogenization
HPMC	Hydroxypropyl methylcellulose
L	Surface length
M_r	Molecular weight
P_∞	Dissolution pressure of an infinitely sized particle
PBS	Phosphate-buffered saline
PDI	Polydispersity index
PEG	Polyethylene glycol
P_r	Dissolution pressure of a particle with radius r
PVA	Polyvinyl alcohol
PVP	Polyvinyl pyrrolidone
QA	Quality attribute
QbD	Quality by design
R	Gas constant
r	Radius
RSM	Response surface methodology
RT	Room temperature
SDS	Sodium dodecyl sulfate

SLS	Sodium lauryl sulfate
T	Absolute temperature
TPGS	D-α-tocopherol polyethylene glycol 1000 succinate
V	Molar volume of particle
V	Flow velocity
v	Velocity
V_A	Attraction potential
VAP peptide	Vesicle-associated membrane protein-associated protein peptide
vdWs	van der Waals
V_R	Repulsion potential
V_T	Total potential energy
γ_{SL}	Interfacial tension between the solid particle and the liquid medium
ΔG	Gibbs free energy
ΔH	Enthalpy change
ΔS	Entropy change
ρ_s/ρ_0	Density of dispersed particle/density of dispersion medium
ε	Dielectric constant
κ	Ion concentration
ξ	Zeta potential

4.1 Introduction

Improved solubility and bioavailability of drug nanocrystal formulations are based on the small particle size. However, small particle size lowers the physical stability of these systems, and they are prone to aggregation [1]. Accordingly, efficient colloidal stabilization is crucial for drug nanocrystal-based formulations. In these systems, physical stability (stable particle size) is reached by using different kinds of stabilizers [2]. However, no universal stabilizer has been found so far, and identifying a suitable stabilizer to enhance the particle size stability of drug nanocrystals is challenging [3].

Drug nanocrystals can be manufactured by top-down or bottom-up techniques [4]. The optimization of the manufacturing process is related to the technique, and different kinds of factorial design and quality by design approaches have been utilized in many studies to improve the understanding of the process [5, 6]. Besides the production technique, stabilizer(s) selection and amount used also require optimization [7, 8]. Target particle size (and particle size deviation) and stabilizer selection are important factors, while these have an impact on solubility and drug release behaviour of drug nanocrystal formulations. It is not only the increased surface area of smaller drug particles that affects drug release, but also one should remember that when the particle size is below 1 μm, the saturated solubility is related to the particle size [9, 10]. The higher saturated solubility, besides the larger surface area of the drug particles and thinner diffusion layer thickness, is according to the Noyes–Whitney equation, increasing the drug release rate.

The formulation of drug nanocrystals requires a well-structured optimized process, with the exact understanding of the physicochemical phenomena behind the small particle size. In this review, first, the production techniques and screening studies of drug nanocrystals are reviewed, followed by the process optimization, including factorial design and design of experiments concepts. Then, drug solubility and dissolution, including dissolution models, are presented. Finally, the stability issues related to drug nanocrystals are discussed.

4.2 Production Techniques for Drug Nanocrystals and Formulation Development

Drug nanocrystals can be produced through a variety of methods, which are broadly categorized into two main approaches: top-down and bottom-up methods [11–13]. In the top-down method, the particle size of bulk drug material is mechanically reduced to the nanometer scale. In contrast, the bottom-up methods build nanocrystals from molecular drug solutions through crystallization processes. The pharmaceutical industry most often applies both approaches in combination to improve process efficiency and product stability [11, 14]. In Table 4.1, the different kinds of approaches for the production of drug nanocrystals are briefly presented.

4.2.1 Top-Down Methods

The most established top-down methods for producing drug nanocrystals are milling (including wet media milling and dry milling) and high-pressure homogenization. In wet media milling (also referred to as bead milling or media milling), the drug powder is suspended in a liquid medium and ground using high-energy collisions with milling beads, such as zirconium oxide beads [7, 54]. Stirred milling creates intense shear forces through bead–drug/drug–drug particle collisions as well as collisions with the vessel walls, gradually reducing the particle size of the drug powder. Dry milling can also be applied, but this is typically leading to amorphous drug forms, as in this case solvent (often water) is not added to the drug powder, which is as a plasticizer stabilizing crystal structures [55, 56]. Although milling is one of the most widely used methods for drug nanocrystal production due to its simplicity and scalability, this technique presents certain typical challenges, such as prolonged processing times, high energy input, and contamination risks [57, 58].

High-pressure homogenization (HPH) involves forcing a drug suspension through a narrow gap at high pressure, thereby generating intense forces such as cavitation, shear, and particle–particle collisions [59, 60]. Especially cavitation plays a central role in reducing the particle size because when cavitation bubbles collapse, they create high-energy microjets that break down solid drug particles. This method is compatible with continuous manufacturing, but it can generate heat, which may limit its use with thermosensitive drug compounds [61].

Table 4.1 Examples for the production of different kinds of drug nanocrystal systems

Drug	Stabilizer/matrix former	Technique	Delivery route	Particle properties: size (nm), PDI, Z-potential (mV)	Suspension or solid dispersion	Dissolution/ permeation, stability, and other notes	References
Aceclofenac	Pectin	Antisolvent precipitation	Transdermal	148 ± 15 nm 0.170 21.2 mV	Nanosuspension	Nanosuspension showed 2.87 times (the radiant tail flicking model) and 2.73 (paw licking model) better analgesic effect than the marketed formulation.	[15]
Amitriptyline hydrochloride	–	Spray freezing into vapor	Oral	90 nm na na	Solid dispersion	Completely dissolved in 5 min, while pure API had dissolved 89% after 65 min.	[16]
Aprepitant	Hydroxyl propyl chitosan	Wet media milling	Oral	151 ± 14 nm <0.35 63.5 ± 0.3 mV	Nanosuspension	30.1 ± 2.2% of pure API had dissolved after 30 min, while almost 90% of nanosuspension had dissolved in 5 min. The relative bioavailability of the nanosuspension was 147.7% compared to Emend®.	[17]

(continued)

Table 4.1 (continued)

Drug	Stabilizer/matrix former	Technique	Delivery route	Particle properties: size (nm), PDI, Z-potential (mV)	Suspension or solid dispersion	Dissolution/ permeation, stability, and other notes	References
Cabazitaxel	TPGS, lipid membrane, *p*-hydroxybenzoic acid, VAP peptides	Thin film hydration	Parenteral	140 nm na 8.2 mV	Nanosuspension	Stability in PBS at 4 °C and FBS 37 °C at least 24 h. Lipid coating provided enhanced stability and longer circulation time.	[18]
Candesartan cilexetil	Poloxamer 407, HPMC E5	Antisolvent precipitation	Oral	159 ± 8 nm 0.177 ± 0.043 −23.7 ± 1.0 mV	Solid dispersion/ nanosuspension	Nanocrystal dissolution rate was 96.02 ± 3.05% vs. 4.09 ± 1.67 and. 14.33 ± 3.45, for pure API and commercial formulation, respectively. Stable up to 3 months.	[19]
Carvedilol	Poloxamer 188, Eudragit® RS PO	Lyophilization	Oral	180 nm 0.35 na	Solid dispersion	Dissolution efficiency in sink condition 85.3 ± 3.0% vs. pure API 51.5 ± 9.6%. Stability at 40 °C up to 4 weeks, at 60 °C up to 2 weeks.	[20]

Carvedilol	Poloxamer 188 and Eudragit® RS PO	Rotary vacuum-drying	Oral	197 nm 0.38 na	Solid dispersion	Dissolution efficiency in sink condition 79.8 ± 1.8% vs pure API 51.5 ± 9.6%. Stability at 40 °C up to 4 weeks, at 60 °C up to 2 weeks.	[20]
Curcumin	Chitosan, sodium alginate, CAP	Ultrasound-assisted antisolvent crystallization and layer-by-layer coating techniques	Oral	421 ± 14 nm 0.21 ± 0.01 −47 ± 3 mV	Suspension	pH-dependent dissolution.	[21]
Curcumin and beclomethasone dipropionate	Poloxamer 188	Wet media milling	Inhalation	221 ± 7 nm 0.25 ± 0.02 na	Nanosuspension	Stabile up to 90 days at RT.	[22]
Cyclosporine	HPMC 3 cps, SDS	Wet media milling	Oral	561.2 ± 31.3 nm 0.374 ± 0.037 −24.4 ± 0.9 mV	Nanosuspension	Saturation solubility for coarse API powder was 6.48 ± 0.88 μg/mL and for nanosuspension 29.41 ± 7.56 μg/mL. Stable up to 1 month at RT.	[23]
Deferasirox	–	Rapid expansion of supercritical solution (RESS)	Oral	175 nm na na	Solid dispersion		[24]

(continued)

Table 4.1 (continued)

Drug	Stabilizer/matrix former	Technique	Delivery route	Particle properties: size (nm), PDI, Z-potential (mV)	Suspension or solid dispersion	Dissolution/ permeation, stability, and other notes	References
Deferasirox	PVA	Wet media milling	Ocular	237.1 ± 6.8 nm na, −18.3 ± 2.6 mV	Nanosuspension	Coarse API solubility was 37.18 ± 4.08 μg/mL^{-1}, while for nanosuspension it was 146.1 ± 5.6 μg/mL^{-1}. Stabile up to 14 days at 4 °C, with minimal increase in particle size and up to 14 days at RT, with more noticeable particle size increase.	[25]
Docetaxel	Sodium acetate	Lyophilization	Injection	161 ± 11 nm na na	Solid dispersion	Less toxic in higher doses than simulated clinical docetaxel with Tween 80/EtOH as carrier.	[26]
Everolimus	PVA, HPMC, Poloxamer 407	Injection	Ocular	156.47 ± 1.10 nm 0.091 ± 0.005 −8.99 mV	Nanosuspension	Compared to micelles with API, nanosuspension had higher release, permeability, and	[27]

						bioavailability. Nanosuspension remained stable at 4 °C for up to 180 days and relatively stable at 40 °C in darkness up to 30 days (stability was not evaluated longer).	
Fenofibrate	PVP K30 and Poloxamer 188	Lyophilization	Oral	246.6 ± 2.6 nm na −23.6 ± 0.6 mV	Solid dispersion	Multiple times higher dissolution rate vs. commercial tablet and pure API. Stable up to 6 months.	[28]
Fenofibrate	Poloxamer 188	Lyophilization	Oral	460 ± 20 nm 0.468 ± 0.01 35.1 ± 1.1 mV	Solid dispersion	The solubility of API in 1% SLS was 23.35 ± 1.54 µg/mL, while nanosuspension solubility was 107.29 ± 1.9 µg/mL. Particle size increased slightly after 180 days at 5 ° C. After 90 days at RT, particle size and PDI increased slightly. 180 days at	[29]

(continued)

Table 4.1 (continued)

Drug	Stabilizer/matrix former	Technique	Delivery route	Particle properties: size (nm), PDI, Z-potential (mV)	Suspension or solid dispersion	Dissolution/ permeation, stability, and other notes	References
						RT, significant increase in size and PDI was observed.	
Fenofibrate	HPMC E15, SDS, glycerin	Wet media milling, film coating	Oral	201 nm na na	Nanosuspension to film	Slight difference was found between nanosized and micronized films regarding solubility; however, in a time span of 40 min, 27% of API was released from nanosized film, while approximately 5% from the micronized film.	[30]
Fluticasone propionate	TPGS, cholesterol	Wet media milling, high-pressure homogenization	Inhalation	192 ± 3.87 nm na −18.7 ± 0.42 mV	Lipo-nanosuspension	Stabile up to 3 months at RT.	[31]
Ganciclovir	Eudragit® RL 100 and PVA	Nanoprecipitation	Ocular	260.8 ± 2.05 nm 0.158	Nanosuspension	Prolonged drug release and enhanced corneal permeability	[32]

				± 0.02 −10 ± 6.45 mV		compared to API solution.	
Genistein	Lecithin, cholesterol, stearamine, TPGS	Wet media milling, thin film hydration	Inhalation	206.2 ± 15.3 nm 0.21 ± 0.01 −21.1 ± 4.1 mV	Nanosuspension	Drug lung retention improved, and macrophage clearance decreased, if coated with lipid layer.	[33]
Glibenclamide	–	Cryo-milling	Oral	550 ± 23 nm na na	Solid dispersion	Cryo-milled API had faster and higher solubilisation than conventionally milled API or native API crystals. Tendency to crystallize was observed, in addition to agglomeration, caused probably due to lack of surfactants.	[34]
Harmine	TPGS, deacetylated gellan gum	High-pressure homogenization, spray-drying	Intranasal	179.21 ± 0.01 nm na −41.4 ± 1.08 mV	Solid dispersion/ gel	Substantial increase in dissolution and bioavailability compared to unformulated API.	[35]
Hydrochlorothiazide	Poly (lactic-co-glycolic) acid, PVA	Emulsion solvent evaporation, lyophilization	Oral	270 nm na, −27.9 mV	Solid dispersion		[36]

(continued)

Table 4.1 (continued)

Drug	Stabilizer/matrix former	Technique	Delivery route	Particle properties: size (nm), PDI, Z-potential (mV)	Suspension or solid dispersion	Dissolution/ permeation, stability, and other notes	References
Hydrocortisone acetate	Poloxamer 188, chitosan chloride	High-pressure homogenization	Oral	716 nm 0.35 na	Nanosuspension	In delayed release test tablets from nanosuspension had released 93% of drug after 10 h, while micronized API tablet released 76%.	[37]
Ibuprofen	Tween 80	Antisolvent crystallization	Injection, intramuscular	300–400 nm 0.1–0.2 −5.44 to −22.9 mV	Nanosuspension	70% of pure API was dissolved after 60 min, while 100% of nanosuspension was released after 40 min. Stable up to 3 months.	[38]
Imiquimod	Sorbitan monostearate, polysorbate 80, copaiba oil	Spontaneous emulsification	Intravaginal	190 ± 8 nm 0.11 ± 0.01 na	Nanoemulsion	Could inhibit the formation of cervical cancer colonies.	[39]
Indomethacin	Polyethylene glycol, poly (*N*-isopropylacrylamide)	Emulsion-templated lyophilization	Oral	480 nm 0.22 −17 ± 4 mV	Solid dispersion	Stable after 8 months in a desiccator.	[40]
Indomethacin	TPGS	Wet media milling, spray-drying	Oral	193 ± 2 nm na na	Nanosuspension/ solid dispersion	Nanosuspension and solid dispersion were compared, and	[41]

						dissolution speed was slowed with dried powder; however, solubility was not affected. Nanosuspension relative particle size compared to fresh solution (193 nm) after 3 months at 4 °C 97 ± 2%, at RT 100 ± 1%, at 40 °C 140 ± 2%.	
Itraconazole	Poloxamer 407	Wet media milling	Ophthalmic	378 ± 31 nm 0.35 ± 0.03 na	Nanosuspension/gel	Drug release after 24 h. Nanosuspension 94.17 ± 16.32% vs. API 28.98 ± 5.04%.	[42]
Ivacaftor and colistin	N-(methylpolyyoxyethylene oxycarbonyl)-1,2-distaroyl-*sn*-glycero-3-phosphoethanolamine sodium salt)	Antisolvent precipitation-ultrasonication and ultrasonic spray-freeze-drying	Inhalation	199.3 ± 4.3 nm 0.100 ± 0.076 na	Solid dispersion	Stabile >120 h in suspension. Exhibited low cytotoxicity. Enhanced antimicrobial activity against colistin-resistant *P. aeruginosa*.	[43]
Ivermectin	Poloxamer 188	Wet media milling, spray-drying	Oral	198.4 ± 0.6 nm	Nanosuspension	Nanosuspension solubility was 47 times higher in	[44]

(continued)

Table 4.1 (continued)

Drug	Stabilizer/matrix former	Technique	Delivery route	Particle properties: size (nm), PDI, Z-potential (mV)	Suspension or solid dispersion	Dissolution/ permeation, stability, and other notes	References
				0.184 na		water and 4.8 in simulated intestinal fluid compared to pure API. No significant change was observed after storage for 100 days at 4 °C or at 25 °C.	
Maraviroc	Poly(vinyl alcohol) and dioctyl sulfosuccinate sodium salt	Emulsion-templated lyophilization	Oral	728 nm 0.3 −25.3 mV	Solid dispersion	Stabile at RT up to 20 months. Has potential for once-daily administration.	[45]
Naproxen	Poloxamer 188	Wet media milling	Oral	270 nm na na	Nanosuspension	C_{max} (μg/mL) was 126 ± 4 for unmilled and 187 ± 18 for nanosuspension. Stable for 4 weeks at 4 °C. Nanosuspension caused less irritation than unmilled API	[46]
Oxybutynin	Poloxamer 407 and vitamin E succinate	Solvent–antisolvent	Transdermal	142 ± 2.8 nm	Nanosuspension	Saturation solubility of coarse API was	[47]

		precipitation, high-pressure homogenization		0.120 ± 0.019 −22.37 ± 2.89 mV		7.53 ± 0.16 μg/mL while nanosuspension had saturation solubility of 85.82 ± 2.93 μg/mL. Incorporating nanosuspension into a transdermal gel did not negatively affect nanosuspension properties.	
Prednisolone acetate	Mannitol and PVA	Spray-drying	Ocular	590 ± 165 nm na na	Nanosuspension	After 30 s nanosuspension showed 83% dissolution compared to 49% of API micro suspension. Stable at RT up to 4 weeks. Nanosuspension had 6.2 higher adherence to cornea than API micro suspension.	[48]
Resveratrol	PVA	High-pressure homogenization, spray-drying	Inhalation	586.6 ± 17.9 nm 0.347 ± 0.035 −4.78 ± 0.13 mV	Nanosuspension	Higher stabilizer ratios can influence flowability, morphology, and fine particle fraction.	[49]

(continued)

Table 4.1 (continued)

Drug	Stabilizer/matrix former	Technique	Delivery route	Particle properties: size (nm), PDI, Z-potential (mV)	Suspension or solid dispersion	Dissolution/ permeation, stability, and other notes	References
Rifampicin	Povacoat® Type F	Wet media milling	Inhalation	340.6 ± 5.4 nm 0.2 ± 0.1 −8.8 ± 0.4 mV	Nanosuspension	Close to 90% of nanosuspension was solubilised after 5 min, while at the same time close to 20% of commercial suspension was solubilized. Stable up to 2 years at 4 °C. Twofold increase in API concentration compared to commercial product.	[50]
Rotigotine	PVA	Wet media milling	Transdermal	274.09 ± 7.43 nm 0.17 ± 0.04 −15.24 ± 2.86 mV	Nanosuspension	After 28 days, coarse API had released 75.02 ± 6.36% into dissolution media while nanosuspension had released 95.95 ± 5.74%. Lyophilized stable powder up to 28 days at RT in desiccator.	[51]

Telmisartan	Modified locust bean gum	Lyophilization	Oral	165.8 ± 4.1 nm na na	Solid dispersion	Dissolution efficiency after 120 min 83.07 ± 1.24%.	[52]
Telmisartan	Poloxamer 188	Lyophilization	Oral	425.5 ± 5.1 nm na na	Solid dispersion	Dissolution efficiency after 120 min 55.65 ± 1.14%.	[52]
Triamcinolone acetonide	Soluplus® and sodium deoxycholate	Solvent–antisolvent precipitation	Ocular	243 ± 6.5 nm na na	Nanosuspension	When nano suspension was incorporated into a gel, it showed improved API concentrations in vitreous humour, which were sustained longer. Stable up to 60 days in RT and 2–8 °C	[53]

API active pharmaceutical ingredient, *CAP* cellulose acetate phthalate, *TPGS* D-α-tocopherol polyethylene glycol 1000 succinate, *na* data not available, *FBS* fetal bovine serum, *HPMC* hydroxy propyl methyl cellulose, *PBS* phosphate-buffered saline, *PVA* poly vinyl alcohol, *RT* room temperature 25 °C, *SDS* sodium dodecyl sulfate, *SLS* sodium lauryl sulfate, *VAP peptides* vesicle-associated membrane protein-associated protein peptides

4.2.2 Bottom-Up Methods

Bottom-up methods produce drug nanocrystals from molecular solutions, typically through precipitation-based or atomization-based techniques. A commonly used precipitation-based method is the antisolvent precipitation technique [62, 63]. In this process, the drug is first dissolved in a suitable organic solvent and then mixed with an antisolvent, which triggers supersaturation, nucleation, and crystal formation. Although this method is relatively simple and efficient, it requires precise control over mixing, solvent removal, and particle stabilization to prevent uncontrolled growth or aggregation.

Atomization-based methods, such as electospraying and spray-drying, are primarily used to dry out drug nanosuspensions into nanocrystals to produce certain dosage forms [64, 65]. Spray-drying is a rapid solvent removal method where a drug solution or suspension is atomized into hot air, consequently forming dry drug particles. Electrospraying uses a high-voltage electric field to atomize a drug solution into fine charged droplets, which form solid drug particles as the solvent rapidly evaporates during flight. In spray-drying, the liquid is atomized into fine droplets, which are dried in a hot gas, typically air.

In addition to these methods, freeze-drying is a widely used drying method in the downstream processing of nanocrystal-based formulations, such as drug nanosuspensions [66, 67]. In this method, a drug solution is frozen and lyophilized under conditions that allow some control over crystal growth and morphology, resulting in a dry powder. Freeze-drying is particularly suitable for heat-sensitive drugs and helps preserve the particle size and structure of drug nanocrystals.

4.2.3 Screening Studies and Scaling Up in Formulation Development

At the start of the process and formulation development of drug nanocrystal formulations, important quality attributes (QAs) that need to be selected include process parameters, drug concentration in the nanosuspension, stabilizer, and even the right production/processing technique [68–73]. This part of the early development is usually performed with a very limited amount of drug substance, though it can require a large number of screening experiments for the mentioned attributes. In certain cases, if the material quantity is not an issue, the screening can be accomplished simply by using small-scale lab equipment [69, 74, 75]. Still, there is also specific machinery designed for high-throughput screening. For example, Juhnke et al. [76] used a screening mill hosting 24 miniature milling beakers for screening drug–stabilizer pairs. With this approach, 42 experiments were conducted using only 110 mg of drug.

Successful screening studies have also been performed effectively by applying, for example, wet milling in vials [77, 78] and 96-well plates [79], which indicates that the screening process can at least be started with simple and cost-effective systems. However, one possible challenge in using these adapted designs instead of conventional milling equipment is the possible wear of the vessels, which needs to be considered and minimized.

Rather than performing laborious single-parameter screening studies, the process is preferably performed using a design of experiments to perform effective screening [71, 75, 80]. In the screening phase for bottom-up processes through solvent–antisolvent precipitation, the process can exploit continuous manufacturing with microfluidic devices or use more conventional batch crystallization on a small scale [71, 81]. The screening phase can be performed using variable systems and volumes from microliters to a few hundred milliliters. For other bottom-up processes, like spray-drying, the approach might be to simply start with small lab-scale equipment in the screening phase [82].

After the screening phase and finding the right formulation composition and processing parameters at a small scale, the process will be scaled up to pilot scale and later to commercial scale. As a basic principle, the process can be intensified by volume, by increasing production speed/throughput, or through parallelization. For example, Tuomela et al. [83] showed that similar quality in terms of particle size was still achieved after moving to larger milling vessels in wet ball milling of drug nanosuspensions. Both Romero et al. and Juhnke et al. have also demonstrated in their studies that scaling up from miniaturized screening to laboratory or even small pilot scale was feasible even when the mill design changed [76, 77]. This provides certainty that the process with optimal formulation can be scaled up to different equipment if the needed energy for drug particle breakage in a top-down system is achieved. Same findings were presented by Hagedorn et al. [70] by illustrating scale-up from screening, performed with dual centrifugation, to production scale with agitator bead milling.

In some cases, there can also be a shift from batch process to continuous process when moving from laboratory scale to production [84]. Considering bottom-up processes, and especially the nano- or microfluidic systems, there are also clear possibilities to parallelize the system, which makes the scaling up easier, considering parameter optimization [81]. To highlight the commercial utilization of, for example, microfluidic systems, this was the chosen manufacturing process of the Comirnaty COVID-19 vaccine. The product is not a drug nanocrystal formulation, but the same principle applies. Alongside parallelization, scaling up by increasing the size of the production vessel in batch crystallization or, for example, spray-drying, is naturally an option but might require some parameter optimization.

4.3 Process Optimization Approaches for Nanocrystals

The successful development and manufacturing of drug nanocrystals require a systematic approach to process optimization. This is often achieved by adopting the principles of quality by design (QbD), which aims to build quality into products and processes throughout the entire production chain. A key aspect of QbD is defining critical quality attributes (CQAs), which represent the desired characteristics of the final drug nanocrystal product and serve as the responses to be measured [5].

For drug nanocrystals, typical CQAs include particle size, particle size distribution [polydispersity index (PDI)], zeta potential, dissolution rate, and physical stability [85–87]. Common critical process parameters (CPPs)/factors investigated include concentrations or amounts of stabilizers and surfactants, drug concentration, stirring or homogenization speed, sonication time, and temperature [5, 87–89]. Systematic optimization through the control of these CPPs is crucial for producing nanocrystals with the desired properties and ensuring their long-term stability [90].

4.3.1 Design of Experiment (DoE)

Design of experiment (DoE) is a powerful statistical technique for both variable screening and process optimization in pharmaceutical development, including the field of nanomedicine. Unlike the traditional one-factor-at-a-time approach, which alters only one factor while keeping others constant, DoE enables the simultaneous variation of multiple experimental parameters [89]. This multivariate approach provides insight throughout a high-dimensional parameter space and allows for a better understanding of how independent variables (factors or CPPs) affect the dependent variables (responses or CQAs) and their interactions [87, 89].

The application of DoE is highly efficient and time-saving. It requires a minimal number of experimental runs to find parameter configurations that optimize various responses of interest [87]. Furthermore, DoE is closely aligned with the principles of quality by design (QbD), aiming for a deeper understanding of the manufacturing process and establishing a design space within which quality is assured [5]. A DoE study follows a systematic workflow that begins with clearly defining the research goals and desired responses, followed by identifying potential experimental variables (factors or CPPs) that may influence these responses, based on prior knowledge or preliminary experiments. After defining the problem and parameters, the process involves selecting an appropriate experimental design based on the study goals (e.g., screening or optimization). Then, conducting the experiments according to the design matrix, and statistically analyzing the results to build models and identify optimal conditions [86, 90]. DoE studies are often performed iteratively, typically beginning with a screening phase to identify the most significant variables, followed by an optimization phase using higher-level designs [86].

4.3.2 Factorial Designs and Response Surface Methodology in Drug Nanocrystal Optimization

Optimizing the formulation and production of drug nanocrystals commonly employs a staged approach within the framework of DoE. This sequential methodology involves an initial phase focused on identifying the most influential factors (CPPs), followed by a second phase aimed at determining their optimal settings.

The initial stage primarily utilizes screening designs, such as factorial designs, to efficiently identify the CPPs that significantly affect the target characteristics of drug nanocrystals. This phase is crucial for "screening out" nonsignificant factors to streamline further investigation [82, 87]. These are typically first-order designs that investigate variables at two levels (high and low, often coded as +1 and −1) to determine the linear influence on the responses.

Once influential variables are selected, response surface methodology (RSM) is employed for the optimization phase. RSM is a second-order multivariate technique that utilizes higher-order optimization designs to investigate significant variables at a higher number of levels, enabling the estimation of nonlinear effects [86]. This allows the experimental data to be fitted to polynomial equations (commonly quadratic) to model the response surface and make statistical predictions [89]. Common RSM designs utilized in drug nanocrystal optimization include Box–Behnken design (BBD), central composite design (CCD), and three-level factorial design. RSM models also facilitate multi-response optimization by integrating individual response models, often via a desirability function, to find an overall optimum satisfying multiple performance criteria simultaneously [82]. This staged application of factorial designs for screening and RSM for optimization provides a more accurate and robust understanding of the process and product compared to traditional methods.

Several studies highlight the effective application of factorial designs and RSM for optimizing drug nanocrystal formulations (Table 4.2). These studies systematically evaluate the relationships between critical process parameters (CPPs) and critical quality attributes (CQAs) within the complex framework of drug nanocrystal formulation and manufacturing. The outcomes of such studies help define the design space within which the process can operate reliably to produce a product meeting the defined quality profile.

4.4 Solubility

Many of the active pharmaceutical ingredients (APIs) and newly discovered chemical entities suffer from high lipophilicity coupled with low aqueous solubility and are therefore classified as class II or class IV drugs according to the Biopharmaceutics Classification System (BCS). Challenges related to solubility and dissolution rate directly affect the bioavailability of APIs, thus posing a major dosing challenge for many therapeutics. Therefore, solubility enhancement methods are developed for overcoming these obstacles, including drug nanocrystals [99–101].

When a drug is orally administered, the first step is the dissolution of the drug from the formulation into the gastric environment, followed by absorption through the gastrointestinal epithelium [102]. Nanocrystals enhance the solubility and dissolution rate of APIs owing to the decreased particle size, resulting in a higher concentration gradient compared to traditional systems [100, 102].

Table 4.2 Case studies of drug nanocrystal fabrications optimized via design of experiment (DoE)

Drug	Primary objective	Target responses (CQAs)	Optimization design	Reference
Azilsartan medoxomil	Increase drug solubility and dissolving rate as well as enhance bioavailability	Particle size, % drug release	3^2 factorial design	[85]
Valsartan	Develop chronotherapeutic drug nanocrystals using a modified anti-solvent precipitation	Particle size, polydispersity index (PDI), and zeta potential	Box–Behnken design	[82]
Glibenclamide	Optimize the process conditions for a novel combinative particle size reduction method (freeze-drying and high-pressure homogenization)	Particle size, zeta potential, and polydispersity index	Customized two-factor, five-level DoE	[80]
Ibuprofen	Improve drug solubility and bioavailability of nanocrystals produced by high-pressure homogenization	Mean particle size, polydispersity index (PDI), and zeta potential	2^2 factorial design	[91]
Cefdinir	Development and optimization of drug nanosuspension by solvent evaporation method to increase solubility and dissolution rate	Particle size, % drug release	2^3 factorial design	[92]
Hesperidin	Accelerate the determination of optimal spray drying parameters to produce well redispersible drug nanocrystals and improve drug solubility	Particle size	Two-step DoE approach: screening model design then RSM design	[69]
Lopinavir	Design and characterization of drug nanocrystals prepared by anti-solvent precipitation method for solubility enhancement aiming for dose reduction and minimizing side effects	Particle size, zeta potential, saturation solubility, and % drug content	3^2 full factorial design	[93]

(continued)

Table 4.2 (continued)

Drug	Primary objective	Target responses (CQAs)	Optimization design	Reference
Atenolol	Enhance the permeability and bioavailability of drug nanocrystal formulations prepared by high-pressure homogenization	Particle size, zeta potential, and production yield (%)	3^2 factorial design	[94]
Ritonavir	Optimize drug nanosuspension preparation by microfluidization using polymeric stabilizers	Particle size, particle size distribution, and zeta potential	Full factorial design	[95]
Artemether	Development and optimization of drug nanocrystal formulations prepared by miniaturized wet bead milling	Particle size, polydispersity index (PDI), and zeta potential	Two-step DoE approach: full 2^3 factorial design (screening) then RSM with axial points (optimization)	[96]
Dexibuprofen	Synthesizing stable drug nanocrystal formulations using top-down method	Particle size, polydispersity index (PDI), and zeta potential	Box–Behnken design	[90]
Haloperidol	Development and optimization of drug nanocrystal formulations prepared by the anti-solvent precipitation method	Particle size, zeta potential, and cumulative % release	3^2 factorial design	[97]
Flibanserin	Optimize drug nanocrystal formulations using the antisolvent sono-precipitation technique	Particle size, polydispersity index, and zeta potential	2^3 full factorial design	[88]
Carvedilol	Optimization of drug nanocrystal formulations prepared by anti-solvent precipitation technique	Particle size, zeta potential, and drug release (%)	3^2 factorial design	[98]

Saturation solubility, C_s, is a drug property that depends on the dissolution medium, the physicochemical properties of the APIs, and the temperature of the medium. This dependency is mostly valid for micron-sized powdered APIs, making it possible to enhance the saturation solubility by decreasing the particle size of the drug particles. Particle size, particularly in the nanometer range, plays a crucial role in determining saturation solubility. Consequently, drug nanocrystal formulations are highly suitable for enhancing saturation solubility and significantly improving the dissolution profile of poorly water-soluble APIs [84, 99, 103]. For nanosized particles, the increase in saturation solubility can be mainly explained by the correlation of the Kelvin and Ostwald–Freundlich equations. The Kelvin equation (4.1) explains this phenomenon concerning vapor pressure:

$$\ln \frac{P_r}{P_\infty} = \frac{2\gamma M_r}{rRT\rho} \tag{4.1}$$

where P_r is the dissolution pressure of a particle with the radius of r, P_∞ is the dissolution pressure of an infinitely sized particle, γ is the surface tension, R is the gas constant, T is the absolute temperature, M_r is the molecular weight of the particle with the radius of r, and ρ is the density of the particle [104]. This equation is mainly used to describe the vapor pressure of over the curve of the surface of lipid droplets. In the case for nanosized particles, the Kelvin equation demonstrates the relationship between increasing curvature by the decreased particle size, resulting in increased vapor pressure.

Since the vapor pressure is directly related to the dissolution pressure, transformation and equilibrium between liquid and gas phase can in principle be viewed in the same way for the transformation between solid (drug nanocrystals) and liquid (dispersion media) phase. This can be summarized as follows; when the saturation solubility is achieved, then the dissolution and recrystallization of the molecules are in equilibrium. The decrease in the particle size leads to an increased curvature of the droplet (or crystal), increasing the surface area, resulting in an increased saturation solubility of the API (Figs. 4.1 and 4.2) [84, 103, 105, 106].

The Oswald–Freundlich equation (4.2), equivalent to the Kelvin equation, explains the relationship between particle size and saturation solubility, which is more significant for particles in the nanosize range:

$$\ln \frac{C_s}{C_\alpha} = \frac{2\sigma V}{RT\rho r} \tag{4.2}$$

where C_s is the saturation solubility of the API, C_α is the solubility of larger particles, σ is the interfacial tension, V is the molar volume of particle, R is the gas constant, T is the absolute temperature, ρ is the density of the particle, and r is the radius of the particle [102, 107, 108].

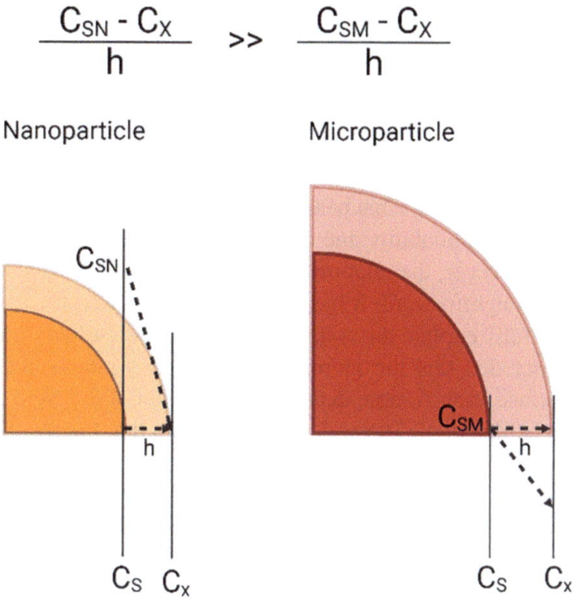

Fig. 4.1 The source of the difference between saturation solubility for micron-sized and nanosized particles in relation to surface curvature and diffusional distance (h) (C_s saturated water on the surface; M micron sized particle; N nanosized particle; C_x concentration at diffusional distance; h diffusional distance). (Ref. [103])

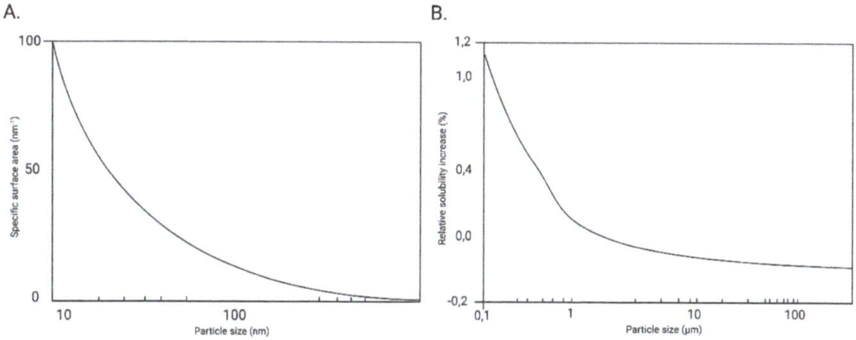

Fig. 4.2 (**a**) Specific surface area as a function of particle size; (**b**) the relationship between saturation solubility and particle size. (Ref. [103])

4.5 Dissolution Rate

The particle size of a drug has a significant impact on the rate of dissolution. This relationship between particle size and drug dissolution can be described by the Nernst–Brunner equation (also known as the Noyes–Whitney equation) (Eq. 4.3):

$$\frac{dm}{dt} = \frac{A \times D}{h} \times (C_S - C_b), \tag{4.3}$$

where dm is the dissolved mass of drug at time dt, A is the surface area, D is the diffusion coefficient, h is the diffusional layer thickness, C_s is the saturated solubility, and C_b is the dissolved concentration.

By reducing particle size, the dissolution rate of a drug is accelerated. This connection is particularly important for drug nanocrystals as the relative surface area increases significantly more with colloidal particles in the submicron region (Fig. 4.2). Drugs that benefit the most from particle size reduction are those that have poor bioavailability due to their slow rate of dissolution.

However, the relationship between drug particle size and dissolution is not quite so straightforward. It has been observed that an increase in surface area alone does not fully explain the measured increase in dissolution rate [109]. Therefore, it can be concluded that the increase in dissolution rate is not solely a result of increased surface area. A more detailed examination of the Nernst–Brunner equation provides an explanation. The equation also shows that the dissolution rate is regulated by the thickness of the diffusion layer (h). In fact, rather than being constant, the diffusion layer is dependent on particle size. The effect on the diffusion layer is explained through the Prandtl equation, where the relationship between L and V is shown to change, resulting in a thinner diffusion layer (Eq. 4.4) [109]:

$$h = k \times \left(\frac{L^{\frac{1}{2}}}{V^{\frac{1}{2}}} \right) \tag{4.4}$$

where h is the diffusion layer thickness, k is a constant, L is the length of the surface in flow direction, and V is the flow velocity [109].

Various studies have presented support for the effect of particle size on the diffusion layer [110–118]. It is well-established that particle size reduction leads to the thinning of the diffusion layer. Consequently, the enhanced dissolution rate observed with nanocrystals can be attributed to an increased surface area, a thinner diffusion layer, and elevated saturation solubility (Fig. 4.2). Moreover, the reduction in diffusion layer thickness and the rise in saturation solubility collectively minimize the concentration gradient ($C_s - C_b$), thereby further promoting the dissolution rate.

It is also noteworthy that the surface area of a drug particle is not constant during dissolution but is a rate-dependent process. This can be accounted for, for example, in the equation derived by Hixson and Crowell [119]. As saturation solubility is particle size-dependent, it dynamically changes during the dissolution process. This phenomenon has been accounted for in the extended Hixson–Crowell model proposed by Ely et al. [120].

It is important to note that drug particle size is, in reality, a particle size distribution. This presents a challenge for dissolution method development. For example, Liu et al. [9] studied the dissolution of itraconazole nanocrystals under non-sink conditions. It has been observed that differences in the dissolution rates of drug nanocrystals with varying particle sizes may become less discernible with increasing sample amounts due to a higher proportion of smaller particles in the added sample. As a result, the smaller particles tend to dissolve preferentially, while the larger ones remain undissolved. In the study conducted by Liu et al. [9], optimal resolution was achieved when the nanosuspension was added in an amount precisely corresponding to the saturation solubility of the drug. This can be explained through the Nernst–Brunner equation, where the dissolution rate is directly dependent on the difference between the saturation solubility of the drug and the dissolved drug concentration ($C_s - C_b$). Thus, careful consideration of the effect of particle size distribution on the results should be made when dissolution testing is conducted. It may be argued that dissolution testing can yield more informative insights into the particle size and size distribution of a sample compared to conventional particle size determination methods [114]. According to the Nernst–Brunner equation, dissolution of a drug particle is preceded by the solvation of drug molecules from the particle surface. Only after this initial solvation step can the molecules diffuse or be convected into the surrounding dissolution medium. While diffusion is generally regarded as the slower, rate-limiting step in the dissolution process [121], Anhalt et al. [122] proposed that solvation may also significantly influence the overall dissolution rate, particularly in the case of nanocrystals.

4.6 Stability of Drug Nanocrystals

4.6.1 Aggregation and Sedimentation

To take advantage of the increased dissolution rate of poorly soluble drugs by reducing the particle size of the drug, the aim for drug nanocrystal formulations is usually to reach a submicron particle size (i.e., <1000 nm). The reduction in the particle size of a drug leads to a higher ratio of surface to volume compared to larger drug particles. Consequently, drug nanocrystals exhibit higher surface curvature and surface molecules that have a higher energy compared to the bulk material. Hence, drug nanocrystals exhibit increased surface energy compared to larger drug particles. This has direct consequences for the chemical and physical stability of drug nanocrystal formulations. Although the high surface energy in drug nanocrystals is one of the major contributors for the improved dissolution rate, it may eventually affect the chemical stability because surface molecules are more reactive due to their higher energy. Because of their high surface energy, drug nanocrystals are also thermodynamically unstable. This leads to physical stability issues primarily in the form of aggregation, sedimentation, and crystal growth (i.e., Ostwald ripening). The tendency of drug nanocrystals to reach a more thermodynamically stable state can be viewed through the change in Gibbs free energy, ΔG, when larger drug particles are divided into smaller particles, which increases the surface area, ΔA (Eq. 4.5) [123]:

$$\Delta G = \gamma_{SL} \times \Delta A, \tag{4.5}$$

where γ_{SL} is the interfacial tension between the solid particle and the liquid medium. As is generally known from thermodynamics, a system tends to minimize its Gibbs free energy to reach a stable state, and equilibrium is reached when $\Delta G = 0$. Thus, according to Eq. (4.3), a stable state can be reached either by reducing the total interfacial area or the interfacial tension between the solid particles and liquid medium. Thus, the driving force to reduce the total interfacial area will lead to aggregation and possibly to sedimentation. Therefore, aggregation and sedimentation are key issues in the stability of drug nanocrystals. Drug nanocrystals exhibit Brownian motion in solution, which consequently leads to random collisions and interactions between particles, thus increasing the probability of aggregation due to the attractive van der Waals forces between the particles [123]. Aggregation of drug nanocrystals reduces the solubility and dissolution rate of drugs due to the increase in particle size and particle size distribution, which can lead to inconsistent dosing. If aggregation occurs in a drug nanocrystal formulation, then larger particles will sediment under the action of gravity. The connection between particle size, medium viscosity, and density difference between the medium and the dispersed phase and the sedimentation rate is described by Stoke's law (Eq. 4.6) [124]:

$$v = \frac{d^2(\rho_s - \rho_0) \times g}{18 \times \mu_0}, \tag{4.6}$$

where v is the settling velocity of a particle, d is the particle diameter, ρ_s and ρ_0 and are the densities of the dispersed particle and the dispersion medium, and g is the gravitational acceleration. As seen from Stoke's law, reducing the particle size, matching particle and medium densities, or increasing medium viscosity can be used to reduce the sedimentation rate of particles. Although sedimentation is a key issue when considering the physical stability of drug nanocrystals, there are surprisingly few studies examining sedimentation issues in aqueous-based nanosuspensions. Wu et al. [125] concluded that the reasons for this are, firstly, that surfactants are generally used in most nanosuspensions to inhibit particle aggregation, which alleviates the sedimentation issues; secondly, that the small nanosized particles significantly reduce the sedimentation rate; and thirdly, that many aqueous nanosuspensions are transformed to dry solid form by spray drying or freeze-drying to circumvent the long-term sedimentation issues [125].

The most common approach to tackle aggregation and possible sedimentation issues for drug nanocrystals is the introduction of stabilizers, that is, surfactants and/or polymers, in the drug nanocrystal formulations. The addition of surfactants or polymers (or a combination of these) that adsorb at the solid–liquid interface of drug nanocrystals can alleviate aggregation tendency of particles by reducing the interfacial tension (e.g., Eq. 4.5). Surfactants and/or polymers adsorbed on drug nanocrystal interfaces also shield the more reactive surface molecules from interacting and reacting with compounds in their surrounding media. The selection of a stabilizer is based on its ability to wet the surface of the particles and offer a barrier to prevent nanoparticles from aggregating [126, 127].

4.6.2 Stabilization Mechanisms

Colloidal dispersions, such as drug nanocrystals, can be stabilized against aggregation through two main mechanisms, that is, electrostatic repulsion and steric stabilization [128, 129]. This can be achieved by adding ionic and nonionic stabilizers to the drug nanocrystal formulations. Ionic stabilizers commonly used in aqueous-based drug nanocrystal formulations include sodium dodecyl sulfate (SDS), lecithin, and docusate sodium, while nonionic surfactants are usually selected from Pluronic® surfactants, polysorbates (like Tween 80), polyethylene glycol (PEG), polyvinyl alcohol (PVA) polyvinylpyrrolidone (PVP), and cellulose polymers such as hydroxypropyl cellulose (HPC) and hydroxypropyl methylcellulose (HPMC) [125].

The classic Derjaguin–Landau–Verwey–Overbeek (DLVO) theory describes the stabilization of colloidal dispersions by electrostatic repulsion [130, 131]. The DLVO theory considers repulsive electrostatic forces and attractive van der Waals (vdWs) forces acting on similar colloidal particles in a medium (Fig. 4.3a).

The repulsive electrostatic forces originate from the electrical double layer (EDL) surrounding the particles in the medium and contribute to preventing the aggregation of colloidal particles. The EDL consists of two layers, that is, the Stern layer consisting of counter ions adsorbed on the particle surface to maintain electrical neutrality of the system, and the Gouy layer, which is the diffusion layer of ions. The total potential energy (V_T) of the particle–particle interaction is a sum of the repulsion potential (V_R) originating from the EDL and the attraction potential (V_A) originating from van der Waals (vdWs) forces. The attraction potential, V_A, is determined by the Hamaker constant (A), particle size, and interparticulate distance (d), while the electrostatic repulsion potential, V_R, depends on particle size, distance between the particles (d), zeta potential (ξ), ion concentration (κ), and the dielectric constant of the medium (ε). Thus, V_R is very sensitive to the ionic strength of the medium because by increasing the ion strength in the medium, the thickness of EDL consequently decreases due to screening of the surface charge [129]. This leads to the susceptibility of the dispersed particles to form aggregates.

The zeta potential (ξ), which is the electric potential determined at the shear plane, that is, at a layer approximately in the middle of the particle surface and the outer edge of the Gouy layer, is a key parameter widely used to predict suspension stability. A rule of thumb is that the higher the absolute value of ξ is, the more stable the suspension is. Usually, a zetapotential of more than +30 mV or less than −30 mV is considered sufficient to electrostatically stabilize a colloidal dispersion.

Nonionic polymeric stabilizers are widely used for steric stabilization of drug nanocrystals. Nonionic stabilizers adsorb on the drug nanocrystals with a segment that strongly interacts with the particle surface, while well-solvated segments of the nonionic stabilizer extend into the surrounding media (Fig. 4.3b). Steric stabilization is very dependent of solvation effects and can be interpreted through the change in Gibbs free energy, ΔG, of the system (Eq. 4.7):

$$\Delta G = \Delta H - T\Delta S, \tag{4.7}$$

where ΔH is the enthalpy change, T is the absolute temperature, and ΔS is the change in entropy. In the case of a good solvent for the segments of the nonionic stabilizers extending into the surrounding media, the segments are well-solvated and cannot interpenetrate each other (Fig. 4.3b). This leads to a reduced number of possible configurations for the segments and a negative entropy consequently increasing the

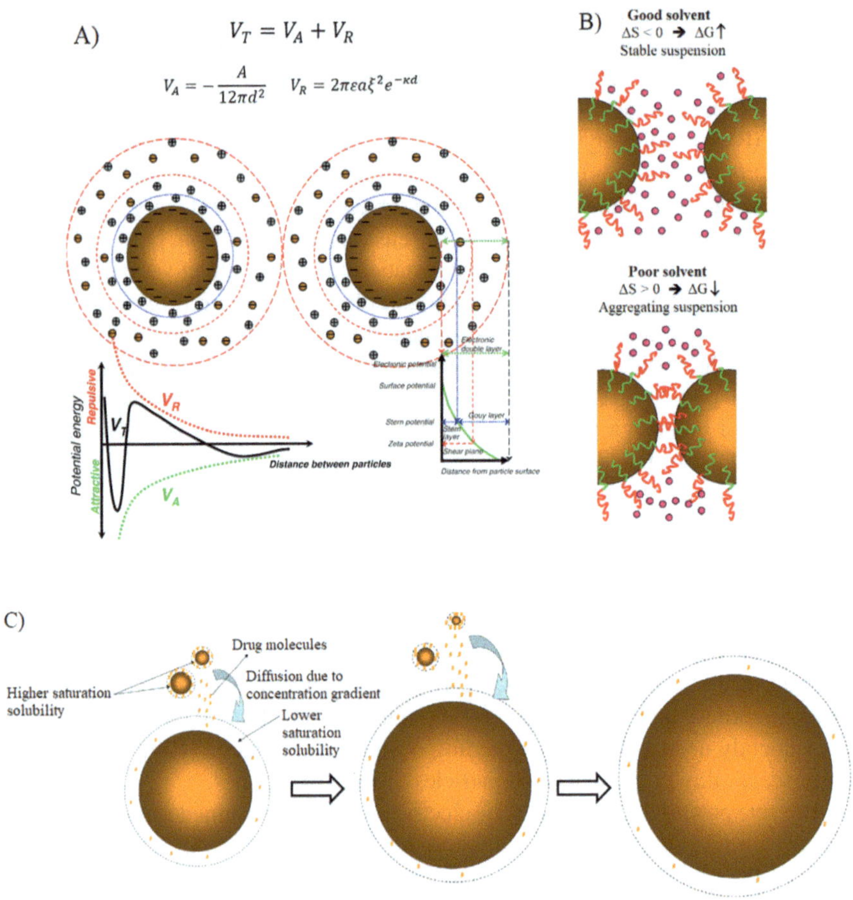

Fig. 4.3 (**a**) Illustration of classical DLVO theory showing the repulsive electrostatic forces (V_R) and attractive van der Waals (vdWs) forces (V_A) acting on similar colloidal particles in a medium. (**b**) Schematic illustration of steric stabilization mechanism according to Gibbs free energy. A good solvent for the stabilizing moiety reduces the number of configurations of the stabilizing tails, resulting in a negative entropy change and an increase in ΔG. In case of a poor solvent, the stabilizing moieties may interpenetrate thermodynamically, which increases the entropy change, leading to a decrease ΔG, thus inducing particle agglomeration. (**c**) Schematic illustration of Ostwald ripening. (Reproduced by permission from Elsevier [125])

change in Gibbs free energy and forming a stable suspension. On the contrary, if the media is a bad solvent for the nonionic stabilizers extending into the surrounding media, then the segments will be able to interpenetrate each other, leading to an increased number of possible configurations. In such a case, the entropy becomes negative, consequently decreasing the Gibbs free energy, which causes the suspension to aggregate. Thus, a good solvation leading to a strong enthalpic interaction between the medium and the stabilizing segment of the stabilizer is a key factor for

Table 4.3 Frequently used stabilizers and their stabilization mechanism for stabilizing nanosuspensions

Stabilization mechanism	Nature/category of stabilizer	Stabilizer
Electrostatic repulsion	Ionic/surfactant	SLS Cetrimonium chloride Dowfax 2A1
	Semisynthetic/polymer	CMC-Na
Electrostatic repulsion/steric hindrance	Natural/polymer	Chitosan
Steric hindrance	Nonionic/surfactant	Poloxamer 188 Poloxamer 407 TPGS Tween 80 Planracare 2000 Saponins
	Amphoteric synthetic/polymer	Lecithin PVP PVA
	Semisynthetic/polymer	HPMC HPC
	Natural/polymer	Sodium alginate
	Natural/food proteins	Whey protein isolate
		Soybean protein isolate β-Lactoglobulin

Adapted from Ref. [135]

successful steric stabilization to prevent particles from aggregating in the medium [132]. Although solvation of the stabilizing segment of the stabilizer is important for steric stabilization, the stabilizing segment also needs to be long enough and dense to minimize the attractive vdWs forces between particles to the extent that the repulsive steric forces dominate in the system [133]. Furthermore, using a combination of the two stabilization mechanisms discussed above by combining an ionic and a nonionic stabilizer could be very beneficial in achieving a stable drug nanocrystal formulation. The use of both ionic and nonionic stabilizers reduces the repulsion between the ionic stabilizer molecules, consequently leading to a closer packing of the stabilizers on the particle surface [134]. Table 4.3 lists frequently used stabilizers and the stabilization mechanism for stabilization of nanosuspensions [135].

4.6.3 Crystal Growth/Ostwald Ripening

Physical stability challenges in drug nanocrystal formulations due to crystal growth, also known as Ostwald ripening, have its origin in that drug solubility increases when particle size decreases, which is also observed from the Oswald–Freundlich equation (4.2). The higher surface free energy of smaller drug particles leads to a higher saturation solubility of smaller particles compared to larger particles (Fig. 4.3c). This creates a drug concentration gradient between small and large particles, which drives drug molecules to diffuse from the higher concentration area surrounding small particles to areas with lower concentration around larger particles. Thus, a supersaturated solution is formed around the larger particles, causing the drug to crystallize onto the larger particles. This transfer of drug molecules from smaller to larger particles continues until all the small particles are dissolved, consequently leading to an increase in particle size and size distribution in drug nanocrystal suspensions. The saturation solubility difference between small and large drug particles and the drug concentration gradients in the medium can be minimized with a narrow particle size distribution, which helps alleviate Ostwald ripening [129]. However, often Ostwald ripening is not a major concern for nanosuspensions with uniform particle size [128, 136]. The inhibition of Ostwald ripening can also be achieved by using stabilizers if they do not themselves enhance the drug solubility [137, 138]. When stabilizers are adsorbed on the particle surface, they can reduce the interfacial tension between the solid and the liquid medium, which prevents Ostwald ripening. Other factors that affect Ostwald ripening are the solubility of the drug, temperature, and mechanical agitation [137].

4.6.4 Chemical Stability

Unlike physical stability issues that are commonly a concern for drug nanocrystal formulations, the chemical stability is specific to the drug that is used as each drug has its specific functional groups and reaction mechanisms that affect chemical stability. For example, chemical groups, such as esters and amides, are susceptible to hydrolytic degradation, while amino functionalities may undergo oxidative degradation [139]. Although chemical stability of nanosuspensions is often not a major concern, because drug nanocrystals are usually dispersed in nanosuspensions with a limited solubility, and therefore the possibility of chemical reactions is not as substantial as that in solution-based formulations [125]. However, the chemical stability of drug molecules with a solubility greater than 1 mg/mL or drugs with low concentrations in drug nanocrystal formulations should be paid extra attention [139]. The chemical stability of drug nanocrystal formulations can be enhanced by transforming the nanosuspensions into dry solid dosage form by using either freeze-drying or spray-drying [126]. Alternatively, increasing the concentration of nanosuspensions can alleviate issues of chemical stability [139].

4.7 Conclusions

Poor solubility is an issue in the drug industry, and drug nanocrystals are one simple route to avoid this challenge. There are different techniques to produce these systems, but before the production, it is important to properly understand the impact of particle size on drug dissolution and solubility. Small particle sizes possess extra demand also for stability issues. Stabilizer selection is very crucial for successful formation of drug nanocrystals, and in the future there may be new stabilizers having more universal use.

References

1. Cerdeira AM, Mazzotti M, Gander B (2010) Miconazole nanosuspensions: influence of formulation variables on particle size reduction and physical stability. Int J Pharm 396(1–2): 210–218
2. Tuomela A, Hirvonen J, Peltonen L (2016) Stabilizing agents for drug nanocrystals: effect on bioavailability. Pharmaceutics 8:16
3. Peppersack C, Flach F, Prziwara P, Damm C, Breitung-Faes S, Peukert W, Kwade A (2023) Conceptual stabilizer selection for nanomilling based on dispersibility parameters. Adv Powder Technol 34:104197
4. Dhaval M, Makwana J, Sakariya E, Dudhat K (2020) Drug nanocrystals: a comprehensive review with current regulatory guidelines. Current Drug Deliv 17(6):470–482
5. Peltonen L (2018) Design space and QbD approach for production of drug nanocrystals by wet media milling techniques. Pharmaceutics 10(3):104
6. Singhal M, Baumgartner A, Turunen E, van Veen B, Hirvonen J, Peltonen L (2020) Nanosuspensions of a poorly soluble investigational molecule ODM-106: impact of milling bead diameter and stabilizer concentration. Int J Pharm 587:119636
7. Liu P, Viitala T, Kartal-Hodzic A, Liang H, Laaksonen T, Hirvonen J, Peltonen L (2015) Interaction studies between indomethacin nanocrystals and PEO/PPO copolymer stabilizers. Pharm Res 32(2):628–639
8. Böck NC, Sundermann J, Koziolek M, Keller BL, Mäder K (2025) Impact of stabilizers on particle size and dispersion behavior in biorelevant media in solid nanocrysal formulations. Eur J Pharm Biopharm 8:114651
9. Liu P, De Wulf O, Laru J, Heikkilä T, van Veen B, Kiesvaara J, Hirvonen J, Peltonen L, Laaksonen T et al (2013) Dissolution studies of poorly soluble drug nanosuspensions in non-sink conditions. AAPS PharmSciTech 14(2):748–756
10. Sarnes A, Østergaard J, Jensen SS et al (2013) Dissolution study of nanocrystal powders of a poorly soluble drug by UV imaging and channel flow methods. Eur J Pharm Sci 50(3–4): 511–519
11. Möschwitzer JP (2013) Drug nanocrystals in the commercial pharmaceutical development process. Int J Pharm 453(1):142–156
12. Gigliobianco MR, Casadidio C, Censi R, Di Martino P (2018) Nanocrystals of poorly soluble drugs: drug bioavailability and physicochemical stability. Pharmaceutics 10(3):134
13. Peltonen L, Hirvonen J (2018) Drug nanocrystals—versatile option for formulation of poorly soluble materials. Int J Pharm 537(1–2):73–83
14. Salazar J, Müller RH, Möschwitzer JP (2014) Combinative particle size reduction technologies for the production of drug nanocrystals. J Pharm (Cairo) 2014(1):265754
15. Nagpal S, Kumari P, Saini K, Kakkar V (2022) Pain management with topical aceclofenac nanosuspension in vitro in-vivo and proof of concept studies. Curr Drug Ther 17(4):289–304
16. Hu Y, Ma C, Sun M, Guo C, Shen J, Wang J, Nie F, Gao B (2019) Preparation and characterization of nano amitriptyline hydrochloride particles by spray freeze drying. Nanomedicine 14(12):1521–1531
17. Liu J, Li S, Ao W, Li Y, Xiao Y, Bai M (2022) Fabrication of an aprepitant nanosuspension using hydroxypropyl chitosan to increase the bioavailability. Biochem Biophys Res Commun 631:72–77
18. Wu S, Lu L, Zhou J, Ran D, Wang S, Xu Q, Xu W, Wang J, Liu Y, Xie C, Luo Z, Lu W (2022) All-stage targeted therapy for glioblastoma based on lipid membrane coated cabazitaxel nanocrystals. J Control Release 345:685–695
19. Jain S, Reddy VA, Arora S, Patel K (2016) Development of surface stabilized candesartan cilexetil nanocrystals with enhanced dissolution rate, permeation rate across CaCo-2, and oral bioavailability. Drug Deliv Transl Res 6(5):498–510

20. Halder S, Ahmed F, Shuma ML, Azad MAK, Kabir ER (2020) Impact of drying on dissolution behavior of carvedilol-loaded sustained release solid dispersion: development and characterization. Heliyon 6(9):e05026
21. Oshi MA, Lee J, Naeem M, Hasan N, Kim J, Kim HJ, Lee EH, Jung Y, Yoo JW (2020) Curcumin nanocrystal/pH-responsive polyelectrolyte multilayer core-shell nanoparticles for inflammation-targeted alleviation of ulcerative colitis. Biomacromolecules 21(9):3571–3581
22. Casula L, Lai F, Pini E, Valenti D, Sinico C, Cardia MC, Marceddu S, Ailuno G, Fadda AM (2021) Pulmonary delivery of curcumin and beclomethasone dipropionate in a multicomponent nanosuspension for the treatment of bronchial asthma. Pharmaceutics 13(8):1300
23. Pınar SG, Canpınar H, Tan Ç, Çelebi N (2022) A new nanosuspension prepared with wet milling method for oral delivery of highly variable drug cyclosporine A: development, optimization and in vivo evaluation. Eur J Pharm Sci 171:106123
24. Asghari I, Esmaeilzadeh F (2012) Formation of ultrafine deferasirox particles via rapid expansion of supercritical solution (RESS process) using Taguchi approach. Int J Pharm 433(1–2):149–156
25. Faizi HS, Nasiri MI, Wu Y, Mishra D, Donnelly RF, Minhas MU, Vora LK, Singh Thakur RR (2024) Deferasirox nanosuspension loaded dissolving microneedles for ocular drug delivery. Int J Pharm 664:124614
26. Ngo AN, Thomas D, Murowchick J, Ayon NJ, Jaiswal A, Youan BBC (2018) Engineering fast dissolving sodium acetate mediated crystalline solid dispersion of docetaxel. Int J Pharm 545(1–2):329–341
27. Tang Z, Yin L, Zhang Y, Yu W, Wang Q, Zhan Z (2019) Preparation and study of two kinds of ophthalmic nano-preparations of everolimus. Drug Deliv 26(1):1235–1242
28. Ibrahim AH, Ibrahim HM, Ismael HR, Samy AM (2018) Optimization and evaluation of lyophilized fenofibrate nanoparticles with enhanced oral bioavailability and efficacy. Pharm Dev Technol 23(4):358–369
29. Ige PP, Baria RK, Gattani SG (2013) Fabrication of fenofibrate nanocrystals by probe sonication method for enhancement of dissolution rate and oral bioavailability. Colloids Surf B: Biointerfaces 108:366–373
30. Sievens-Figueroa L, Bhakay A, Jerez-Rozo JI, Pandya N, Romañach RJ, Michniak-Kohn B, Iqbal Z, Bilgili E, Davé RN (2012) Preparation and characterization of hydroxypropyl methyl cellulose films containing stable BCS class II drug nanoparticles for pharmaceutical applications. Int J Pharm 423(2):496–508
31. Fu TT, Cong ZQ, Zhao Y, Chen WY, Liu CY, Zheng Y, Yang FF, Liao YH (2019) Fluticasone propionate nanosuspensions for sustained nebulization delivery: an in vitro and in vivo evaluation. Int J Pharm 572:118839
32. Peter S, Mathews MM, Saju F, Paul S (2023) Development, optimization and in vitro characterization of Eudragit-ganciclovir nanosuspension or treating herpes simplex keratitis. J Pharm Innov 18(3):1328–1337
33. He Y, Liu C, Han R, Liang Y, Mak JCW, Zhu Y, Li H, Zheng Y (2023) Reducing systemic absorption and macrophages clearance of genistein by lipid-coated nanocrystals for pulmonary delivery. Chin Chem Lett 34(1):107484
34. Martena V, Censi R, Hoti E, Malaj L, Di Martino P (2013) Preparation of glibenclamide nanocrystals by a simple laboratory scale ultra cryo-milling. J Nanopart Res 15(6):1712
35. Huang G, Xie J, Shuai S, Wei S, Chen Y, Guan Z, Zheng Q, Yue P, Wang C (2021) Nose-to-brain delivery of drug nanocrystals by using Ca2+ responsive deacetylated gellan gum based in situ-nanogel. Int J Pharm 594:120182
36. Chourasiya V, Bohrey S, Pandey A (2015) Hydrochlorothiazide containing PLGA nanoparticles: design, characterization, in-vitro drug release and release kinetic study. Polymer Sci Ser B 57(6):645–653
37. Möschwitzer J, Müller RH (2006) Spray coated pellets as carrier system for mucoadhesive drug nanocrystals. Eur J Pharm Biopharm 62(3):282–287

38. Haghighizadeh A, Mahdavi H, Rajabi O (2023) Continuous and size-controlled preparation of ibuprofen nanosuspension by antisolvent crystallization method using hollow fiber membrane. J Pharm Innov 18(1):195–204
39. Frank LA, Gazzi RP, Mello PA, Chaves P, Peña F, Beck RCR, Buffon A, Pohlmann AR, Guterres SS (2020) Anti-HPV nanoemulsified-imiquimod: a new and potent formulation to treat cervical cancer. AAPS PharmSciTech 21(2):54
40. Wais U, Jackson AW, Zuo Y, Xiang Y, He T, Zhang H (2016) Drug nanoparticles by emulsion-freeze-drying via the employment of branched block copolymer nanoparticles. J Control Release 222:141–150
41. van Eerdenbrugh B, Froyen L, van Humbeeck J, Martens JA, Augustijns P, van den Mooter G (2008) Drying of crystalline drug nanosuspensions—the importance of surface hydrophobicity on dissolution behavior upon redispersion. Eur J Pharm Sci 35(1–2):127–135
42. Permana AD, Utami RN, Layadi P, Himawan A, Juniarti N, Anjani QK, Utomo E, Mardikasari SA, Arjuna A, Donnelly RF (2021) Thermosensitive and mucoadhesive in situ ocular gel for effective local delivery and antifungal activity of itraconazole nanocrystal in the treatment of fungal keratitis. Int J Pharm 602:120623
43. Yu S, Pu X, Ahmed MU, Yu HH, Mutukuri TT, Li J, Zhou QT (2021) Spray-freeze-dried inhalable composite microparticles containing nanoparticles of combinational drugs for potential treatment of lung infections caused by Pseudomonas aeruginosa. Int J Pharm 610:121160
44. Lopez-Vidal L, Parodi P, Actis MR, Camacho N, Real DA, Paredes AJ, Irazoqui FJ, Real JP, Palma SD (2024) Formulation and optimization of pH-sensitive nanocrystals for improved oral delivery. Drug Deliv Transl Res 14(5):1301–1318
45. Savage AC, Tatham LM, Siccardi M, Scott T, Vourvahis M, Clark A, Rannard SP, Owen A (2019) Improving maraviroc oral bioavailability by formation of solid drug nanoparticles. Eur J Pharm Biopharm 138:30–36
46. Liversidge GG, Conzentino P (1995) Drug particle size reduction for decreasing gastric irritancy and enhancing absorption of naproxen in rats. Int J Pharm 125(2):309–313
47. Sheng Y, Zhang S, Ling J, Hu C, Zhang Z, Lv H (2022) Oxybutynin nanosuspension gel for enhanced transdermal treatment for overactive bladder syndrome. Pharm Dev Technol 27(4): 459–468
48. Nandwani Y, Kaur A, Bansal AK (2021) Generation of ophthalmic nanosuspension of prednisolone acetate using a novel technology. Pharm Res 38(2):319–333
49. Liu Q, Guan J, Sun Z, Shen X, Li L, Jin L, Mao S (2019) Influence of stabilizer type and concentration on the lung deposition and retention of resveratrol nanosuspension-in-microparticles. Int J Pharm 569:118562
50. Melo KJC, Henostroza MAB, Löbenberg R, Bou-Chacra NA (2020) Rifampicin nanocrystals: towards an innovative approach to treat tuberculosis. Mater Sci Eng C 112:110895
51. Li Y, Wang J, Vora LK, Sabri AH, Bin McGuckin MB, Paredes AJ, Donnelly RF (2024) Dissolving microarray patches loaded with a rotigotine nanosuspension: a potential alternative to Neupro® patch. J Control Release 372:304–317
52. Singh G, Sharma S, Gupta GD (2017) Extensive diminution of particle size and amorphization of a crystalline drug attained by eminent technology of solid dispersion: a comparative study. AAPS PharmSciTech 18(5):1770–1784
53. Khan MS, Ravi PR, Dhavan DS (2023) Design, optimization, in vitro and in vivo evaluation of triamcinolone acetonide nanocrystals loaded in situ gel for topical ocular delivery. Colloids Surf B: Biointerfaces 231:113539
54. Bitterlich A, Laabs C, Krautstrunk I, Dengler M, Juhnke M, Grandeury A, Bunjes H, Kwade A (2015) Process parameter dependent growth phenomena of naproxen nanosuspension manufactured by wet media milling. Eur J Pharm Biopharm 92:171–179
55. Modhave D, Saraf I, Karn A, Paudel A (2020) Understanding concomitant physical and chemical transformations of simvastatin during dry ball milling. AAPS PharmSciTech 21(5):1–10

56. Holzapfel K, Rades T, Leopold CS (2023) Co-amorphous systems consisting of indomethacin and the chiral co-former tryptophan: solid-state properties and molecular mobilities. Int J Pharm 636:122840
57. Jarvis M, Krishnan V, Mitragotri S (2018) Nanocrystals: a perspective on translational research and clinical studies. Bioeng Transl Med 4(1):5
58. Lestari MLAD, Müller RH, Möschwitzer JP (2019) The scalability of wet ball milling for the production of nanosuspensions. Pharm Nanotechnol 7(2):147–161
59. Zhang D, Tan T, Gao L, Zhao W, Wang P (2007) Preparation of azithromycin nanosuspensions by high pressure homogenization and its physicochemical characteristics studies. Drug Dev Ind Pharm 33(5):569–575
60. Sharma P, Denny WA, Garg S (2009) Effect of wet milling process on the solid state of indomethacin and simvastatin. Int J Pharm 380(1–2):40–48
61. Ye X, Patil H, Feng X, Tiwari RV, Lu J, Gryczke A, Kolter K, Langley N, Majumdar S, Neupane D, Mishra SR, Repka MA (2015) Conjugation of hot-melt extrusion with high-pressure homogenization: a novel method of continuously preparing nanocrystal solid dispersions. AAPS PharmSciTech 17(1):78
62. Cho E, Cho W, Cha K-H, Park J, Kim M-S, Kim J-S, Park HJ, Hwang S-J (2010) Enhanced dissolution of megestrol acetate microcrystals prepared by antisolvent precipitation process using hydrophilic additives. Int J Pharm 396:91–98
63. Sinha B, Müller RH, Möschwitzer JP (2013) Bottom-up approaches for preparing drug nanocrystals: formulations and factors affecting particle size. Int J Pharm 453(1):126–141
64. Arzi RS, Sosnik A (2018) Electrohydrodynamic atomization and spray-drying for the production of pure drug nanocrystals and co-crystals. Adv Drug Deliv Rev 131:79–100
65. Wang M, Rutledge GC, Myerson AS, Trout BL (2012) Production and characterization of carbamazepine nanocrystals by electrospraying for continuous pharmaceutical manufacturing. J Pharm Sci 101(3):1178–1188
66. Ma YQ, Zhang ZZ, Li G, Zhang J, Xiao HY, Li XF (2014) Solidification drug nanosuspensions into nanocrystals by freeze-drying: a case study with ursodeoxycholic acid. Pharm Dev Technol 21(2):180–188
67. Mohammady M, Mohammadi Y, Yousefi G (2020) Freeze-drying of pharmaceutical and nutraceutical nanoparticles: the effects of formulation and technique parameters on nanoparticles characteristics. J Pharm Sci 109(11):3235–3247
68. Kakran M, Shekogar R, Gopal Sahoo N, Al Shaal L, Li L, Muller RH (2012) Fabrication of quercetin nanocrystals: comparison of different methods. Eur J Pharm Biopharm 80:113–121
69. Wei Q, Keck CM, Müller RH (2018) Solidification of hesperidin nanosuspension by spray drying optimized by design of experiment (DoE). Drug Dev Ind Pharm 44(1):1–12
70. Hagedorn M, Liebich L, Bögershausen A, Massing U, Hoffman S, Mende S, Rischer M (2019) Rapid development of API nano-formulations from screening to production combining dual centrifugation and wet agitator bead milling. Int J Pharm 565:187–198
71. Henriquez L, Bahloul B, Alfareth K, Oyoun F, Frejkova M, Kostka L, Etrych T, Kalshoven L, Guillaume A, Mignet N, Corvis Y (2024) Step-by-step standardization of the bottom-up semi-automated nanocrystallization of pharmaceuticals: a quality by design and design of experiments joint approach. Small 20(25):2306054
72. Li J, Zhou Y, Aisha M, Wu J, Wang H, Huang F, Sun M (2021a) Preparation of loratadine nanocrystal tablets to improve the solubility and dissolution for enhanced oral bioavailability. J Pharm Pharmacol 73:937–946
73. Hladek F, Chvila S, Navratil O, Balouch M, Stepanek F (2022) Systematic investigation of wet-milling kinetics and colloidal stability of pharmaceutical nanocrystals. Crystal Growth Des 22(12):6928–6940
74. Liu P, Rong X, Laru J, Van Veen B, Kiesvaara J, Hirvonen J, Laaksonen T, Peltonen L (2011) Nanosuspensions of poorly soluble drugs: preparation and development by wet milling. Int J Pharm 411(1–2):215–222

75. Oktay AN, Ilbasmis-Tamer S, Karakucuk A, Celebi N (2020) Screening of stabilizing agents to optimize flurbiprofen nanosuspensions using experimental design. J Drug Deliv Sci Technol 57:101690
76. Juhnke M, Berghausen J, Timpe C (2010) Accelerated formulation development for nanomilled active pharmaceutical ingredients using a screening approach. Chem Eng Technol 33(9):1412–1418
77. Romero GB, Keck CM, Muller RH (2016) Simple low-cost miniaturization approach for pharmaceutical nanocrystals production. Int J Pharm 501:236–244
78. Zuo J, Barros de Araujo GL, Stephano MA, Zuo Z, Bou-Chacra NA, Löbenberg R (2020) Design space approach in the development of esculetin nanocrystals by a small-scale wet-bead milling process. J Drug Deliv Sci Technol 55:101486
79. Van Eerdenbrugh B, Stuyven B, Froyen L, Van Humbeeck J, Martens JA, Augustijns P, Van den Mooter G (2009) Downscaling drug nanosuspension production: processing aspects and physicochemical characterization. AAPS PharmSciTech 10(1):44–53
80. Salazar J, Heinzerling O, Müller RH, Möschwitzer JP (2011) Process optimization of a novel production method for nanosuspensions using design of experiments (DoE). Int J Pharm 420: 395–403
81. Chauhan R, Minocha N, Coliaie P, Singh PG, Korde A, Kelkar M, Langston M, Liu C, Nazemifard N, Patience D, Skliar D, Nere NK, Singh MR (2023) Emerging microfluidic platforms for crystallization process development. Chem Eng Res Des 197:908–930
82. Sreeharsha N, Naveen NR, Anitha P, Goudanavar PS, Ramkanth S, Fattepur S, Telsang M, Habeebuddin M, Anwer MK (2022) Development of nanocrystal compressed minitablets for chronotherapeutic drug delivery. Pharmaceuticals 15(3):311
83. Tuomela A, Laaksonen T, Laru J, Antikainen O, Kiesvaara J, Ilkka J, Oksala O, Rönkkö S, Järvinen K, Hirvonen J, Peltonen L (2015) Solid formulations by a nanocrystal approach: critical process parameters regarding scale-ability of nanocrystals for tableting applications. Int J Pharm 485:77–86
84. Malamatari M, Taylor KM, Malamataris S, Douromis D, Kachrimanis K (2018) Pharmaceutical nanocrystals: production by wet milling and applications. Drug Discov Today 23(3): 534–547
85. Shirsath N, Marathe D, Jaiswal P, Zawar L (2022) A 3^2 factorial design approach for formulation and optimization of azilsartan medoxomil nanosuspension for solubility enhancement. Indian J Pharm Educ Res 56(2):S365–S373
86. Williamson EM, Sun Z, Mora-Tamez L, Brutchey RL (2022) Design of experiments for nanocrystal syntheses: a how-to guide for proper implementation. Chem Mater 34(22): 9823–9835
87. Rampado R, Peer D (2023) Design of experiments in the optimization of nanoparticle-based drug delivery systems. J Control Release 358:398–419
88. Naguib MJ, Makhlouf AI (2021) Scalable flibanserin nanocrystal-based novel sublingual platform for female hypoactive sexual desire disorder: engineering, optimization adopting the desirability function approach and in vivo pharmacokinetic study. Drug Deliv 28(1): 1301–1311
89. Williamson EM, Tappan BA, Mora-Tamez L, Barim G, Brutchey RL (2021) Statistical multiobjective optimization of thiospinel CoNi2S4 nanocrystal synthesis via design of experiments. ACS Nano 15(6):9422–9433
90. Sutarna TH, Gozali D, Panantarani C, Alatas F (2023) Application of the behnken design box to optimize the production of dexibuprofen nanocrystals using the top-down method. Seybold Report 18:6
91. Fernandes AN, Ferreira J, Fangueiro A, Santos F, Veiga C, Cabral A, Silva AM, Souto EB (2017) Ibuprofen nanocrystals developed by 2^2 factorial design experiment: a new approach for poorly water-soluble drugs. Saudi Pharm J 25(8):1117–1124

92. Patil OA, Patil IS, Mane RU, Randive DS, Bhutkar MA, Bhinge SD (2018) Formulation optimization and evaluation of cefdinir nanosuspension using 2^3 factorial design. Marmara Pharm J 22(4):587–598
93. Shinde A, Jadhav N, More H, Naikwadi H (2019) Design and characterisation of lopinavir nanocrystals for solubility and dissolution enhancement. Pharm Sci Asia 46:193–205
94. Castañeda L (2019) A facile method for formulation of atenolol nanocrystal drug with enhanced bioavailability. In: Movahedi B (ed) Nanocrystalline materials. IntechOpen, London. Available from: https://doi.org/10.5772/intechopen.88191
95. Karakucuk A, Celebi N, Teksin ZS (2016) Preparation of ritonavir nanosuspensions by microfluidization using polymeric stabilizers: I. A design of experiment approach. Eur J Pharm Sci 95:111–121
96. Macedo LDO, Morales IAC, Barbosa EJ, Stephano MA, de Araujo GL, Bou-Chacra NA (2022) Thermal study, process optimization, and water solubility improvement of a freeze-dried artemether nanosuspension for malaria treatment. J Drug Deliv Sci Technol 78:103915
97. Saritha D, Chandra Bose PS, Osmani RAM, Iriventi P (2024) Development, optimization and in vitro characterization of haloperidol nanocrystals using 3^2 factorial design. Int J Appl Pharm 16(3):187
98. Pujitha R, Daisy CS (2023) Formulation development, optimization and evaluation of carvedilol loaded nanocrystal orodispersıble tablets for the management of hypertension. Int J Pharm Res Appl 8(6):2651–2662
99. McGuckin MB, Wang J, Ghanma R, Qin N, Palma SD, Donnelly R, Paredes AJ (2022) Nanocrystals as a master key to deliver hydrophobic drugs via multiple administration routes. J Control Release 345:334–353
100. Phuna ZX, Panda BP, Shivashekaregowda NKH, Madhavan P (2022) Recent development in nanocrystal based drug delivery for neurodegenerative diseases: scope, challenges, current and future prospects. J Drug Deliv Sci Technol 68:102921
101. Jakubowska E (2024) A short history of drug nanocrystals—methods, milestones and meaning in pharmaceutical technology. J Drug Deliv Sci Technol 102:106400
102. Liu J, Tu L, Cheng M, Feng J, Jin Y (2020) Mechanisms for oral absorption enhancement of drugs by nanocrystals. J D Deliv Sci Technol 56:101607
103. Junghanns JUAH, Müller RH (2008) Nanocrystal technology, drug delivery and clinical applications. Int J Nanomedicine 3(3):295–310
104. Junyaprasert VB, Morakul B (2015) Nanocrystals for enhancement of oral bioavailability of poorly water-soluble drugs. Asian J Pharm Sci 10:13–23
105. de Oliveira Macedo L, Barbosa EJ, Löbenberg R, Bou-Chacra NA (2021) Anti-inflammatory drug nanocrystals: state of art and regulatory perspective. Eur J Pharm Sci 158:105654
106. Parmar PK, Wadhawan J, Bansal AK (2021) Pharmaceutical nanocrystals: a promising approach for improved topical drug delivery. Drug Discov Today 26:2329–2349
107. Fontana F, Figueiredo P, Zhang P, Hirvonen JT, Liu D, Santos HA (2018) Production of pure drug nanocrystals and nano co-crystals by confinement methods. Adv Drug Deliv Rev 131:3–21
108. Jakubowska E, Lulek J (2021) The application of freeze-drying as a production method of drug nanocrystals and solid dispersions—a review. J Drug Deliv Sci Technol 62:102357
109. Bisrat M, Nyström C (1988) Physicochemical aspects of drug release. VIII. The relation between particle size and surface specific dissolution rate in agitated suspensions. Int J Pharm 47(1):223–231
110. Higuchi WI, Hiestand EN (1963) Dissolution rates of finely divided drug powders I: effect of a distribution of particle sizes in a diffusion-controlled process. J Pharm Sci 52(1):67–71
111. Higuchi WI, Rowe EL, Hiestand EN (1963) Dissolution rates of finely divided drug powders II: micronized methylprednisolone. J Pharm Sci 52(2):162–164
112. Hintz RJ, Johnson KC (1989) The effect of particle size distribution on dissolution rate and oral absorption. Int J Pharm 51(1):9–17

113. Andersson SBE, Frenning G, Alderborn G, Gråsjö J (2022) Effect of fluid velocity and particle size on the hydrodynamic diffusion layer thickness. Eur J Pharm Biopharm 180:1–10
114. Galli C (2006) Experimental determination of the diffusion boundary layer width of micron and submicron particles. Int J Pharm 313(1):114–122
115. Hixson AW, Baum SJ (1941) Agitation. Mass transfer coefficients in liquid-solid agitation systems. Ind Eng Chem 33(4):478–485
116. Niebergall PJ, Milosovich G, Goyan JE (1963) Dissolution rate studies II: dissolution of particles under conditions of rapid agitation. J Pharm Sci 52(3):236–241
117. Sheng JJ, Sirois PJ, Dressman JB, Amidon GL (2008) Particle diffusional layer thickness in a USP dissolution apparatus II: a combined function of particle size and paddle speed. J Pharm Sci 97(11):4815–4829
118. Sinko PD, Salehi N, Halseth T, Meyer PJ, Amidon GL, Ziff RM, Amidon GE (2024) Particle size, dose, and confinement affect passive diffusion flux through the membrane concentration boundary layer. Mol Pharm 21(1):201–215
119. Hixson AW, Crowell JH (1931) Dependence of reaction velocity upon surface and agitation. Ind Eng Chem 23(8):923–931
120. Ely DR, Edwin García R, Thommes M (2014) Ostwald–Freundlich diffusion-limited dissolution kinetics of nanoparticles. Powder Technol 257:120–123
121. Mosharraf M, Nyström C (1995) The effect of particle size and shape on the surface specific dissolution rate of microsized practically insoluble drugs. Int J Pharm 122(1):35–47
122. Anhalt K, Geissler S, Harms M, Weigandt M, Fricker G (2012) Development of a new method to assess nanocrystal dissolution based on light scattering. Pharm Res 29(10):2887–2901
123. Berre FL, Chauveteau G, Pefferkorn E (1998) Perikinetic and orthokinetic aggregation of hydrated colloids. J Colloid Interface Sci 199:1–12
124. Sinko PJ, Singh Y (2011) Ch 17: Colloidal dispersions. In: Martin's physical pharmacy and pharmaceutical sciences, 6th edn. Lippincott Williams & Wilkins, Baltimore, p 411
125. Wu L, Zhang J, Watanabe W (2011) Physical and chemical stability of drug nanoparticles. Adv Drug Deliv Rev 63:456–469
126. Eerdenbrugh BV, Mooter GVD, Augustijns P (2008) Top-down production of drug nanocrystals nanosuspension stabilization, miniaturization and transformation into solid products. Int J Pharm 364:64–75
127. Gao L, Zhang D, Chen M (2008) Drug nanocrystals for the formulation of poorly soluble drugs and its application as a potential drug delivery system. J Nanopart Res 10:845–862
128. Rabinow BE (2004) Nanosuspensions in drug delivery. Natl Rev 3:785–796
129. Nutan MTH, Reddy IK (2009) General principles of suspensions. In: Kulshreshtha AK, Singh ON, Wall GM (eds) Pharmaceutical suspensions: from formulation development to manufacturing. Springer, pp 39–66
130. Derjaguin BV, Landau L (1941) Theory of the stability of strongly charged lyophobic sols and of the adhesion of strongly charged particles in solutions of electrolytes. Acta Physicochim 14: 633–662
131. Vervey EJW, Overbeek JTG (1948) Theory of the stability of lyophobic colloids. Elsevier, New York
132. Kim C-J (2004) Surface chemistry and colloids. In: Advanced pharmaceutics: physicochemical principles. CRC Press, Boca Raton, pp 193–256
133. Choi J-Y, Yoo JY, Kwak H-S, Nam BU, Lee J (2005) Role of polymeric stabilizers for drug nanocrystal dispersions. Curr Appl Phys 5:472–474
134. Rosen MJ (2004) Dispersion and aggregation of solid in liquid medium by surfactants. In: Surfactants and interfacial phenomena. Wiley, Hoboken, pp 337–362
135. Li J, Gang Z, Zhang H, Gao J, Zheng A (2021b) Progress in the development of stabilization strategies for nanocrystal preparations. Drug Deliv 28(1):19–36
136. Patravale VB, Date AA, Kulkarni RM (2004) Nanosuspensions a promising drug delivery strategy. J Pharm Pharmacol 56:827–840

137. McClements DJ (2004) Emulsion stability. In: Food emulsions: principles, practices, and techniques, 2nd edn. CRC Press, Boca Raton, pp 185–234
138. Tadros TF (2005) Surfactans in pharmaceutical formulations. In: Applied surfactants: principles and applications. Wiley-VCH, Weinheim, pp 433–502
139. Garad S, Wang J, Joshi Y, Panicucci R (2009) Preclinical development for suspensions. In: Kulshreshtha AK, Singh ON, Wall GM (eds) Pharmaceutical suspensions: from formulation development to manufacturing. Springer, New York, pp 39–66

Nanocrystal Technology in Targeted Drug Delivery

5

Sunny Mukherjee, Nitisha Beniwal, and Aravind Kumar Rengan

Abstract

Nanocrystal technology has emerged as a revolutionary approach in the field of targeted drug delivery, offering significant advancements in the treatment of various diseases. This chapter explores the synthesis, characterization, and application of nanocrystals in enhancing the efficacy and specificity of drug delivery systems. Nanocrystals, with their unique properties such as high surface area-to-volume ratio and tunable size, enable the delivery of poorly soluble drugs, improving their bioavailability and therapeutic potential. The chapter illuminates the various synthesis methods, including top-down and bottom-up approaches, highlighting the importance of particle size control and surface modification in achieving optimal drug delivery performance. A significant focus is placed on the mechanisms of targeted delivery, where nanocrystals can be engineered to recognize and bind specific cellular receptors, thereby enhancing drug accumulation at the desired site while minimizing off-target effects. Case studies in oncology, infectious diseases, and neurological disorders illustrate the practical applications and benefits of nanocrystal-based drug delivery systems. The chapter also addresses the challenges and prospects of this technology, including scalability, regulatory considerations, and potential toxicity. Through a comprehensive examination of current research and clinical advancements, this chapter aims to provide a thorough understanding of how nanocrystal technology is transforming targeted drug delivery and its implications for personalized medicine.

S. Mukherjee · N. Beniwal · A. K. Rengan (✉)
Department of Biomedical Engineering, Indian Institute of Technology Hyderabad, Sangareddy, Telangana, India
e-mail: aravind@bme.iith.ac.in

N. K. Mehra, K. K. Bansal (eds.), *Nanocrystal Technologies in Biomedical Sciences*,
https://doi.org/10.1007/978-981-95-2968-1_5

Keywords

Targeting · Nanocrystal · Drug delivery · Personalized medicine

5.1 Introduction

Nanocrystal technology has emerged as a transformative tool in the field of drug delivery, offering new solutions to the persistent challenges faced by conventional drug formulations [1]. As medicine advances, the demand for highly effective, targeted drug delivery systems that can deliver therapeutic agents with precision to specific tissues or organs has grown exponentially [2]. Traditional drug delivery methods often encounter limitations such as poor bioavailability, limited solubility, and a lack of specificity, leading to suboptimal therapeutic outcomes. Nanocrystals, with their unique physicochemical properties, are helping to address these challenges, offering unprecedented opportunities for the treatment of complex diseases, including cancer, neurological disorders, and infectious diseases [3].

Nanocrystals are finely engineered, submicron-sized particles that exhibit an exceptional surface area-to-volume ratio [4]. This characteristic significantly enhances the dissolution rate of poorly water-soluble drugs, which is a critical factor in improving their bioavailability. By reducing drug molecules to the nanoscale, nanocrystals increase the surface area available for dissolution, enabling faster absorption and more efficient drug delivery [5]. Additionally, their small size and customizable surface properties allow nanocrystals to be functionalized with targeting ligands, making them highly suitable for targeted drug delivery applications (Fig. 5.1) [6, 7].

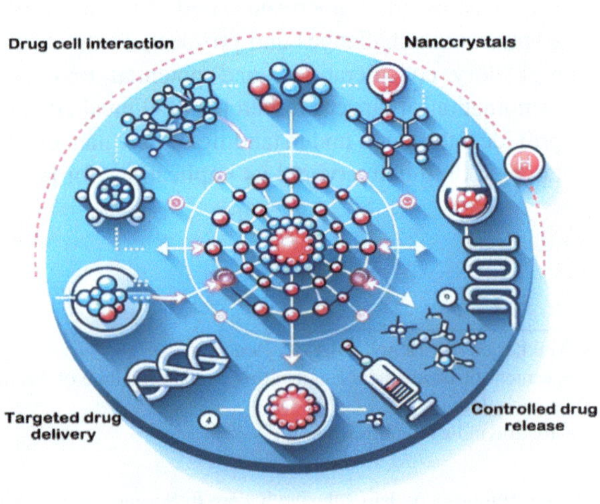

Fig. 5.1 Schematic illustration depicting the central role of nanocrystals in advanced drug delivery systems. Nanocrystals enable enhanced drug–cell interactions, facilitate targeted drug delivery through passive and active mechanisms, and support controlled drug release profiles. Their unique physicochemical properties, including high drug loading, tunable surface chemistry, and nanoscale dimensions, allow for improved solubility, optimized biodistribution, and increased therapeutic efficacy

Nanocrystals in Drug Delivery One of the most significant advantages of nanocrystals is their ability to enhance the bioavailability of poorly soluble drugs, which is a common problem in the pharmaceutical industry. Poor solubility often limits the absorption of drugs in the gastrointestinal tract, reducing their effectiveness [8]. By reducing the size of drugs to the nanometer scale, nanocrystals significantly increase the surface area for dissolution, allowing for faster and more efficient absorption [9]. This has made nanocrystals a promising solution for the delivery of drugs that were previously considered too insoluble for effective therapeutic use. For instance, drugs like paclitaxel (used in cancer treatment) and cyclosporine (used for immunosuppression) have shown improved solubility and bioavailability when formulated as nanocrystals, allowing for lower dosages, reduced side effects, and enhanced therapeutic outcomes [10].

Nanocrystals in Oncology: Revolutionizing Cancer Treatment The application of nanocrystals in oncology has been particularly promising. The unique properties of nanocrystals allow for the delivery of chemotherapeutic agents with improved solubility, bioavailability, and controlled release. One of the most studied examples is paclitaxel-loaded nanocrystals, which have been shown to increase the drug's solubility and facilitate more efficient drug release at the tumor site. Research by Zhang et al. demonstrated that paclitaxel nanocrystals exhibited enhanced tumor targeting and reduced systemic toxicity, leading to improved therapeutic outcomes in cancer treatment. Furthermore, the ability of nanocrystals to overcome biological barriers, such as the blood–brain barrier, holds great potential for treating central nervous system cancers and other neurological disorders [11].

Targeted Drug Delivery with Nanocrystals A major advantage of nanocrystals is their capacity to enable targeted drug delivery, ensuring that drugs are delivered precisely to the site of action, thus reducing the exposure of healthy tissues and minimizing side effects. Traditional drug delivery systems often result in systemic distribution of the drug, which can lead to unwanted interactions and toxicity. However, nanocrystals can be engineered with specific surface modifications that allow them to bind selectively to target cells or tissues [4].

Targeting can be achieved through two main strategies: passive targeting and active targeting.

- *Passive Targeting*: This approach takes advantage of the natural characteristics of nanocrystals, such as their small size and high surface area, which enable them to accumulate in certain tissues, such as tumors. This process relies on the enhanced permeability and retention (EPR) effect, where nanoparticles accumulate in tumor tissues due to the leaky vasculature. This passive mechanism has been successfully employed in cancer treatment, where nanocrystals have been shown to accumulate in tumor sites, providing sustained drug release and improving therapeutic outcomes [12].
- *Active Targeting*: Active targeting involves modifying the surface of nanocrystals with ligands, such as antibodies, peptides, or small molecules, that can

specifically bind to receptors on target cells. This strategy offers more precision than passive targeting, allowing for the selective delivery of drugs to specific cell types. For example, research by Colombo et al. in 2022 demonstrated that paclitaxel-loaded nanocrystals functionalized with folic acid could target folate receptor-positive cancer cells, enhancing drug uptake and improving therapeutic efficacy [13].

Nanocrystals in Personalized Medicine As personalized medicine continues to evolve, nanocrystals are playing an increasingly crucial role in delivering drugs tailored to individual patients' needs. Personalized medicine aims to treat patients based on their specific molecular and genetic profiles, and nanocrystal-based drug delivery systems are well-suited to this approach. By using nanocrystals, clinicians can deliver drugs directly to the tumor or diseased tissue, reducing side effects and improving the precision of treatment. This is particularly important in cancer therapy, where treatment strategies are becoming increasingly tailored to the molecular signatures of tumors. In cancer immunotherapy, nanocrystals are being used to deliver immune checkpoint inhibitors, such as PD-1 inhibitors, directly to tumor cells. This approach has shown promise in enhancing the immune system's ability to target and destroy cancer cells. Research by Reda et al. [14] found that nanocrystals loaded with immune checkpoint inhibitors significantly increased the therapeutic efficacy of immunotherapy while reducing systemic toxicity. An overview of these processes is illustrated in the schematic in Fig. 5.2, which highlights the synthesis of nanocrystals, their functionalization with targeting ligands, and their application in delivering drugs to specific tissues or organs with precision.

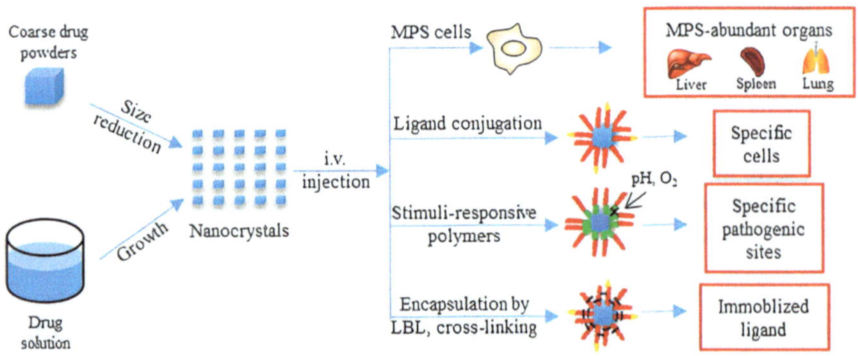

Fig. 5.2 Schematic representation of nanocrystal preparation and targeting mechanisms. The process begins with size reduction of coarse drug powders or nanocrystal growth from a drug solution, followed by intravenous (I.V.) injection. Targeting is achieved through ligand conjugation for specific cells, stimuli-responsive polymers for targeting specific pathogenic sites, and encapsulation using layer-by-layer (LBL) cross-linking for immobilized ligand delivery. Key organs abundant in mononuclear phagocyte system (MPS) cells, such as the liver, spleen, and lungs, are also depicted as primary accumulation sites. (Reproduced with copyright permission from Ref. [10])

Nanocrystal-based drug delivery systems have been the subject of extensive research and clinical trials, demonstrating significant improvements in drug solubility, pharmacokinetics, and reduced toxicity. Clinical studies investigating paclitaxel-loaded nanocrystals in cancer treatment have shown enhanced drug accumulation at the tumor site, leading to better therapeutic outcomes compared to conventional formulations. As research continues to advance, nanocrystals are expected to play a pivotal role in the development of more targeted and personalized treatment options for a wide range of diseases [15]. By enhancing drug solubility, bioavailability, and targeted delivery, nanocrystals are paving the way for more effective and personalized treatments [16]. With ongoing research and clinical success, nanocrystals hold the potential to revolutionize the way we treat diseases, from cancer to neurological disorders, offering a new paradigm in precision medicine [17]. The primary objective of this chapter is to provide an in-depth examination of nanocrystal technology, focusing on its synthesis, characterization, and application in targeted drug delivery [18]. The chapter will discuss the various synthesis methods used to prepare nanocrystals, exploring their advantages and limitations [19, 20]. In addition, advanced characterization techniques that enable the evaluation of nanocrystal properties, such as size, morphology, and surface characteristics, will be reviewed [21]. Furthermore, the chapter will delve into the mechanisms by which nanocrystals can be engineered to target specific cells or tissues, a critical step in enhancing the therapeutic efficacy of drug delivery systems [22]. Finally, we will explore the practical applications of nanocrystals in the treatment of diseases such as cancer, neurological disorders, and infectious diseases, drawing insights from case studies and clinical trials.

5.2 Synthesis of Nanocrystals: Techniques and Challenges

Nanocrystal synthesis methods are broadly categorized into top-down and bottom-up approaches, each offering distinct advantages for engineering particles with properties optimized for targeted drug delivery. These strategies directly influence critical parameters such as particle size, surface charge, and crystallinity, which govern biodistribution, cellular uptake, and therapeutic precision [23]. Figures 5.3 and 5.4 present schematic overviews of these methods, illustrating their significance in linking the synthesis process to functional outcomes.

5.2.1 Top-Down Methods: Precision Through Mechanical Energy

Top-down techniques fragment bulk drug materials into nanoscale particles while preserving crystalline integrity. High-pressure homogenization (HPH), a gold-standard method, subjects drug suspensions to pressures of 500–2500 bar, generating nanocrystals (100–300 nm) via cavitation and shear forces. This process ensures narrow size distributions (PDI <0.2), critical for passive targeting via the enhanced permeability and retention (EPR) effect in tumor tissues [18]. Ball milling,

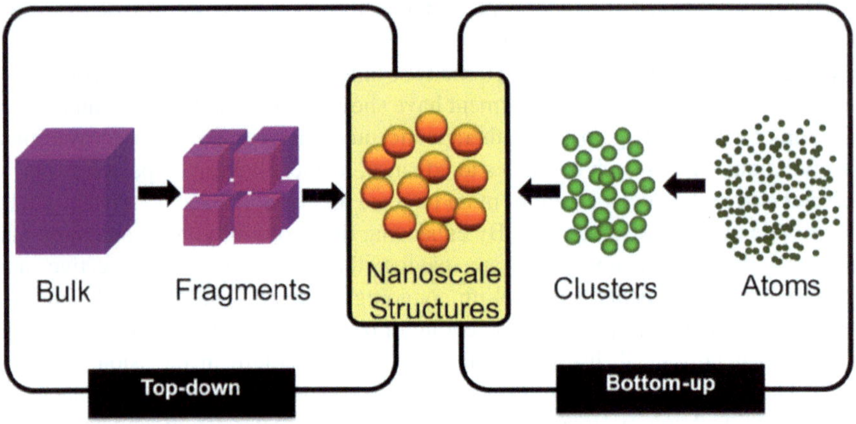

Fig. 5.3 Schematic comparison of top-down and bottom-up approaches for synthesizing nanocrystals in cancer drug therapy. (Reproduced with copyright permission from Ref. [29])

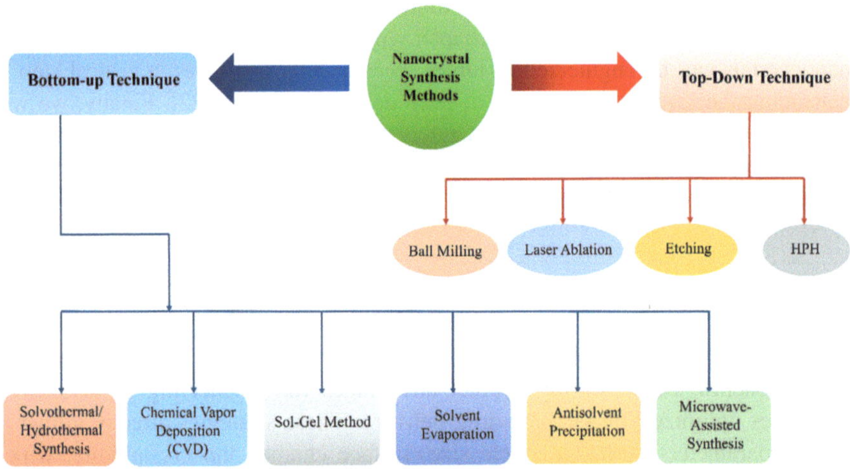

Fig. 5.4 Schematic representation of nanocrystal synthesis methods, comparing bottom-up techniques such as sol–gel method, solvent evaporation, and microwave-assisted synthesis with top-down techniques including ball milling, laser ablation, etching, and high-pressure homogenization (HPH). This highlights the diverse strategies for achieving nanoscale structures in drug delivery applications

another top-down approach, employs high-energy collisions between zirconium oxide beads and drug particles to produce nanocrystals stabilized by polymers like hydroxypropyl methylcellulose (HPMC) [24]. These methods are scalable and ideal for poorly soluble drugs, such as paclitaxel, where maintaining crystallinity (>80%) ensures controlled dissolution and sustained release at target sites. Laser ablation and etching complement these techniques, offering contamination-free nanocrystals (e.g., via pulsed lasers) or chemically refined surfaces for ligand conjugation. Collectively, top-down methods excel in producing particles with tunable sizes (50–500 nm) and surface charges (zeta potential: −20 to −40 mV), enabling evasion of immune clearance and prolonged circulation (Fig. 5.3).

5.2.2 Bottom-Up Methods: Controlled Assembly for Functional Surfaces

Bottom-up strategies construct nanocrystals from molecular precursors, allowing precise control over size, morphology, and surface functionality [25]. Solvent evaporation and antisolvent precipitation are widely used: a drug dissolved in a volatile solvent (e.g., acetone) is mixed with an antisolvent (water), inducing supersaturation and nucleation. By adjusting solvent-to-antisolvent ratios (1:10–1:20) and stabilizers (e.g., TPGS for P-gp inhibition), nanocrystals of 50–200 nm are achieved, with surfaces primed for active targeting ligands like folate or transferrin [26]. Microwave-assisted synthesis accelerates nucleation through dielectric heating, producing iron oxide nanocrystals (10–30 nm) in minutes with high crystallinity and biocompatibility [27]. The sol–gel method, though more common for inorganic nanoparticles, enables hybrid nanocrystals by hydrolyzing metal alkoxides (e.g., titanium isopropoxide) into gel networks, later calcined into crystalline structures [28]. These bottom-up approaches prioritize surface modularity, enabling functionalization with antibodies or peptides for receptor-mediated uptake. By aligning synthesis parameters (e.g., pressure in HPH, stabilizer chemistry in precipitation) with targeting goals, researchers engineer nanocrystals that balance scalability, stability, and specificity. Figures 5.3 and 5.4 visually map these relationships, emphasizing how mechanical energy (top-down) and thermodynamic control (bottom-up) shape nanocrystals for precise drug delivery.

- *Sol–Gel Method*: The sol–gel process is a versatile technique that involves transitioning from a liquid solution (sol) to a solid gel to form nanocrystals. Metal alkoxides or metal salts are typically used as precursors, which are hydrolyzed and polymerized to form nanoscale crystal structures [30]. This technique is especially effective for creating metal oxide nanocrystals and composites, which are often used in catalysis, sensors, and drug delivery systems [28].
- *Antisolvent Precipitation*: This method involves dissolving the drug in a solvent and then adding a nonsolvent (antisolvent), causing the drug to precipitate out of the solution as nanocrystals [31]. The size of the nanocrystals can be controlled by adjusting the rate of solvent and antisolvent addition, as well as other factors such as temperature and stirring speed. This method is scalable and cost-effective, but controlling the uniformity of the crystals can be challenging [32].
- *Microwave-Assisted Synthesis*: Microwave heating can be used to accelerate the formation of nanocrystals by enhancing the molecular interactions between drug molecules. This method allows for the rapid formation of nanocrystals and can

produce highly uniform particles [27]. It has been shown to be especially useful for synthesizing nanocrystals with specific surface characteristics or functional groups, which can improve drug delivery performance [33].

5.3 Characterization of Nanocrystals: Techniques and Analysis

Thorough characterization of nanocrystals is essential to ensure their suitability for targeted drug delivery, as properties such as particle size, surface charge, and crystallinity directly influence biological fate and therapeutic efficacy. Dynamic light scattering (DLS) is routinely employed to determine hydrodynamic diameter and polydispersity index (PDI), with optimal nanocrystals for passive tumor targeting typically ranging from 100 to 300 nm and exhibiting a narrow PDI (<0.2) [34]. Zeta potential measurements assess surface charge, where values beyond ± 20 mV generally indicate sufficient electrostatic stabilization to prevent aggregation and reduce rapid clearance by the reticuloendothelial system [23].

X-ray diffraction (XRD) and differential scanning calorimetry (DSC) are used to confirm the crystalline state of nanocrystals, which is crucial for maintaining controlled drug release profiles and ensuring batch-to-batch consistency [35]. Scanning electron microscopy (SEM) and transmission electron microscopy (TEM) provide detailed insights into particle morphology and surface structure, which can affect cellular uptake and biodistribution [36, 37]. Additionally, Fourier-transform infrared spectroscopy (FTIR) [38] and X-ray photoelectron spectroscopy (XPS) [39] are valuable for verifying surface modifications, such as ligand or antibody conjugation, that enable active targeting.

Figure 5.5 provides a summary of the key characterization techniques, demonstrating the contribution of each analytical method to the evaluation of critical quality attributes of nanocrystals. This comprehensive characterization framework ensures that nanocrystals are optimized not only for stability and reproducibility but also for precise and effective targeted drug delivery.

5.4 Mechanisms of Targeted Drug Delivery

Targeted drug delivery is a sophisticated approach designed to deliver therapeutic agents directly to the site of action, thereby enhancing their efficacy and reducing systemic side effects [3]. By exploiting specific characteristics of the target tissues, cells, or organs, targeted drug delivery systems (TDDS) improve the precision of drug release, allowing for better therapeutic outcomes [17]. The mechanisms involved in targeted drug delivery are diverse, ranging from passive targeting through enhanced permeability and retention (EPR) effects to more precise active targeting using ligands that bind to specific receptors. We explore the major mechanisms of targeted drug delivery next [3].

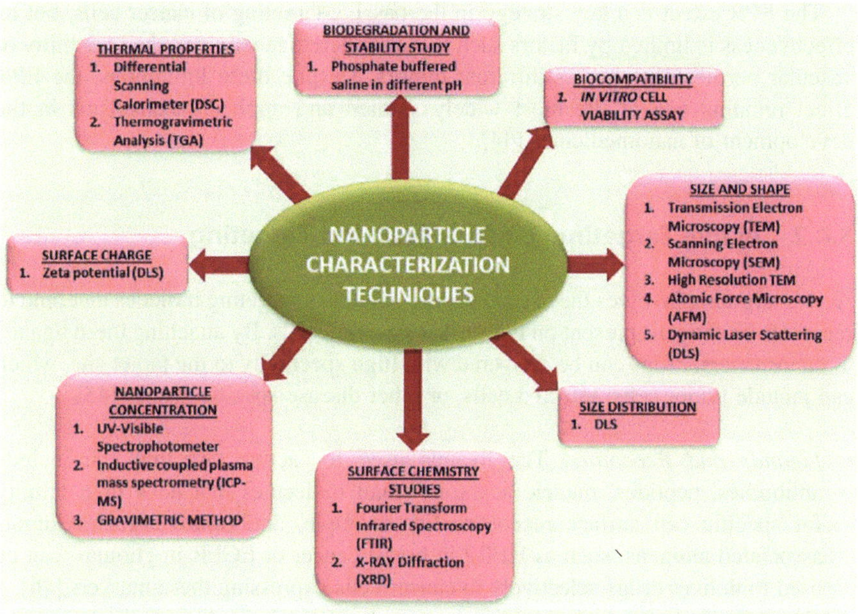

Fig. 5.5 Summary of characterization techniques for nanocrystals in targeted drug therapy, illustrating methods such as transmission electron microscopy (TEM) for particle morphology, dynamic light scattering (DLS) for size distribution, Fourier-transform infrared spectroscopy (FTIR) for surface functionalization, high-performance liquid chromatography (HPLC) for drug loading and release profiling, and stability assays to evaluate formulation robustness, all essential for optimizing nanocrystal performance and therapeutic efficacy [40]

5.4.1 Passive Targeting: Enhanced Permeability and Retention (EPR) Effect

The EPR effect is a phenomenon that occurs in tumor tissues and inflamed areas, where blood vessels exhibit abnormal architecture. These vessels are typically leaky, allowing nanoparticles or nanocarriers to accumulate more easily in the tumor tissue. This increased permeability, combined with the impaired lymphatic drainage system in tumors, results in the enhanced retention of therapeutic agents in the tumor site [41].

- *Tumor Vasculature*: Tumors often exhibit irregular and leaky blood vessels with wide fenestrations that allow large molecules or nanoparticles to pass through. The leaky nature of these vessels, coupled with the poor drainage of the lymphatic system in tumors, leads to the accumulation of drug-loaded nanocrystals [42].
- *Nanocarriers in EPR*: Various types of nanocarriers, such as liposomes, micelles, and polymeric nanoparticles, capitalize on the EPR effect to enhance the accumulation of drugs in tumor tissues. The size of the nanocarrier plays a crucial role in determining its ability to penetrate the tumor vasculature and remain trapped in the tumor [43].

The EPR effect is a key strategy in the passive targeting of cancer cells, but its effectiveness is limited by factors such as tumor heterogeneity and the variability of vascular permeability across different tumors. Despite these limitations, the EPR effect remains one of the most widely studied and applied mechanisms in the development of nanomedicines [44].

5.4.2 Active Targeting: Ligand-Mediated Targeting

Active targeting involves the use of specific ligands or targeting moieties that bind to receptors or antigens present on the surface of target cells. By attaching these ligands to nanocarriers, drugs can be delivered with high specificity to the target site, which can include tumor cells, infected cells, or other disease-specific tissues [45].

- *Ligands and Receptors*: The ligands used for active targeting can include antibodies, peptides, nucleic acids, or small molecules that have high affinity for specific cell surface receptors. For example, antibodies targeting tumor-associated antigens, such as HER2 in breast cancer or EGFR in gliomas, can be used to deliver drugs selectively to cancer cells expressing these markers [46].
- *Drug Conjugates*: Active targeting can also be achieved through the use of antibody–drug conjugates (ADCs), where the antibody acts as a delivery vehicle, guiding the drug directly to the target cells. ADCs have gained prominence in cancer therapy as they combine the specificity of antibodies with the potency of cytotoxic drugs [47].

Active targeting provides a more precise approach to drug delivery compared to passive methods. However, the successful application of this strategy depends on the identification of suitable targets and the development of stable and efficient drug conjugates [48, 49].

5.4.3 Stimuli-Responsive Targeting

Stimuli-responsive drug delivery systems respond to specific changes in the local environment, such as pH, temperature, light, or enzymatic activity, to trigger the release of drugs at the target site. These systems offer a level of control and flexibility, as the release of the therapeutic agent is triggered by external or internal stimuli.

- *pH-Sensitive Delivery*: pH-sensitive nanocarriers are engineered to exploit the acidic microenvironment characteristic of tumors and inflamed tissues, enabling site-specific drug release. For example, poly(acrylic acid)-polyhedral oligomeric silsesquioxane (PAA-POSS) nanocrystals loaded with doxorubicin have demonstrated a pronounced pH-dependent release profile, with approximately

75% of the doxorubicin payload released within 24 h at pH 5.0 (tumor lysosomal conditions), compared to only 40% at physiological pH 7.4. The hydrophobic POSS core contributes to high drug loading and enhanced stability in the bloodstream, and in vivo experiments have shown a 2.3-fold increase in tumor accumulation compared to non-pH-sensitive controls, as well as a notable reduction in cardiotoxicity [50]. Another example involves albumin-coated PLGA nanocrystals modified with acid-transforming masked agents (ATRAM), which exhibit up to 79% drug release at pH 5.0 compared to less than 10% at pH 7.4 due to protonation-induced surface charge reversal, significantly enhancing cellular uptake in breast cancer cells [51].

- *Enzyme-Sensitive Systems*: Enzyme-sensitive nanocarriers are designed to respond to specific enzymes that are overexpressed in disease sites, such as tumors or inflamed tissues, thereby achieving controlled and localized drug release. A notable case study involves hyaluronic acid (HA)-coated mesoporous silica nanocrystals loaded with doxorubicin. In the presence of tumor-associated hyaluronidase, the HA coating is degraded, exposing biotin targeting ligands and resulting in a significant increase in drug release—up to 65% within 12 h. In animal models, these enzyme-responsive nanocrystals achieved a 45% reduction in tumor volume, outperforming nonresponsive controls [52]. Another advanced example features tocopheryl polyethylene glycol succinate (TPGS) nanoparticles designed with matrix metalloproteinase-2 (MMP-2) cleavable peptide linkers. When administered to glioblastoma models, the overexpression of MMP-2 in the tumor microenvironment triggered the degradation of the peptide linkers, resulting in the efficient release of temozolomide. This approach led to a 90% drug release in glioma cells and extended the median survival of glioma-bearing mice from 22 to 53 days [53].
- *Thermal or Light-Responsive Systems*: Thermal and light-responsive nanocarriers provide spatial and temporal control over drug release, making them highly attractive for precision medicine applications. One compelling example is the use of gold nanorod-embedded polymersomes loaded with doxorubicin. These nanocarriers are designed to respond to near-infrared (NIR) light, which induces localized heating and triggers rapid, pH-dependent drug release. Upon NIR irradiation, these systems achieved up to 80% release of doxorubicin within 20 min, and in vivo studies demonstrated complete tumor regression in mice after three rounds of NIR treatment, with no observable systemic toxicity. Similarly, thermosensitive poly(N-isopropylacrylamide) (PNIPAM) nanogels loaded with curcumin have shown a fourfold increase in drug release at hyperthermic temperatures (42 °C) compared to normal body temperature. When combined with localized heating in breast cancer models, these nanogels reduced tumor growth by 70% relative to free curcumin, underscoring the efficacy of thermal-responsive systems [54]. Stimuli-responsive targeting is particularly useful in applications such as cancer therapy, where localized drug release can enhance the therapeutic effect while minimizing side effects.

5.4.4 Nanoparticle-Mediated Targeting via Cellular Uptake

Cellular uptake of nanoparticles plays a crucial role in targeted drug delivery. Nanoparticles can enter cells via various endocytosis pathways, including clathrin-mediated endocytosis, caveolae-mediated endocytosis, and macropinocytosis. The size, charge, and surface properties of nanoparticles influence the efficiency of cellular uptake, allowing for the design of nanocarriers that can selectively target specific cell types [55].

- *Surface Modifications for Cellular Targeting*: Surface functionalization of drug nanocrystals with targeting ligands significantly enhances cellular recognition and uptake, enabling precise delivery to diseased tissues. Beyond folic acid conjugation that can be selectively internalized by cancer cells that overexpress folate receptors [56], transferrin-modified docetaxel nanocrystals exemplify this strategy: docetaxel nanocrystals (DTX-NCs) coated with apo-transferrin (Tf) via adsorption demonstrated a 2.3-fold increase in cellular uptake by A549 lung cancer cells compared to non-targeted nanocrystals, attributed to Tf receptor-mediated endocytosis. Similarly, antibody-conjugated nanocrystals leverage covalent bonding methods like carbodiimide or maleimide chemistry to attach monoclonal antibodies (e.g., anti-EGFR or HER2) directly to nanocrystal surfaces, enhancing tumor-specific binding and reducing off-target effects [57]. RGD peptide-decorated nanocrystals further illustrate this principle, where cyclic arginylglycylaspartic acid (cRGD) ligands target integrin receptors overexpressed in glioblastoma, improving cellular internalization by 40% in U87MG glioma models [58]. These modifications exploit receptor–ligand interactions to bypass biological barriers (e.g., blood–brain barrier) and amplify therapeutic efficacy while maintaining the nanocrystals' inherent advantages of high drug loading and carrier-free design.
- *Targeting* via *Specific Pathways*: Targeting nanoparticles to specific cellular pathways or intracellular compartments can further enhance their therapeutic efficacy [59, 60]. Nanocarriers can be designed to accumulate in organelles such as the mitochondria or nucleus, improving the therapeutic effect on specific intracellular targets [61]. By manipulating the size, surface properties, and functionalization of nanoparticles, targeted drug delivery can be achieved at both the cellular and subcellular levels.

5.4.5 Magnetic Targeting

Magnetic targeting utilizes magnetic fields to guide drug-loaded nanoparticles to specific sites within the body. Magnetic nanoparticles (MNPs), typically made of iron oxide or other magnetic materials, can be loaded with drugs and guided to target tissues through external magnetic fields. Once injected into the bloodstream, MNPs are directed by an external magnet placed over the target area. This method can be used to increase the local concentration of therapeutic agents at the target site, particularly in tissues with high vascularity, such as tumors or inflamed tissues.

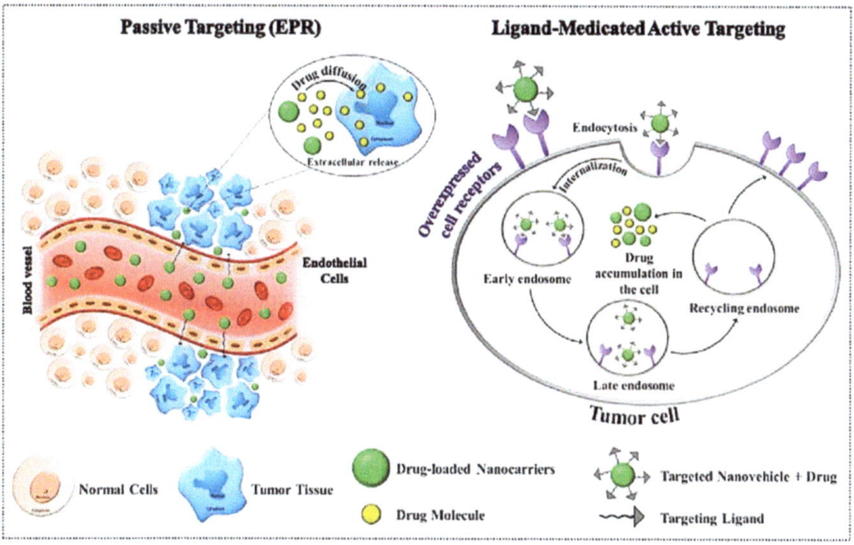

Fig. 5.6 Visualization of nanocrystals engineered for targeted drug delivery, highlighting their enhanced solubility, surface functionalization for precise tumor targeting, and controlled drug release mechanisms that improve therapeutic efficacy while minimizing systemic side effects. (Reproduced with copyright permission from Ref. [63])

Magnetic targeting has been explored in cancer therapy, where MNPs can enhance tumor targeting by providing an additional external force. It has also been proposed for use in treating inflammatory diseases, where inflammation sites can be targeted using MNPs. It is an innovative approach that can provide a noninvasive means of directing therapeutic agents to specific sites, although its application is still largely in the experimental stage.

Targeted drug delivery mechanisms offer exciting possibilities for improving the specificity, efficacy, and safety of pharmaceutical therapies. The choice of mechanism, whether passive targeting via the EPR effect, active targeting using ligands, stimuli-responsive systems, cellular uptake, or magnetic targeting, depends on the nature of the disease, the drug being delivered, and the desired therapeutic outcome (Fig. 5.6). By tailoring the delivery system to exploit these mechanisms, researchers are steadily advancing the field of nanomedicine and paving the way for more personalized and effective treatments [62].

5.5 Applications of Nanocrystals in Drug Delivery

Nanocrystals are a promising tool in the field of drug delivery, owing to their unique properties such as high surface area, increased solubility, and enhanced bioavailability. These properties make nanocrystals highly effective for the delivery of poorly water-soluble drugs. Their applications span various therapeutic areas, from cancer treatment to gene therapy, and they can be tailored to target specific tissues or cells, further enhancing the precision and efficacy of treatment. We explore several key applications of nanocrystals in drug delivery next.

5.5.1 Enhanced Oral Bioavailability of Poorly Soluble Drugs

One of the primary challenges in pharmaceutical development is the limited bioavailability of drugs that have poor water solubility. Nanocrystals significantly improve the solubility and dissolution rate of these compounds by reducing the particle size, which increases the surface area and enhances the rate at which the drug dissolves in the gastrointestinal tract [4].

- *Improvement in Solubility and Absorption*: By reducing the particle size to the nanometer range, nanocrystals offer a larger surface area for dissolution, leading to faster absorption. Drugs like paclitaxel, which is poorly soluble in water, have shown improved bioavailability when formulated as nanocrystals [64].
- *Techniques for Formulation*: Several techniques, including wet milling, high-pressure homogenization, and solvent evaporation, are used to prepare nanocrystals. These methods help achieve a consistent particle size distribution that maximizes the drug's solubility and bioavailability [18].

This application of nanocrystals is particularly important for oral formulations, where enhancing solubility can dramatically improve the therapeutic effectiveness of drugs that would otherwise be ineffective due to poor absorption.

5.5.2 Targeted Cancer Therapy

Cancer is one of the most common diseases where nanocrystals have found significant application. By modifying nanocrystals with targeting ligands, such as antibodies or peptides, they can be directed specifically to tumor cells, reducing the toxic effects of chemotherapy on healthy tissues. The high surface area of nanocrystals allows for the efficient attachment of targeting agents, enabling precise targeting and release of the drug at the tumor site [65].

- *Tumor-Targeting* via *EPR Effect*: As discussed earlier, the EPR effect is a key mechanism in cancer therapy. Nanocrystals can accumulate more readily in tumors due to the leaky vasculature characteristic of tumor blood vessels. By incorporating therapeutic agents into nanocrystals, drugs can be delivered more specifically to tumor cells, reducing systemic toxicity [66].
- *Combination Therapy*: Nanocrystal-based combination therapies, defined by their high drug loading and carrier-free crystalline structure, have shown significant promise in cancer treatment. For example, paclitaxel and niclosamide nanocrystals, prepared via antisolvent precipitation and co-loaded into a

thermosensitive hydrogel, achieved sustained dual-drug release and resulted in 80% tumor growth inhibition in breast cancer models, demonstrating synergistic suppression of cancer stem cells [57]. Similarly, SN-38 nanocrystals coated with a lipid membrane enabled high drug loading and effective delivery across the blood–brain barrier, extending survival in glioma-bearing mice by combining chemotherapy with vasculogenic mimicry destruction [67]. Additionally, crystalline nanodrugs composed of 5-aminolevulinic acid, Fe^{3+}, and curcumin enabled the combination of photodynamic therapy and chemotherapy, achieving 97% tumor suppression in murine models through controlled, spatiotemporal drug release [68]. Ursolic acid and paclitaxel self-assembled nanocrystals further exemplify this strategy, providing enhanced plasma half-life, reduced hepatotoxicity, and synergistic cytotoxicity in liver and breast cancer models [69]. Collectively, these examples highlight how true nanocrystal-based combination therapies can offer potent, targeted, and safer cancer treatments by leveraging dual-drug synergy and efficient, carrier-free delivery. The combination of high drug-loading capacity and targeted delivery makes nanocrystals a promising approach for enhancing the efficacy of cancer therapies, particularly those with limited effectiveness or significant side effects [43].

5.5.3 Treatment of Cardiovascular Diseases

Nanocrystals are also being explored for the treatment of cardiovascular diseases, particularly for the delivery of drugs that target atherosclerosis, thrombosis, and heart failure. For instance, statins, which are used to lower cholesterol levels, can benefit from nanocrystal formulations by improving their solubility and bioavailability [70].

- *Targeting Atherosclerotic Plaques*: Nanocrystals functionalized with specific ligands offer a promising strategy for targeting atherosclerotic plaques by binding selectively to endothelial cell markers or plaque-associated proteins, thereby enhancing drug delivery directly to diseased vascular sites while minimizing systemic side effects [71]. A notable case study demonstrated this approach using gadolinium-doped amorphous calcium carbonate nanocrystals (CC-Aln NPs) surface-modified with alendronate, a ligand that targets hydroxyapatite deposits commonly found in calcified atherosclerotic plaques. In apolipoprotein E knockout mouse (ApoE-deficient mouse), these functionalized nanocrystals exhibited a 2.7-fold higher accumulation in plaques compared to nontargeted nanoparticles, as confirmed by MRI and synchrotron X-ray fluorescence imaging, and showed a significant increase in signal intensity in aortic arches with minimal off-target retention. This targeted delivery not only allowed for earlier and more precise detection of vulnerable plaques but also enabled a 60% reduction in the required drug dose while maintaining therapeutic efficacy, underscoring the potential of ligand-functionalized nanocrystals to improve the safety and effectiveness of treatments for atherosclerosis [72].

- *Controlled Release for Cardiovascular Drugs*: Nanocrystals can provide controlled release of cardiovascular drugs, such as antithrombotic agents, ensuring a steady concentration of the drug in the bloodstream over extended periods [73]. Nanocrystal-based formulations have demonstrated significant potential in optimizing the controlled release of cardiovascular drugs, as exemplified by a recent study on nimodipine nanocrystals for managing subarachnoid hemorrhage-induced vasospasm. Nimodipine, a calcium channel blocker, traditionally suffers from low oral bioavailability (<10%) and requires frequent dosing (every 4 h), often leading to systemic hypotension. To address this, researchers developed nimodipine nanocrystals (159 nm) using a combination of micro-precipitation and high-pressure homogenization. The nanocrystals achieved a 3.5-fold faster dissolution rate than conventional formulations (Nimotop®), enabling sustained plasma concentrations over 24 h with a 40% reduction in peak drug levels in preclinical models. This controlled release profile minimized blood pressure fluctuations—a critical advantage given that 25% of patients on standard nimodipine experience hypotensive episodes while maintaining therapeutic efficacy in preventing cerebral vasospasm. The nanocrystals' large surface area facilitated gradual dissolution aligned with first-order kinetics, contrasting sharply with the rapid burst release of traditional formulations. This case underscores how nanocrystal technology can balance efficacy and safety in cardiovascular therapies requiring stable drug levels, offering a template for improving antithrombotic agents like rivaroxaban or apixaban, where precise dose control is essential to avoid bleeding complications [4]. Nanocrystals represent a novel approach to improving the delivery and efficacy of cardiovascular treatments, particularly for drugs that face challenges in solubility and bioavailability [74].

5.5.4 Gene and Nucleic Acid Delivery

Gene therapy involves the delivery of genetic material, such as DNA, RNA, or small interfering RNA (siRNA), to specific cells to treat diseases at the genetic level [75]. Nanocrystals are increasingly being studied for their ability to deliver these nucleic acids due to their small size, high surface area, and ease of surface modification [76].

Nanocrystals have emerged as versatile platforms for targeted gene and nucleic acid delivery, addressing challenges such as nuclease degradation and inefficient cellular uptake. For instance, a self-assembled cellulose nanocrystal/zinc/DNA nanocomplex (CZD) utilized ligand-to-metal charge transfer (LMCT) between zinc ions and phosphate groups in cellulose nanocrystals (CNCs) and DNA to form carrier-free nanoparticles without chemical conjugation. This system achieved 85% transfection efficiency in HEK293 cells 1.7-fold higher than polyethyleneimine (PEI) vectors while maintaining low cytotoxicity (<10% cell death at 100 μg/mL) and protecting DNA from enzymatic degradation for over 24 h. The CNC's anionic surface allowed modular functionalization with folate ligands, enabling selective uptake in folate receptor-positive cancer cells (e.g., HeLa) via receptor-mediated endocytosis. In another study, bacterial cellulose nanocrystals (BCNCs) derived from *Komagataeibacter xylinus* were crosslinked with glutaraldehyde to enhance mechanical stability and modified with cationic polymers (e.g., methacrylamide) for siRNA delivery [77]. The cationic BCNCs exhibited 65.3% nucleic acid adsorption efficiency via electrostatic interactions and sustained siRNA release over 72 h, achieving 60% KRAS gene silencing in colorectal cancer cells (HT-29) with minimal cytotoxicity (cell viability >90% at 0.1 mg/mL). In vivo, these BCNCs reduced tumor volume by 45% in murine xenograft models through targeted accumulation and lysosomal escape facilitated by their needle-like morphology (25 ± 10 nm diameter, 626 ± 172 nm length) [78]. Both systems exemplify how nanocrystals combine high payload capacity, biocompatibility, and tunable surface chemistry to overcome biological barriers, enabling precise gene therapy applications. Gene delivery via nanocrystals offers an exciting opportunity for treating genetic disorders, cancers, and other conditions that require the modification or silencing of specific genes [79].

5.5.5 Delivery of Biologic Drugs

Biologic drugs, including proteins, monoclonal antibodies, and vaccines, are often large molecules that face challenges in terms of stability, solubility, and targeted delivery. Nanocrystals can help address these challenges by providing a stable carrier system for biologic drugs, protecting them from degradation and enhancing their delivery to the target site [80].

- *Stabilization of Biologic Drugs*: Nanocrystals can protect biologic drugs from degradation due to environmental factors like temperature and pH changes, ensuring that the active ingredient remains intact during transportation and administration [81].
- *Targeted Delivery of Monoclonal Antibodies*: Monoclonal antibodies, which are used for cancer treatment and autoimmune diseases, can be loaded onto nanocrystals and targeted to specific cells or tissues [82]. Monoclonal antibodies (mAbs) can be integrated with nanocrystal-based systems to achieve precise therapeutic delivery, particularly in cancer and autoimmune diseases. By conjugating mAbs to nanocrystals, these platforms enhance targeting specificity while minimizing systemic side effects. For example, anti-CD20 monoclonal antibodies were coupled with doxorubicin-loaded carbon nanoparticles (ACNP-DOX-DSPE-PEG2000-anti-CD20) to target CD20-positive B-cell lymphoma cells (Raji cells). The nanocrystals demonstrated selective uptake via antibody–antigen binding, leading to a 3.5-fold increase in intracellular doxorubicin accumulation compared to nontargeted nanoparticles. Confocal microscopy and flow cytometry confirmed receptor-mediated endocytosis, resulting in 80% tumor cell apoptosis in vitro while sparing CD20-negative cells [83]. Another case study

highlights trastuzumab-conjugated paclitaxel nanorods (ADNs) for HER2-positive breast cancer. The antibody-functionalized nanorods improved tumor-specific binding, achieving 90% cytotoxicity in HER2-overexpressing SK-BR-3 cells versus 20% in HER2-negative cells. In vivo models showed a 60% reduction in tumor volume compared to free paclitaxel, attributed to prolonged circulation and reduced off-target accumulation [84]. These examples illustrate how nanocrystal-mAb hybrids leverage both passive (enhanced permeability and retention effect) and active targeting mechanisms, enabling higher therapeutic indices and reduced systemic toxicity [85].

Nanocrystals enable more effective delivery of biologic drugs, ensuring their stability, solubility, and targeted action [86].

5.5.6 Delivery of Anti-infective Agents

Nanocrystals have also shown promise in the delivery of anti-infective agents, including antibiotics and antiviral drugs. These drugs often face challenges such as poor solubility, rapid metabolism, or limited tissue penetration. Nanocrystals can improve the solubility and bioavailability of these agents, allowing for more effective treatment of infections.

- *Targeted Delivery to Infected Tissues*: Nanocrystals can be designed to target infected tissues or cells, such as those involved in bacterial or viral infections. This targeted approach can reduce the need for high doses and minimize side effects, which are common with conventional anti-infective treatments [80].
- *Enhanced Drug Penetration*: Nanocrystals can enhance the penetration of drugs into tissues where infections are localized, such as in the case of intracellular pathogens. By improving drug bioavailability and targeting, nanocrystals can provide a more efficient means of combating infections [87].

The application of nanocrystals in anti-infective drug delivery is especially important in the face of rising antimicrobial resistance as it offers a way to improve the effectiveness of existing drugs.

5.6 Clinical Applications

Nanocrystals have shown promising results in various preclinical and clinical studies, especially in overcoming challenges related to the solubility and bioavailability of poorly water-soluble drugs. The following case studies highlight the potential of nanocrystal technology in drug delivery and its clinical applications across various therapeutic areas.

5.6.1 Rapamune® (Sirolimus) Nanocrystals

One of the first commercial applications of nanocrystals in drug delivery was the development of Rapamune® (sirolimus) nanocrystals, which was formulated to improve the bioavailability of sirolimus, an immunosuppressive drug. Sirolimus has poor water solubility, limiting its therapeutic effectiveness [88].

The nanocrystal formulation of sirolimus was developed to improve solubility and absorption. By reducing the particle size to the nanoscale, the formulation significantly increased the dissolution rate and, consequently, the bioavailability of the drug. Clinical studies showed that the nanocrystal formulation improved the pharmacokinetics of sirolimus, allowing for more consistent drug absorption and potentially reducing the risk of organ rejection in transplant patients [89]. The improved bioavailability also allowed for lower doses, which may reduce the incidence of side effects [90].

5.6.2 Kyorin Pharmaceutical's Nanocrystal Drug Delivery System

Another notable example of nanocrystals in clinical use comes from *Kyorin Pharmaceutical*, which has developed a nanocrystal-based drug delivery system for various poorly soluble drugs, including an anti-inflammatory agent, S-2496 [4]. The aim was to enhance the therapeutic efficacy of the drug while minimizing side effects [80].

S-2496, an anti-inflammatory drug, was reformulated into nanocrystals to increase its solubility. This formulation achieved a significant increase in the drug's bioavailability, leading to a higher concentration in the bloodstream following oral administration. Early-phase clinical trials of the nanocrystal formulation showed enhanced absorption and improved clinical outcomes in patients suffering from chronic inflammation. The formulation allowed for a reduction in dosing frequency while maintaining therapeutic effects, offering better patient compliance and fewer side effects [91].

5.6.3 Nanocrystals for Anticancer Drug Delivery

Nanocrystals have been extensively studied for their application in cancer therapy, where they offer advantages in overcoming the poor solubility of chemotherapy drugs and improving targeted drug delivery [92]. One of the most notable studies is the formulation of paclitaxel nanocrystals, a widely used anticancer agent [11]. Paclitaxel, a hydrophobic drug, has low solubility in water, which limits its efficacy when administered via traditional routes. By converting paclitaxel into nanocrystals, the drug's solubility was significantly enhanced, allowing for better absorption and controlled release. Preclinical studies and early-phase clinical trials of paclitaxel nanocrystals have shown promising results in terms of enhanced drug bioavailability and improved tumor targeting. By delivering higher concentrations of the drug to the

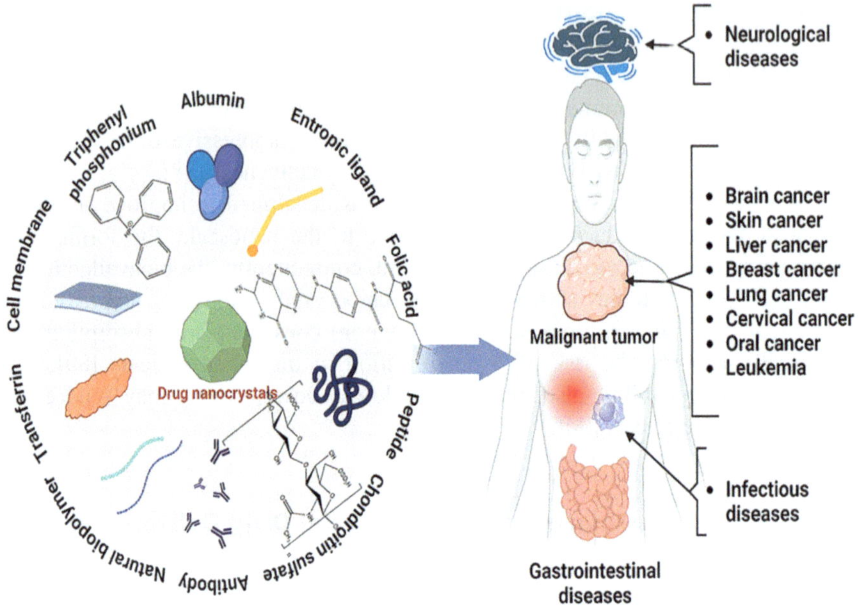

Fig. 5.7 Illustration of specialized nanocrystals engineered for cancer drug therapy, highlighting their precise size and surface modifications that enable targeted delivery to tumor cells. The image showcases the uniform morphology of the nanocrystals, their functionalized surfaces for drug attachment, and their interaction with cancerous tissues, demonstrating their potential to enhance treatment efficacy while minimizing side effects. (Reproduced with copyright permission from Ref. [92])

tumor site while reducing systemic exposure, paclitaxel nanocrystals may improve therapeutic outcomes and reduce the side effects often associated with chemotherapy [93]. This advancement, along with a myriad of others, is depicted in Fig. 5.7, showcasing the engineered nanocrystals and their targeted delivery to cancerous tissues.

5.6.4 Nanocrystals for Oral Insulin Delivery

Another exciting clinical application of nanocrystals is in the oral delivery of insulin, which traditionally requires injection. Nanocrystal formulations of insulin have been developed to overcome the challenge of insulin's poor bioavailability when administered orally due to its large molecular size and instability in the gastrointestinal tract. Insulin nanocrystals were designed to increase the drug's solubility, stability, and absorption from the gastrointestinal tract. The formulation was created using various stabilizing agents and surfactants to protect the insulin molecules and enhance their bioavailability. Early-stage clinical trials have demonstrated that oral insulin nanocrystals can achieve effective blood glucose control, potentially offering a noninvasive alternative to insulin injections. These formulations hold promise for improving the quality of life for diabetic patients by providing an easier and more convenient means of insulin delivery [94].

5.6.5 Nanocrystals for Antibiotic Delivery

Nanocrystals have also shown promise in the delivery of antibiotics, particularly for the treatment of infections caused by multidrug-resistant bacteria. One such example is the use of ciprofloxacin nanocrystals, which was formulated to address the challenges of poor solubility and bioavailability of this antibiotic.

Ciprofloxacin nanocrystals were developed to increase the solubility and bioavailability of the antibiotic, improving its effectiveness when delivered to infected tissues. The formulation also helped enhance the antibiotic's ability to penetrate bacterial cell walls. Clinical studies involving ciprofloxacin nanocrystals have demonstrated enhanced antibacterial activity and better tissue penetration, providing a promising strategy for combating antibiotic resistance [95]. These case studies highlight the diverse applications of nanocrystals in drug delivery, as summarized in Table 5.1.

These studies highlight the diverse clinical applications of nanocrystal-based drug delivery systems, showcasing their potential to enhance the solubility, bioavailability, and targeted delivery of a wide range of therapeutic agents. From improving the efficacy of immunosuppressants like sirolimus to enabling noninvasive insulin

Table 5.1 Summary of various nanocrystal formulations and their testing stages

Drug name	Mode of synthesis	Carrier system	Testing phase	Reference
Rapamune® (sirolimus)	High-pressure homogenization/ media milling	Polyethylene glycol (PEG)	Clinical studies (improved pharmacokinetics and bioavailability)	[88]
S-2496 (anti-inflammatory)	High-pressure homogenization/ bead milling	Polysorbate 80	Early-phase clinical trials (enhanced absorption and clinical outcomes)	[91]
Paclitaxel (anticancer)	Wet milling/ high-pressure homogenization	Polyvinyl alcohol (PVA)	Preclinical and early-phase clinical trials (enhanced solubility and tumor targeting)	[93]
Insulin (oral delivery)	Solvent evaporation or antisolvent precipitation	Chitosan	Early-stage clinical trials (effective blood glucose control)	[94]
Ciprofloxacin (antibiotic)	High-pressure homogenization/ wet milling	Hydroxypropyl methylcellulose (HPMC)	Clinical studies (enhanced antibacterial activity and tissue penetration)	[95]

delivery and combating multidrug-resistant infections, nanocrystals offer promising solutions to some of the most pressing challenges in modern medicine. As research advances and clinical data accumulate, the clinical application of nanocrystals is expected to expand, providing more effective and patient-friendly drug delivery options in the future [96].

5.7 Challenges and Limitations of Nanocrystal-Based Drug Delivery

While nanocrystals offer numerous advantages in drug delivery, several challenges and limitations remain, which can impact their development and clinical application. These challenges primarily involve issues related to formulation, manufacturing, stability, and regulatory hurdles. We discuss some of the most prominent challenges and limitations associated with nanocrystal-based drug delivery systems next.

5.7.1 Stability Concerns and Aggregation

One of the primary concerns with nanocrystals is their stability, especially in aqueous environments. Due to their small size and large surface area, nanocrystals are prone to surface energy-driven phenomena such as aggregation, which can lead to the formation of larger particles. This aggregation can reduce the effective surface area and compromise the bioavailability of the drug [97].

- *Aggregation During Storage*: Nanocrystals tend to aggregate over time due to their increased surface energy, which results in clumping. This can lead to a loss of their desirable small size, thereby affecting their dissolution rate and bioavailability [98].
- *Prevention Strategies*: To mitigate aggregation, nanocrystals are often stabilized with surfactants, polymers, or other excipients that can prevent particle aggregation. However, the use of stabilizers may introduce other issues, such as increased toxicity or interference with drug release profiles [99].

Addressing the stability of nanocrystals is crucial to ensuring their effectiveness in drug delivery, particularly for long-term storage and clinical use.

5.7.2 Limited Drug-Loading Capacity

Although nanocrystals have a high surface area, the drug-loading capacity of nanocrystals is often limited. This can be particularly problematic when attempting to deliver high doses of drugs, or when working with drugs that have low solubility. A low drug-loading capacity requires the administration of larger volumes of the nanocrystal formulation to achieve therapeutic concentrations, which may not always be feasible or desirable [100].

- *Impact on Dosing and Administration*: To achieve sufficient therapeutic levels, nanocrystals with low drug loading may require higher volumes of administration, which can be inconvenient for patients, particularly for intravenous formulations. Additionally, this may lead to issues related to dosing accuracy and patient compliance [101].
- *Overcoming Low Drug Loading*: Advances in formulation techniques, such as the use of co-encapsulation strategies or the development of core–shell nanocrystals, are being explored to improve the drug-loading capacity of nanocrystals. These approaches can allow for a higher payload of drugs while maintaining particle stability and enhancing therapeutic efficacy [102].

Despite these challenges, ongoing research into improving drug-loading capacity holds promise for enhancing the clinical applicability of nanocrystals.

5.7.3 Manufacturing and Scalability

The manufacturing of nanocrystals presents significant challenges, particularly when it comes to scaling up production for commercial purposes. Techniques like high-pressure homogenization and wet milling are commonly used to produce nanocrystals, but these processes can be time-consuming and difficult to scale efficiently. This limits the ability to produce nanocrystal formulations on a large scale, which can be problematic for meeting the demands of clinical trials or commercial market [103].

- *Process Optimization*: High-pressure homogenization and other nanocrystal production methods need to be optimized for large-scale manufacturing. However, achieving consistency in particle size and drug loading during large-scale production is a significant challenge [104].
- *Cost Considerations*: The complexity of nanocrystal manufacturing also raises concerns about the cost-effectiveness of these systems. High production costs can be a barrier to the widespread adoption of nanocrystal-based formulations, especially compared to traditional drug delivery systems [105].

Despite these hurdles, advances in manufacturing technologies and process scaling are helping to make the large-scale production of nanocrystals more feasible.

The development and clinical translation of nanocrystal-based drug delivery systems face distinct regulatory and safety challenges compared to conventional formulations. Regulatory agencies like the FDA require comprehensive safety evaluations to address concerns about nanocrystals' unique physicochemical properties, which may influence biodistribution, biocompatibility, and long-term toxicity. A critical issue is the potential for nanocrystals to accumulate in tissues over time, raising toxicity risks that necessitate rigorous studies on their interactions with biological systems and immune responses. For instance, the particulate nature of nanocrystals can lead to phagocytic clearance and unintended immune activation, requiring tailored pharmacokinetic and toxicological assessments. Additionally, stabilizers and surface modifiers used in formulations must be evaluated for safety, as their presence could exacerbate biocompatibility issues. The classification of nanocrystals under nanotechnology guidelines subjects them to heightened scrutiny, often prolonging approval timelines and increasing development costs. Harmonizing international regulatory standards and advancing methods for molecular safety assessments such as standardized in vitro/in vivo models and predictive computational tools are essential to streamline approvals while ensuring patient safety. These efforts must balance innovation with robust risk mitigation, particularly as nanocrystal formulations evolve toward targeted therapies and personalized medicine.

While regulatory challenges exist, the ongoing refinement of nanocrystal formulations and comprehensive safety evaluations are helping to improve their acceptance in clinical settings.

5.7.4 Limited Understanding of In-Vivo Behavior

Despite the significant progress in nanocrystal formulation and characterization, the in vivo behavior of nanocrystals is still not fully understood. Once administered, nanocrystals can interact with biological systems in complex ways, and these interactions can vary depending on factors like the size, surface charge, and shape of the nanocrystals. These factors can influence the distribution, clearance, and overall therapeutic efficacy of the nanocrystal-based drugs [106].

- *Biodistribution and Clearance*: The size and surface characteristics of nanocrystals play a critical role in determining their biodistribution and clearance from the body. Nanocrystals can accumulate in various organs, including the liver and spleen, and understanding these processes is crucial for optimizing drug delivery [107].
- *Cellular Uptake and Release*: Nanocrystals can be taken up by cells through mechanisms like endocytosis, but the efficiency of this process is influenced by factors such as surface modifications and the type of targeting ligands used. Additionally, the release of the drug from the nanocrystal in the targeted tissue is another important consideration for ensuring efficacy [105].

Ongoing research is working to clarify the in vivo behavior of nanocrystals, which will help optimize their use in clinical applications.

While nanocrystal-based drug delivery systems hold significant promise in enhancing the solubility, bioavailability, and targeted delivery of drugs, several challenges remain that must be addressed for their widespread clinical adoption. Stability concerns, limited drug-loading capacity, manufacturing and scalability issues, regulatory hurdles, and uncertainties regarding in vivo behavior are key obstacles that researchers and pharmaceutical companies must overcome. Addressing these challenges will be crucial for realizing the full potential of nanocrystals in drug delivery, but ongoing advancements in formulation techniques, manufacturing processes, and safety evaluations suggest that these obstacles are surmountable in the near future [108].

5.8 Conclusion

The integration of nanocrystal technology into drug delivery systems represents a remarkable advancement in modern pharmaceutical science, addressing critical challenges that have long plagued the development and efficacy of therapeutic agents. Nanocrystal-based drug delivery systems represent a transformative advancement in pharmaceutical science, offering significant improvements over traditional drug formulations. By harnessing the unique properties of nanocrystals such as enhanced solubility, increased bioavailability, and the ability to tailor drug release these systems address many longstanding challenges in drug delivery, particularly for poorly soluble compounds and complex therapeutic agents.

A defining strength of nanocrystal technology lies in its capacity for targeted drug delivery. Through strategic surface modifications and the integration of nanocarriers, nanocrystals can be engineered to home in on specific tissues or cells. This targeted approach not only maximizes therapeutic efficacy but also minimizes systemic side effects, a critical advantage in the treatment of diseases like cancer, where precision in drug delivery is paramount. Clinical applications highlighted in recent studies underscore the real-world impact of nanocrystal formulations. Whether improving the effectiveness of anticancer agents, enabling oral delivery of biologics such as insulin, or enhancing the treatment of multidrug-resistant infections, nanocrystals are already demonstrating their versatility and potential to elevate patient care across diverse therapeutic areas. Despite these promising developments, several challenges remain. Achieving reliable large-scale production, ensuring long-term safety and stability, and addressing potential toxicity are ongoing areas of research. Continued innovation in the design, functionalization, and manufacturing of nanocrystals will be essential to fully realize their potential as targeted drug delivery vehicles.

Nanocrystal-based drug delivery systems offer a promising path toward more effective, personalized, and patient-friendly therapies. As research and technology evolve, nanocrystals are poised to play an increasingly central role in the future landscape of targeted drug delivery and therapeutic optimization.

References

1. Ezike TC, Okpala US, Onoja UL, Nwike CP, Ezeako EC, Okpara OJ, Okoroafor CC, Eze SC, Kalu OL, Odoh EC, Nwadike UG, Ogbodo JO, Umeh BU, Ossai EC, Nwanguma BC (2023) Advances in drug delivery systems, challenges and future directions. Heliyon 9:e17488. https://doi.org/10.1016/j.heliyon.2023.e17488
2. Sim S, Wong NK (2021) Nanotechnology and its use in imaging and drug delivery (review). Biomed Reports 14:42. https://doi.org/10.3892/br.2021.1418

3. Patra JK, Das G, Fraceto LF, Campos EVR, Rodriguez-Torres MDP, Acosta-Torres LS, Diaz-Torres LA, Grillo R, Swamy MK, Sharma S, Habtemariam S, Shin HS (2018) Nano based drug delivery systems: recent developments and future prospects. J Nanobiotechnology 16:1–33. https://doi.org/10.1186/s12951-018-0392-8
4. Gigliobianco MR, Casadidio C, Censi R, Di Martino P (2018) Nanocrystals of poorly soluble drugs: drug bioavailability and physicochemical stability. Pharmaceutics 10:134. https://doi.org/10.3390/pharmaceutics10030134
5. Ding Y, Zhao T, Fang J, Song J, Dong H, Liu J, Li S, Zhao M (2024) Recent developments in the use of nanocrystals to improve bioavailability of APIs. Wiley Interdiscip Rev Nanomedicine Nanobiotechnology 16:e1958. https://doi.org/10.1002/wnan.1958
6. Chang T-L, Zhan H, Liang D, Liang JF (2015) Nanocrystal technology for drug formulation and delivery. Front Chem Sci Eng 9:1–14. https://journal.hep.com.cn/fcse
7. Ran Q, Wang M, Kuang W, Ouyang J, Han D, Gao Z, Gong J (2022) Advances of combinative nanocrystal preparation technology for improving the insoluble drug solubility and bioavailability. Crystals 12:1–21. https://doi.org/10.3390/cryst12091200
8. Zhang X, Xia M, Su X, Yuan P, Li X, Zhou C, Wan Z, Zou W (2021) Photolytic degradation elevated the toxicity of polylactic acid microplastics to developing zebrafish by triggering mitochondrial dysfunction and apoptosis. J Hazard Mater 413:125321. https://doi.org/10.1016/j.jhazmat.2021.125321
9. Rossier B, Jordan O, Allémann E, Rodríguez-Nogales C (2024) Nanocrystals and nanosuspensions: an exploration from classic formulations to advanced drug delivery systems. Drug Deliv Transl Res 14:3438–3451. https://doi.org/10.1007/s13346-024-01559-0
10. Lu Y, Li Y, Wu W (2016) Injected nanocrystals for targeted drug delivery. Acta Pharm Sin B 6:106–113. https://doi.org/10.1016/j.apsb.2015.11.005
11. Zhang H, Hu H, Zhang H, Dai W, Weng X, Wang X, Zhang Q (2015) Effects of PEGylated paclitaxel nanocrystals on breast cancer and its lung metastasis. Nanoscale 7:10790–10800. https://doi.org/10.1039/c4nr07450e
12. Zhang XD, Chen J, Luo Z, Wu D, Shen X, Song SS, Sun YM, Liu PX, Zhao J, Huo S, Fan S, Fan F, Liang XJ, Xie J (2014) Enhanced tumor accumulation of Sub-2 nm gold nanoclusters for cancer radiation therapy. Adv Healthc Mater 3:133–141. https://doi.org/10.1002/adhm.201300189
13. Colombo E, Coppini DA, Maculan S, Seneci P, Santini B, Testa F, Salvioni L, Vanacore GM, Colombo M, Passarella D (2022) Folic acid functionalization for targeting self-assembled paclitaxel-based nanoparticles. RSC Adv 12:35484–35493. https://doi.org/10.1039/d2ra06306a
14. Reda M, Ngamcherdtrakul W, Nelson MA, Siriwon N, Wang R, Zaidan HY, Bejan DS, Reda S, Hoang NH, Crumrine NA, Rehwaldt JPC, Bindal A, Mills GB, Gray JW, Yantasee W (2022) Development of a nanoparticle-based immunotherapy targeting PD-L1 and PLK1 for lung cancer treatment. Nat Commun 13:4261. https://doi.org/10.1038/s41467-022-31926-9
15. Kumar M, Goswami P, Jha A, Manjit M, Satpute AP, Koch B, Mishra B (2024) Formulation and evaluation of cetuximab functionalized phospholipid modified nanocrystals of paclitaxel for non-small cell lung cancer therapy. Sci Rep 14:1–18. https://doi.org/10.1038/s41598-024-80283-8
16. Hristova-Panusheva K, Xenodochidis C, Georgieva M, Krasteva N (2024) Nanoparticle-mediated drug delivery systems for precision targeting in oncology. Pharmaceuticals 17:1–25. https://doi.org/10.3390/ph17060677
17. Kaur A, Parmar P, Jadhav S, Bansal A (2022) Advances in nanocrystals as drug delivery systems. In: Nanoparticle therapeutics: production technologies, types of nanoparticles, and regulatory aspects. Academic Press, London, pp 413–454. https://doi.org/10.1016/B978-0-12-820757-4.00011-9
18. Joshi K, Chandra A, Jain K, Talegaonkar S (2019) Nanocrystalization: an emerging technology to enhance the bioavailability of poorly soluble drugs. Pharm Nanotechnol 7:259–278. https://doi.org/10.2174/2211738507666190405182524

19. Das KP, Chandra J (2022) Nanoparticles and convergence of artificial intelligence for targeted drug delivery for cancer therapy: current progress and challenges. Front Med Technol 4:1–14. https://doi.org/10.3389/fmedt.2022.1067144
20. Chenthamara D, Subramaniam S, Ramakrishnan SG, Krishnaswamy S, Essa MM, Lin F-H, Qoronfleh MW (2019) Therapeutic efficacy of nanoparticles and routes of administration. Biomater Res 23:20. https://doi.org/10.1186/s40824-019-0166-x
21. Mourdikoudis S, Pallares RM, Thanh NTK (2018) Characterization techniques for nanoparticles: comparison and complementarity upon studying nanoparticle properties. Nanoscale 10:12871–12934. https://doi.org/10.1039/c8nr02278j
22. Ulbrich K, Holá K, Šubr V, Bakandritsos A, Tuček J, Zbořil R (2016) Targeted drug delivery with polymers and magnetic nanoparticles: covalent and noncovalent approaches, release control, and clinical studies. Chem Rev 116:5338–5431. https://doi.org/10.1021/acs.chemrev.5b00589
23. Rodriguez-Loya J, Lerma M, Gardea-Torresdey JL (2024) Dynamic light scattering and its application to control nanoparticle aggregation in colloidal systems: a review. Micromachines 15:24. https://doi.org/10.3390/mi15010024
24. Sherif El-Eskandarany M, Al-Hazza A, Al-Hajji LA, Ali N, Al-Duweesh AA, Banyan M, Al-Ajmi F (2021) Mechanical milling: a superior nanotechnological tool for fabrication of nanocrystalline and nanocomposite materials. Nano 11:2484. https://doi.org/10.3390/nano11102484
25. Baig N, Kammakakam I, Falath W, Kammakakam I (2021) Nanomaterials: a review of synthesis methods, properties, recent progress, and challenges. Mater Adv 2:1821–1871. https://doi.org/10.1039/d0ma00807a
26. Nikumbhe B, Bhaskar R, Ola M, Pagar P, Sawarkar P (2024) The evolution of nanocrystalline drug delivery system. J Drug Deliv Ther 14:214–222. https://doi.org/10.22270/jddt.v14i6.6628
27. Gawande MB, Shelke SN, Zboril R, Varma RS (2014) Microwave-assisted chemistry: synthetic applications for rapid assembly of nanomaterials and organics. Acc Chem Res 47:1338–1348. https://doi.org/10.1021/ar400309b
28. Schubert U (2015) Chemistry and fundamentals of the sol-gel process. In: The sol-gel handbook. Wiley-VCH, Weinheim, pp 1–28. https://doi.org/10.1002/9783527670819.ch01
29. Rawat RS (2015) Dense plasma focus – from alternative fusion source to versatile high energy density plasma source for plasma nanotechnology. J Phys Conf Ser 591:12021. https://doi.org/10.1088/1742-6596/591/1/012021
30. Emrie DB (2024) Sol-gel synthesis of nanostructured mesoporous silica powder and thin films. J Nanomater 4:1–16. https://doi.org/10.1155/2024/6109770
31. Wu CY, Wang W (2022) Application of antisolvent precipitation method for formulating excipient-free nanoparticles of psychotropic drugs. Pharmaceutics 14:819. https://doi.org/10.3390/pharmaceutics14040819
32. Ghosh P, Rasmuson A, Hudson SP (2023) Impact of additives on drug particles during liquid antisolvent crystallization and subsequent freeze-drying. Org Process Res Dev 27:2020–2034. https://doi.org/10.1021/acs.oprd.3c00204
33. Naganthran A, Verasoundarapandian G, Khalid FE, Masarudin MJ, Zulkharnain A, Nawawi NM, Karim M, Abdullah CAC, Ahmad SA (2022) Synthesis, characterization and biomedical application of silver nanoparticles. Materials (Basel) 15:1–43. https://doi.org/10.3390/ma15020427
34. Bhattacharjee S (2016) DLS and zeta potential – what they are and what they are not? J Control Release 235:337–351. https://doi.org/10.1016/j.jconrel.2016.06.017
35. Holder CF, Schaak RE (2019) Tutorial on powder x-ray diffraction for characterizing nanoscale materials. ACS Nano 13:7359–7365. https://doi.org/10.1021/acsnano.9b05157
36. Malatesta M (2021) Transmission electron microscopy as a powerful tool to investigate the interaction of nanoparticles with subcellular structures. Int J Mol Sci 22:12789. https://doi.org/10.3390/ijms222312789

37. Buhr E, Senftleben N, Klein T, Bergmann D, Gnieser D, Frase C, Bosse H (2009) Characterization of nanoparticles by scanning electron microscopy in transmission mode. Meas Sci Technol 20:84025. https://doi.org/10.1088/0957-0233/20/8/084025
38. Sowa JK, Cadena DM, Mehmood A, Levine BG, Roberts ST, Rossky PJ (2024) IR spectroscopy of carboxylate-passivated semiconducting nanocrystals: simulation and experiment. J Phys Chem C 128:8724–8731. https://doi.org/10.1021/acs.jpcc.4c01988
39. Khanjanzadeh H, Behrooz R, Bahramifar N, Gindl-Altmutter W, Bacher M, Edler M, Griesser T (2018) Surface chemical functionalization of cellulose nanocrystals by 3-aminopropyltriethoxysilane. Int J Biol Macromol 106:1288–1296. https://doi.org/10.1016/j.ijbiomac.2017.08.136
40. Dhilip Kumar SS, Abrahamse H (2020) Advancement of nanobiomaterials to deliver natural compounds for tissue engineering applications. Int J Mol Sci 21:1–27. https://doi.org/10.3390/ijms21186752
41. Prabhakar U, Maeda H, Jain RK, Sevick-Muraca EM, Zamboni W, Farokhzad OC, Barry ST, Gabizon A, Grodzinski P, Blakey DC (2013) Challenges and key considerations of the enhanced permeability and retention effect for nanomedicine drug delivery in oncology. Cancer Res 73:2412–2417. https://doi.org/10.1158/0008-5472.CAN-12-4561
42. Shi J, Kantoff PW, Wooster R, Farokhzad OC (2017) Cancer nanomedicine: progress, challenges and opportunities. Nat Rev Cancer 17:20–37. https://doi.org/10.1038/nrc.2016.108
43. Sharifi M, Cho WC, Ansariesfahani A, Tarharoudi R, Malekisarvar H, Sari S, Bloukh SH, Edis Z, Amin M, Gleghorn JP, ten Hagen TLM, Falahati M (2022) An updated review on EPR-based solid tumor targeting nanocarriers for cancer treatment. Cancers (Basel) 14:2868. https://doi.org/10.3390/cancers14122868
44. Nel A, Ruoslahti E, Meng H (2017) New insights into "permeability" as in the enhanced permeability and retention effect of cancer nanotherapeutics. ACS Nano 11:9567–9569. https://doi.org/10.1021/acsnano.7b07214
45. Yan S, Na J, Liu X, Wu P (2024) Different targeting ligands-mediated drug delivery systems for tumor therapy. Pharmaceutics 16:1–19. https://doi.org/10.3390/pharmaceutics16020248
46. Khongorzul P, Ling CJ, Khan FU, Ihsan AU, Zhang J (2020) Antibody-drug conjugates: a comprehensive review. Mol Cancer Res 18:3–19. https://doi.org/10.1158/1541-7786.MCR-19-0582
47. Jiang Z, Guan J, Qian J, Zhan C (2019) Peptide ligand-mediated targeted drug delivery of nanomedicines. Biomater Sci 7:461–471. https://doi.org/10.1039/c8bm01340c
48. Gocheva G, Ivanova A (2019) A look at receptor-ligand pairs for active-targeting drug delivery from crystallographic and molecular dynamics perspectives. Mol Pharm 16:3293–3321. https://doi.org/10.1021/acs.molpharmaceut.9b00250
49. Yoo J, Park C, Yi G, Lee D, Koo H (2019) Active targeting strategies using biological ligands for nanoparticle drug delivery systems. Cancers (Basel) 11:640. https://doi.org/10.3390/cancers11050640
50. Kim WJ, Lee EH, Kwon YJ, Ye SK, oh Kim K (2022) Targeted drug release system based on pH-responsive PAA-POSS nanoparticles. RSC Adv 12:18209–18214. https://doi.org/10.1039/d2ra01141g
51. Palanikumar L, Al-Hosani S, Kalmouni M, Nguyen VP, Ali L, Pasricha R, Barrera FN, Magzoub M (2020) pH-responsive high stability polymeric nanoparticles for targeted delivery of anticancer therapeutics. Commun Biol 3:1–17. https://doi.org/10.1038/s42003-020-0817-4
52. Wang X (2016) Matrix metalloproteinases-2/9-sensitive peptide-conjugated polymer micelles for site-specific release of drugs and enhancing tumor accumulation: preparation and in vitro and in vivo evaluation. Int J Nanomedicine 11:1643–1661
53. Awan UA, Raza A, Ali S, Saeed RF, Akhtar N (2021) Doxorubicin-loaded gold nanorods: a multifunctional chemo-photothermal nanoplatform for cancer management. Beilstein J Nanotechnol 12:295–303. https://doi.org/10.3762/bjnano.12.24

54. Santhamoorthy M, Kim SC (2023) Dual pH- and thermo-sensitive poly (N-isopropylacrylamide-co-allylamine) nanogels for curcumin delivery: swelling–deswelling behavior and phase transition mechanism. Gels 9:536. https://doi.org/10.3390/gels9070536
55. Kamaly N, Yameen B, Wu J, Farokhzad OC (2016) Degradable controlled-release polymers and polymeric nanoparticles: mechanisms of controlling drug release. Chem Rev 116:2602–2663. https://doi.org/10.1021/acs.chemrev.5b00346
56. Sousa De Almeida M, Susnik E, Drasler B, Taladriz-Blanco P, Petri-Fink A, Rothen-Rutishauser B (2021) Understanding nanoparticle endocytosis to improve targeting strategies in nanomedicine. Chem Soc Rev 50:5397–5434. https://doi.org/10.1039/d0cs01127d
57. de La Pomélie D, Santé-Lhoutellier V, Sayd T, Théron L, Gatellier P (2019) Using a dynamic artificial digestive system to investigate heme iron nitrosylation during gastro-intestinal transit. Food Chem 281:231–235. https://doi.org/10.1016/j.foodchem.2018.12.094
58. Ly PD, Ly KN, Phan HL, Nguyen HHT, Duong VA, Nguyen HV (2024) Recent advances in surface decoration of nanoparticles in drug delivery. Front Nanotechnol 6:1–20. https://doi.org/10.3389/fnano.2024.1456939
59. Rajesh S, Lillard JWL Jr (2000) Nanoparticle-based targeted drug delivery. Exp Mol Pathol 86:215–223. https://doi.org/10.1016/j.yexmp.2008.12.004
60. Gao H, Yang Z, Zhang S, Cao S, Shen S, Pang Z, Jiang X (2013) Ligand modified nanoparticles increases cell uptake, alters endocytosis and elevates glioma distribution and internalization. Sci Rep 3:2534. https://doi.org/10.1038/srep02534
61. Sanità G, Carrese B, Lamberti A (2020) Nanoparticle surface functionalization: how to improve biocompatibility and cellular internalization. Front Mol Biosci 7:587012. https://doi.org/10.3389/fmolb.2020.587012
62. Parhi R (2017) Cross-linked hydrogel for pharmaceutical applications: a review. Adv Pharm Bull 7:515–530. https://doi.org/10.15171/apb.2017.064
63. Masoudifar R, Pouyanfar N, Liu D, Ahmadi M, Landi B, Akbari M, Moayeri-Jolandan S, Ghorbani-Bidkorpeh F, Asadian E, Shahbazi MA (2022) Surface engineered metal-organic frameworks as active targeting nanomedicines for mono- and multi-therapy. Appl Mater Today 29:101646. https://doi.org/10.1016/j.apmt.2022.101646
64. Haddad R, Alrabadi N, Altaani B, Li T (2022) Paclitaxel drug delivery systems: focus on nanocrystals' surface modifications. Polymers (Basel) 14:658. https://doi.org/10.3390/polym14040658
65. Park J, Choi Y, Chang H, Um W, Ryu JH, Kwon IC (2019) Alliance with EPR effect: combined strategies to improve the EPR effect in the tumor microenvironment. Theranostics 9:8073–8090. https://doi.org/10.7150/thno.37198
66. Islam W, Kimura S, Islam R, Harada A, Ono K, Fang J, Niidome T, Sawa T, Maeda H (2021) Epr-effect enhancers strongly potentiate tumor-targeted delivery of nanomedicines to advanced cancers: further extension to enhancement of the therapeutic effect. J Pers Med 11:487. https://doi.org/10.3390/jpm11060487
67. Koseki Y, Ikuta Y, Taemaitree F, Saito N, Suzuki R, Dao ATN, Onodera T, Oikawa H, Kasai H (2021) Fabrication of size-controlled SN-38 pure drug nanocrystals through an ultrasound-assisted reprecipitation method toward efficient drug delivery for cancer treatment. J Cryst Growth 572:126265. https://doi.org/10.1016/j.jcrysgro.2021.126265
68. Fang F, Wang S, Song Y, Sun M, Chen WC, Zhao D, Zhang J (2023) Continuous spatiotemporal therapy of a full-API nanodrug via multi-step tandem endogenous biosynthesis. Nat Commun 14:1–14. https://doi.org/10.1038/s41467-023-37315-0
69. Wang J, Zhao H, Zhi K, Yang X (2020) Exploration of the natural active small-molecule drug-loading process and highly efficient synergistic antitumor efficacy. ACS Appl Mater Interfaces 12:6827–6839. https://doi.org/10.1021/acsami.9b18443
70. Yang F, Xue J, Wang G, Diao Q (2022) Nanoparticle-based drug delivery systems for the treatment of cardiovascular diseases. Front Pharmacol 13:1–14. https://doi.org/10.3389/fphar.2022.999404

71. Liu H, Pietersz G, Peter K, Wang X (2022) Nanobiotechnology approaches for cardiovascular diseases: site-specific targeting of drugs and nanoparticles for atherothrombosis. J Nanobiotechnol 20:1–23. https://doi.org/10.1186/s12951-022-01279-y
72. Martínez-Parra L, Piñol-Cancer M, Sanchez-Cano C, Miguel-Coello AB, Di Silvio D, Gomez AM, Uriel C, Plaza-García S, Gallego M, Pazos R, Groult H, Jeannin M, Geraki K, Fernández-Méndez L, Urkola-Arsuaga A, Sánchez-Guisado MJ, Carrillo-Romero J, Parak WJ, Prato M, Herranz F, Ruiz-Cabello J, Carregal-Romero S (2023) A comparative study of ultrasmall calcium carbonate nanoparticles for targeting and imaging atherosclerotic plaque. ACS Nano 17:13811–13825. https://doi.org/10.1021/acsnano.3c03523
73. Smith BR, Edelman ER (2023) Nanomedicines for cardiovascular disease. Nat Cardiovasc Res 2:351–367. https://doi.org/10.1038/s44161-023-00232-y
74. Farokhi S, Razavi ZS, Mavaei M, Shadab A, Afkhami H, Sardarabadi H (2024) New perspectives on arteriosclerosis treatment using nanoparticles and mesenchymal stem cells. Discov Appl Sci 6:411. https://doi.org/10.1007/s42452-024-06113-8
75. Mendes BB, Conniot J, Avital A, Yao D, Jiang X, Zhou X, Sharf-Pauker N, Xiao Y, Adir O, Liang H, Shi J, Schroeder A, Conde J (2022) Nanodelivery of nucleic acids. Nat Rev Methods Prim 2:24. https://doi.org/10.1038/s43586-022-00104-y
76. Mukalel AJ, Riley RS, Zhang R, Mitchell MJ (2019) Nanoparticles for nucleic acid delivery: applications in cancer immunotherapy. Cancer Lett 458:102–112. https://doi.org/10.1016/j.canlet.2019.04.040
77. Singhsa P, Narain R, Manuspiya H (2018) Bacterial Cellulose Nanocrystals (BCNC) preparation and characterization from three bacterial cellulose sources and development of functionalized BCNCs as nucleic acid delivery systems. ACS Appl Nano Mater 1:209–221. https://doi.org/10.1021/acsanm.7b00105
78. Munawaroh HSH, Anwar B, Yuliani G, Murni IC, Arindita NPY, Maulidah GS, Martha L, Hidayati NA, Chew KW, Show PL (2023) Bacterial cellulose nanocrystal as drug delivery system for overcoming the biological barrier of cyano-phycocyanin: a biomedical application of microbial product. Bioengineered 14:2252226. https://doi.org/10.1080/21655979.2023.2252226
79. Jiao L, Sun Z, Sun Z, Liu J, Deng G, Wang X (2024) Nanotechnology-based non-viral vectors for gene delivery in cardiovascular diseases. Front Bioeng Biotechnol 12:1–13. https://doi.org/10.3389/fbioe.2024.1349077
80. Jarvis M, Krishnan V, Mitragotri S (2019) Nanocrystals: a perspective on translational research and clinical studies. Bioeng Transl Med 4:5–16. https://doi.org/10.1002/btm2.10122
81. Dhaval M, Makwana J, Sakariya E, Dudhat K (2020) Drug nanocrystals: a comprehensive review with current regulatory guidelines. Curr Drug Deliv 17:470–482. https://doi.org/10.2174/1567201817666200512104833
82. Wathoni N, Puluhulawa LE, Joni IM, Muchtaridi M, Mohammed AFA, Elamin KM, Milanda T, Gozali D (2022) Monoclonal antibody as a targeting mediator for nanoparticle targeted delivery system for lung cancer. Drug Deliv 29:2959–2970. https://doi.org/10.1080/10717544.2022.2120566
83. Jiang S, Wang X, Zhang Z, Sun L, Pu Y, Yao H, Li J, Liu Y, Zhang Y, Zhang W (2016) CD20 monoclonal antibody targeted nanoscale drug delivery system for doxorubicin chemotherapy: an in vitro study of cell lysis of CD20-positive Raji cells. Int J Nanomedicine 11:5505–5518. https://doi.org/10.2147/IJN.S115428
84. Abedin MR, Powers K, Aiardo R, Barua D, Barua S (2021) Antibody–drug nanoparticle induces synergistic treatment efficacies in HER2 positive breast cancer cells. Sci Rep 11:1–17. https://doi.org/10.1038/s41598-021-86762-6
85. Goodall S, Jones ML, Mahler S (2015) Monoclonal antibody-targeted polymeric nanoparticles for cancer therapy – future prospects. J Chem Technol Biotechnol 90:1169–1176. https://doi.org/10.1002/jctb.4555

86. Turner CT, McInnes SJP, Voelcker NH, Cowin AJ (2015) Therapeutic potential of inorganic nanoparticles for the delivery of monoclonal antibodies. J Nanomater 2015:309602. https://doi.org/10.1155/2015/309602
87. Bin Liew K, Janakiraman AK, Sundarapandian R, Khalid SH, Razzaq FA, Ming LC, Khan A, Kalusalingam A, Ng PW (2022) A review and revisit of nanoparticles for antimicrobial drug delivery. J Med Life 2022:328–335. https://doi.org/10.25122/jml-2021-0097
88. Haeri A, Osouli M, Bayat F, Alavi S, Dadashzadeh S (2018) Nanomedicine approaches for sirolimus delivery: a review of pharmaceutical properties and preclinical studies. Artif Cells Nanomedicine Biotechnol 46:1–14. https://doi.org/10.1080/21691401.2017.1408123
89. Shen L-J, Wu F-LL (2007) Nanomedicines in renal transplant rejection? Focus on sirolimus. Int J Nanomedicine 2:25–32. https://doi.org/10.2147/nano.2007.2.1.25
90. de Oliveira Macedo L, Masiero JF, Bou-Chacra NA (2024) Drug nanocrystals in oral absorption: factors that influence pharmacokinetics. Pharmaceutics 16:1141. https://doi.org/10.3390/pharmaceutics16091141
91. Lu L, Xu Q, Wang J, Wu S, Luo Z, Lu W (2022) Drug nanocrystals for active tumor-targeted drug delivery. Pharmaceutics 14:797. https://doi.org/10.3390/pharmaceutics14040797
92. Lhaglham P, Jiramonai L, Jia Y, Huang B, Huang Y, Gao X, Zhang J, Liang XJ, Zhu M (2024) Drug nanocrystals: surface engineering and its applications in targeted delivery. IScience 27: 111185. https://doi.org/10.1016/j.isci.2024.111185
93. Liu F, Park JY, Zhang Y, Conwell C, Liu Y, Bathula SR, Huang L (2010) Targeted cancer therapy with novel high drug-loading nanocrystals. J Pharm Sci 99:3542–3551. https://doi.org/10.1002/jps.22112
94. Hunt NJ, Lockwood GP, Heffernan SJ, Daymond J, Ngu M, Narayanan RK, Westwood LJ, Mohanty B, Esser L, Williams CC, Kuncic Z, McCourt PAG, Le Couteur DG, Cogger VC (2024) Oral nanotherapeutic formulation of insulin with reduced episodes of hypoglycaemia. Nat Nanotechnol 19:534–544. https://doi.org/10.1038/s41565-023-01565-2
95. Yayehrad AT, Wondie GB, Marew T (2022) Different nanotechnology approaches for ciprofloxacin delivery against multidrug-resistant microbes. Infect Drug Resist 15:413–426. https://doi.org/10.2147/IDR.S348643
96. Nwabuife JC, Omolo CA, Govender T (2022) Nano delivery systems to the rescue of ciprofloxacin against resistant bacteria "E. coli; P. aeruginosa; Saureus; and MRSA" and their infections. J Control Release 349:338–353. https://doi.org/10.1016/j.jconrel.2022.07.003
97. Li J, Wang Z, Zhang H, Gao J, Zheng A (2021) Progress in the development of stabilization strategies for nanocrystal preparations. Drug Deliv 28:19–36. https://doi.org/10.1080/10717544.2020.1856224
98. Tuomela A, Hirvonen J, Peltonen L (2016) Stabilizing agents for drug nanocrystals: effect on bioavailability. Pharmaceutics 8:16. https://doi.org/10.3390/pharmaceutics8020016
99. Udabe J, Martin-Saldaña S, Tao Y, Picchio M, Beloqui A, Paredes AJ, Calderón M (2024) Unveiling the potential of surface polymerized drug nanocrystals in targeted delivery. ACS Appl Mater Interfaces 16:47124–47136. https://doi.org/10.1021/acsami.4c07669
100. Bhattacharjee S (2022) Craft of co-encapsulation in nanomedicine: a struggle to achieve synergy through reciprocity. ACS Pharmacol Transl Sci 5:278–298. https://doi.org/10.1021/ACSPTSCI.2C00033
101. Jiang B, Mu M, Zhou Y, Zhang J, Li W (2024) Nanoparticle-empowered core–shell microcapsules: from architecture design to fabrication and functions. Small 20:1–19. https://doi.org/10.1002/smll.202311897
102. Shen S, Wu Y, Liu Y, Wu D (2017) High drug-loading nanomedicines: progress, current status, and prospects. Int J Nanomedicine 12:4085–4109. https://doi.org/10.2147/IJN.S132780
103. Raghava Srivalli KM, Mishra B (2016) Drug nanocrystals: a way toward scale-up. Saudi Pharm J 24:386–404. https://doi.org/10.1016/j.jsps.2014.04.007
104. Bosetti R, Jones SL (2019) Cost-effectiveness of nanomedicine: estimating the real size of nano-costs. Nanomedicine 14:1367–1370. https://doi.org/10.2217/nnm-2019-0130

105. Thakkar S, Shah V, Misra M, Kalia K (2017) Nanocrystal based drug delivery system: conventional and current scenario. Recent Pat Nanotechnol 11:130–145. https://doi.org/10.2174/1872210510666161014122439
106. Gao L, Liu G, Ma J, Wang X, Zhou L, Li X (2012) Drug nanocrystals: in vivo performances. J Control Release 160:418–430. https://doi.org/10.1016/j.jconrel.2012.03.013
107. Lu Y, Qi J, Dong X, Zhao W, Wu W (2017) The in vivo fate of nanocrystals. Drug Discov Today 22:744–750. https://doi.org/10.1016/j.drudis.2017.01.003
108. Patel A, Dharamsi A (2021) Nanocrystals- a substantial platform for drug delivery applications. Curr Nanomedicine 11:202–212. https://doi.org/10.2174/2468187312666211221124154

Future Outlook for Diverse Type of Nanocrystal Based Drug Delivery System

6

Sandeep Kumar, Kritika Sharma, Pradeep Kumar, Sonali Batra, and Sumit Sharma

Abstract

The future of nanocrystal-based drug delivery systems holds immense potential for revolutionizing pharmaceutical formulations, particularly in enhancing drug solubility, stability, and bioavailability. Continuous advancements in nanocrystal engineering are expected to enable more precise control over particle size, surface characteristics, and drug release kinetics, making them highly adaptable for diverse therapeutic applications. Hybrid nanocrystal systems integrating polymers, lipids, or inorganic materials are likely to emerge, offering synergistic benefits such as targeted delivery and extended circulation time. The evolution of surface modification techniques, including functionalization with ligands, peptides, or antibodies, will further enhance the specificity and efficacy of nanocrystals for precision medicine, particularly in oncology, neurodegenerative disorders, and infectious diseases.

Keywords

Nanocrystal · Bioavailability · Targeting · Stability · High-pressure homogenization

S. Kumar · K. Sharma · S. Sharma (✉)
School of Pharmaceutical Sciences, Delhi Pharmaceutical Sciences and Research University, New Delhi, India

P. Kumar
Department of Botany, Swami Shraddhanand College, University of Delhi, Delhi, India

S. Batra
Department of Pharmaceutical Sciences, Indira Gandhi University, Meerpur, Rewari, Haryana, India

N. K. Mehra, K. K. Bansal (eds.), *Nanocrystal Technologies in Biomedical Sciences*,
https://doi.org/10.1007/978-981-95-2968-1_6

6.1 Introduction

Nanotechnology is the science, engineering, and application of materials and devices at the nanoscale, typically between 1 and 100 nm and today nanotechnology is encountered all around our daily lives, whether it be in the production of computer chips, the increasing field of biotechnology, simply cosmetic research and products, for medical use (e.g., mini robots to clean arteries), the food industry, or the application of drugs for pharmaceutical use. In drug delivery and therapeutic applications, the technology to nanonize (i.e., reduce in size below 1000 nm) is one of the key factors for modern drug therapy, now and in the years to come [1]. Nanotechnology-based systems for drug delivery are needed because they offer various benefits over traditional drug delivery methods, including augmented drug solubility along with targeted drug delivery, bioavailability, crossing biological hurdles (e.g., blood-brain barrier), controlled and sustained release, reduced toxicity and side effects, enhanced stability and protection, personalized medicine applications, and many more.

Since the beginning of the 1990s, elan nanosystems propagated the use of nanocrystals instead of microcrystals for oral bioavailability enhancement, but also to use nanocrystals suspended in water (nanosuspensions) for intravenous or pulmonary drug delivery [2]. Drug nanocrystals (NCs) are pure solid drug particles within the nanometer scale (10–1000 nm), surrounded by a stabilizer layer formed during the crystallization of drug molecules. As a colloidal dispersion system, they offer increased saturation solubility and drug loading capacity due to their large specific surface area, with only minimal amounts of surfactants required for stabilization. Nanocrystals of drugs comprising active pharmaceutical ingredients and required stabilizers (e.g. Polymeric, non-ionic, and ionic stabilizers) enhance the bioavailability of poorly water-soluble drugs or new drug candidates. They can be delivered through oral, intravenous, and other administration routes [3]. Another key feature of drug nanocrystals is that they consist entirely of pure drug without any carrier material, unlike polymeric nanoparticles. When dispersed in a liquid medium, they form what is known as "nanosuspensions," as opposed to "microsuspensions" or "macrosuspensions." To maintain stability, these dispersed particles typically require stabilization through polymeric or surfactants stabilizers. The dispersion medium can include water, aqueous solutions, or nonaqueous solvents such as liquid polyethylene glycol or oils [1]. Nanocrystals are conventionally produced by through top-down or bottom-up approaches. These fabrication techniques are discussed in later part of this chapter.

The smaller particle sizes modify the physicochemical characteristics, which also leads to higher kinetic solubility. In addition, a rise in nanocrystal particle curvature leads to a higher dissolution pressure, which ultimately increases solubility. Poorly water-soluble drugs exhibit improved dissolution rates when formulated as nanocrystals, primarily due to their increased surface area and enhanced apparent solubility. The resulting higher concentration gradient across biological membranes facilitates more efficient passive diffusion. Furthermore, the reduced particle size leads to a higher surface-to-volume ratio, promoting stronger adhesion of the nanocrystals to biological membranes. This property enhances retention time at the absorption site and improves the permeability of the active pharmaceutical ingredient. These attributes have been effectively leveraged in the development of oral drug delivery systems to enhance bioavailability [4].

6.1.1 Attributes of Nanocrystal

Nanocrystal DDS have attracted intense interest being consequence of their properties that boost the solubility, dissolution, bioavailability, and therapeutic effectiveness of poorly soluble drugs. Such properties of these systems are:

1. *Enhanced Solubility and Dissolution Rate*
   Approximately 40% of drugs on the market and 90% of those in development stage have poor aqueous solubility, according to the biopharmaceutical classification system (BCS). Aqueous insoluble drugs present significant challenges in terms of solubility and bioavailability, which must be addressed in the pharmaceutical industry. Low water solubility can hinder medical development and cause clinical issues, including variability in patient responses, difficulty maintaining a safe therapeutic index, increased costs, and potential toxicity or inefficacy. Nanocrystal technology is a promising solution for improving the aqueous solubility of poorly soluble drugs, with potential for industrial applications [5].
2. *Improved Bioavailability*
   Bioavailability of drugs with low solubility is a significant challenge in pharmaceutical science. Due to inadequate bioavailability, patients require higher drug dosages to achieve therapeutic effects. Increasing dosage can lead to negative side effects, harm patients' physical and mental health, and reduce medication compliance. The reduced particle size of nanocrystals allows for rapid dissolution, leading to higher plasma drug concentrations. This enhances drug absorption and therapeutic outcomes [6].
3. *Targeted Delivery and Combination Therapy*
   Nanodrug delivery systems encapsulate drugs either internally or by modifying the carrier surface. Nanomaterials' unique properties, including surface effects, catalytic properties, and chemical reactivity, can be tailored to specific pathological sites. Precision delivery of drugs at the disease locus prevents premature release and improves therapeutic efficacy while minimizing toxic effects on healthy tissues. Generally, targeted delivery includes both passive and active targeting strategies. Passive targeting combines the intrinsic properties of the nanodrug delivery system with the physiological or pathological characteristics of the target site to ensure effective drug accumulation. Active targeting involves the nanodrug delivery system's surface interacting with specific molecules or proteins at the desired site (e.g., ligand-receptor interactions, lectin-sugar interactions, antibody-antigen interactions), resulting in selective drug

concentration in targeted tissues or cells. Active targeting is categorized as biologically directed, in-vivo stimulus-directed, ex-vivo stimulus-directed, and biologically directed [7].

4. *Stability Considerations*
   Nanocrystals of different sizes have different affinities and functionalities, with 660 nm crystals promoting follicular accumulation and 250 nm crystals facilitating rapid dissolution. Exposure to light, oxygen, moisture, and enzymatic degradation, along with adverse physiological conditions, can cause drugs to decompose or alter before reaching their intended target site. Premature degradation significantly reduces drug therapeutic efficacy. Improving drug stability is crucial for pharmaceuticals to perform optimally. Nanocarriers, with their small size, large surface area, and ease of modification, offer a more stable and secure environment for drugs. This sequestration reduces pre-target degradation and inactivation issues, improving drug stability during delivery [8].
5. *Controlled Release Profiles*
   Nanocrystal formulations can be engineered for reducing dosing frequency, controlled drug release and improving patient adherence. This can be achieved by optimizing the physicochemical properties of the nanocrystals and their stabilizers [1].

6.2 Advancement in Nanocrystal Fabrication and Classification

NCs are usually synthesized using top-down techniques (by mechanically fragmenting or attriting a crystalline powder) or bottom-up techniques (by modified crystallization or anti-solvent precipitation). Although top-down approaches are extensively employed in industrial settings for nanocrystal production, they are associated with inherent limitations that underscore the need for the development and exploration of alternative methodologies. Nevertheless, bottom-up methods have a greater chance of yielding small-sized drug NCs with lower energy consumption, although they are not yet completely established for commercial applications.

6.2.1 Top-Down Techniques

Top-down techniques in pharmaceutical manufacturing involve breaking down bulk materials into smaller, nanoscale particles to enhance drug solubility, bioavailability, and therapeutic efficacy. Unlike bottom-up approaches that build structures atom by atom, top-down techniques start with larger drug compounds and refine them into desired forms. The pharmaceutical industry increasingly relies on these methods for innovative drug delivery systems and nanomedicines.

6.2.1.1 Milling

Wet ball milling (WBM) is utilized in preparing most NC products on sale in the market. Wet ball milling technique crushes the particles by utilizing milling balls within a milling container and employs mechanical attrition in wetting them using an aqueous surfactant solution and shearing. Wet ball milling produces smaller particles, possibly as small as a few hundred micrometers; nevertheless, NCs can be produced by altering the conventional milling procedure. Furthermore, the manufacturing process can be reproducible. Lower crystallinity, longer operation time, high energy input, and metal milling pearl or ball erosion-induced contamination are some of the major drawbacks of this production process. On the other hand, microparticles can be generated through jet milling without the application of organic solvents. Because of the extremely short production time and lack of organic solvents, one of the main benefits of using solvent-free jet milling is the prevention of toxicity [9].

6.2.1.2 High-Pressure Homogenization (HPH)

High-pressure homogenization (HPH) the second most significant method of making drug nanocrystals, is a high-energy disintegration technique in which high pressure is employed to decrease the particle size of API in liquid medium with surfactants. The primary mechanisms of size reduction of HPH are through cavitation forces, shear forces, and collision. The HPH process is relatively simple and closely resembles the WBM technique. Initially, a coarse drug is dispersed in a suitable medium containing surfactants and/or stabilizers to form a premix, typically using high-speed stirring. This premix is then subjected to multiple homogenization cycles using a HPH to reduce the particle size and produce a stable nanosuspension. The ideal number of homogenization periods depends on drug delivery technologies, the initial material size, machine maximum pressure, and hardness of drug. High pressure homogenizers can be used at any scale, from milliliters to large production batches [10].

6.2.1.3 Laser Ablation (LA)

In LA, materials are expelled by projecting a laser beam onto a solid target. The expelled materials create nanoparticles in the surrounding liquid. Laser-induced fragmentation is employed for transforming mixed suspensions of microparticles into nano-sized particles. Particle size is affected by suspension properties, laser intensity, and scanning speed. Laser irradiation is classified as femtosecond, picosecond, or nanosecond based on its pulse duration. Laser ablation does not require the use of organic solvents, making it a cleaner technique; however, the application of extremely high laser energy may induce changes in the crystalline state and lead to partial oxidative degradation of the drug. LA has been used in preparing curcumin, megestrol acetate, and paclitaxel nanoparticles [5].

6.2.1.4 Ultrasound

Ultrasound utilizes acoustic waves to break up drug particles into smaller units of drug particles. Ultrasound can promote nucleation by acoustic cavitation, within the rapid dispersal and solution of the drug. Ultrasound is highly reproducible and easy to implement in a laboratory setting. Ultrasound-assisted precipitation of nanocrystals influences growth, agglomeration, nucleation, and mixing phenomena. The NC size is characterized by cavitation depth, horn immersion depth, horn length, and ultrasound treatment [5].

6.2.2 Bottom-Up Techniques

6.2.2.1 High Gravity-Controlled Precipitation (HGCP)

This is a precipitation-based technique that utilizes gravitational control to promote the formation of smaller and more uniform drug nanocrystals. Rotational speed, volumetric flow rate, and concentration of reactants are some of the factors that influence the size of particles. The HGCP method may not be suitable for industrial applications due to local oversaturation of the feed stream at the turbulent edge during mixing, resulting in continuous nucleation [12].

6.2.2.2 Microfluidization

The microfluidization process uses carefully engineered microchannels to impact liquids at high pressure levels of as much as 1700 bar (25,000 psi). The compressed-air-driven pump compounds active chemicals and lipids at high speeds through microchannels and produces stable 50–100 nm nano-sized delivery systems.

6.2.3 Combination Technique

Combination processes such as HPH and milling, spray-drying, and additionally treated by homogenization or milling. Precipitation preceded by HPH and freeze-drying preceded by milling are examples of general combination processes used to increase production and product benefits [11].

6.3 Emerging Types of Pharmaceutical Nanocrystals

6.3.1 Organic Nanocrystals

Organic nanocrystals are crystals of an organic substance of nanometer order. Due to their intriguing properties for fundamental research and the anticipated wide range of applications, organic nanocrystals are the nanomaterials that have garnered the most attention recently [12]. Organic nanocrystals are used to increase aqueous solubility of drugs with low solubility. Serving as carrier-free, nano-sized delivery systems, they improve saturation solubility, adhesion to cell membranes, and dissolution rates. Various production techniques for nanocrystals are available, like reprecipitation method, microwave irradiation method, milling, supercritical fluid crystallization method, and hybrid methods like nanoedge™, and smart crystal [13].

6.3.2 Stimuli-Responsive Nanocrystalline Formulation

Biomedical and pharmaceutical areas have been transformed by nanotechnology, specially with stimuli-responsive NCs. By reacting to pH, temperature, or enzymes, these "smart" drug delivery devices provide reduced adverse effects, enhancing therapeutic efficacy controlled and targeted drug release. NCs circumvent the drawbacks of traditional drug delivery thanks to their high payload capacity, biocompatibility, and biodegradability. They improve therapy results and patient compliance by enabling precise medication administration to sick locations in response to internal or external signals [14].

6.4 Target-Specific Nanocrystal Delivery Systems

6.4.1 Brain Targeted Drug Nanocrystal

Brain drug targeting aims to deliver therapeutics that cross or bypass the brain's protective barriers against foreign substances. The blood-brain barrier (BBB) protects the central nervous system both anatomically and physiologically. The BBB has three interfaces: endothelial cells in capillaries, the cerebrospinal fluid barrier (CSF and ISF), and the arachnoid epithelium. There is no effective treatment for nearly all brain diseases. Developing new therapies for neurological and neurodegenerative diseases remains a significant challenge. One of the main challenges in drug delivery is selectively directing the drug to the target site while avoiding side effects in other districts. Nanotechnological formulations can deliver drugs to the brain through various transport mechanisms, such as paracellular, transcellular, carrier-mediated, receptor-mediated, and adsorptive transcytosis, as previously mentioned. Nanopharmaceutics transport is influenced by various physicochemical properties, including particle size, lipophilicity/hydrophilicity, surface charge, hydrogen bonds, protein binding, shape, and functional coating [15]. Among the nano systems studied for brain delivery, NCs have demonstrated significant advantages, particularly their ability to provide a potent drug therapeutic concentration useful for achieving therapeutic action in brain diseases [16].

Zingale et al. highlighted that curcumin, a naturally occurring polyphenol derived from *Curcuma longa,* exhibits numerous pharmacological properties, including antioxidant, anti-inflammatory, and neuroprotective effects. However, its therapeutic potential is significantly compromised by factors such as poor aqueous solubility, rapid metabolism, and low bioavailability. To address these drawbacks, nanocrystal formulation has emerged as a promising drug delivery approach, especially for targeting the brain. In this context, curcumin nanocrystals were developed using a combination of nanoprecipitation and high-pressure homogenization techniques, yielding particles with an average size in the 100–200 nm range. The precipitation involved dissolving curcumin in an organic solvent and rapidly introducing it to an aqueous solution containing stabilisers like poloxamer and polyvinylpyrrolidone (PVP), followed by repeated homogenization cycles to refine particle size and ensure uniform dispersion. A DoE methodology was employed to investigate how formulation parameters-such as drug nanocrystal attributes systematically. Key properties like particle size distribution, surface charge (zeta potential), crystallinity, and drug loading efficiency were thoroughly evaluated. The optimized nanocrystals demonstrated significantly enhanced cellular uptake, approximately four times greater than that of free curcumin, attributed to the small particle size and large surface area, facilitating improved membrane interaction. Moreover, stabilizers played a crucial role in maintaining dispersion stability, reducing aggregation, and improving mucosal adhesion, which contributed to increased drug retention in brain-targeted regions such as the olfactory bulb. These findings suggest that nanocrystal-based delivery systems offer an effective strategy for enhancing curcumin's therapeutic potential in the treatment of neurological disorders through targeted nasal administration [16].

6.4.2 Skin-Targeted Drug Nanocrystal

Nanocarriers enable effective transdermal delivery of water-insoluble drugs by enhancing dissolution and absorption through the skin, reducing systemic side effects [17]. For example, cortisone nanocrystals improve bioavailability, making them a viable alternative to oral administration. Lohan et al. found that dexamethasone nanocrystals penetrated the skin more effectively than conventional creams [18]. Similarly, antioxidant formulations in creams or gels enhance the bioavailability of compounds like lutein and quercetin. Diosmin nanocrystals have been incorporated into wafers for diabetic ulcer treatment, benefiting from improved adhesion and drug absorption. NC-based patches and follicular suspensions offer sustained release, with the hair follicle acting as a reservoir. Hydrating excipients further enhance skin penetration. Low-viscosity NC formulations facilitate efficient absorption of anti-inflammatory drugs like diclofenac, reducing gastric side effects. Additionally, micro-needles create microchannels that enhance NC penetration, as demonstrated by Pireddu et al., who found this approach improved diclofenac delivery without causing NSAID-related gastric injuries [19, 20].

6.4.3 Pulmonary Targeted Drug Nanocrystal

Nanocrystals (NCs) have gained attention for pulmonary drug delivery due to advantages like high vascularization, minimal enzymatic activity, and bypassing hepatic metabolism. NCs overcome challenges in conventional inhalation therapies, such as poor drug release, particle size heterogeneity, and macrophage clearance. Their small size (<300 nm) allows better lung retention and penetration through mucus. Studies show that PEG-coated NCs enhance mucus diffusion, while curcumin NCs improve bioavailability. Dry powder inhalation formulations optimize lung deposition, and targeted NCs using hyaluronic acid or mannose offer cancer-specific drug delivery. These strategies enhance therapeutic efficacy while reducing systemic toxicity [21].

6.4.4 Ocular Targeted Drug Nanocrystal

Ocular drug delivery faces significant challenges due to the eye's unique anatomy and physiology, limiting drug absorption to only 5% of the administered dose. Constraints such as limited conjunctival sac volume, nasolacrimal drainage, eye barriers, and blinking hinder drug retention. Nanocarrier (NC) technology offers solutions, particularly for inflammatory diseases and glaucoma, by enhancing drug bioavailability and prolonging retention time. Strategies include using mucoadhesive stabilizers like cetylpyridinium chloride and benzalkonium chloride, formulating in situ gels, and optimizing particle size. Smaller particles (<200 nm) and suitable stabilizers improve drug absorption and retention, enhancing both short-term efficacy and sustained drug release [22].

6.4.5 Intracellular Targeting Approaches

To precisely deliver medications into cells, researchers use "carriers." These carriers open certain cell doors (receptors) using unique "keys" (ligands or peptides). This maximizes the drug's impact and minimizes adverse effects by enabling it to reach the appropriate location within the cell, such as the nucleus or mitochondria [23].

The strategies for intracellular targeting include:

1. Receptor-mediated endocytosis: Affixing ligands (such as short peptides, aptamers, or antibodies) to the drug carrier that selectively attach to target cells' overexpressed receptors to initiate internalization via endocytosis pathways [24].
2. Intracellular targeting with engineered proteins: Some possible approaches to ensuring the co-localization of transduced protein drugs and their targets in locations other than the cytoplasm include intracellular targeting programs that employ signal sequences as a method of localizing provided proteins within subcellular compartments [25]. Chali et al. reported that Atovaquone (Ato), a hydrophobic drug recognized for its antiprotozoal activity and its ability to inhibit STAT3 signalling, shows considerable promise and poor bioavailability, limiting its ability to reach intracellular targets effectively. To overcome these limitations, researchers developed two distinct nanocarrier systems, namely ovalbumin-based nano capsules (Ato-nCap) and human serum albumin-coated nanocrystals (Ato-nCry). The Ato-nCap formulation was prepared using a water-in-oil emulsion technique. In this process, Atovaquone was first solubilized in dimethyl sulfoxide (DMSO) and added to an aqueous mixture containing ovalbumin and a block copolymer surfactant. The emulsion was then dispersed into cyclohexane under high-speed stirring, followed by sonication to generate nanodroplets. Crosslinking of the ovalbumin shell using 2,4-toluene diisocyanate (TDI) formed nanocapsules with a distinct core-shell structure, averaging about 300 nm in hydrodynamic diameter. Once transferred into water, the diffusion of DMSO led to the precipitation of Ato within the core, successfully encapsulating the drug. In a separate strategy, Ato-nCry were produced through solvent

evaporation. Here, Ato was first dissolved in chloroform, creating an oil phase that was emulsified into an aqueous SDS solution. Evaporation of chloroform facilitated the formation of Ato crystals within the droplets, which were subsequently coated with human serum albumin (HSA). This protein coating not only improved the nanocrystal's stability and compatibility but also aided in enhancing cellular uptake while reducing immune clearance. Both nanoformulations exhibited promising uptake in vitro; however, Ato-nCry showed markedly better drug release and intracellular distribution, effectively reaching various immune cell types, including T-cells, B-cells, and cells of the myeloid lineage. Flow cytometry data revealed that Ato-nCry were internalized across a broader range of immune cells compared to Ato-nCap, which mainly targeted myeloid cells. Furthermore, Ato-nCry significantly suppressed STAT3 phosphorylation in both IL-10 stimulated dendritic cells and melanoma cell lines, pointing to their strong therapeutic potential. On the other hand, the limited drug release from Ato-nCap resulted in reduced efficacy. These results underline the critical role of nanocarrier design in optimizing intracellular drug delivery and support the utility of HSA-coated nanocrystals as a promising platform for targeted cancer therapy [26].

6.5 Smart Nanocrystal Platforms for Controlled Release

Smart nanocrystal platforms enable controlled drug delivery, which maximizes therapeutic effectiveness and minimizes side effects. They are designed nanostructures that respond to specific stimuli, enabling targeted delivery and controlled release kinetics. The recent advances have been promising in a variety of pharmaceutical applications [27].

6.5.1 pH-Responsive Release Mechanisms

A range of chemical stimulus (redox, pH, ionic) sensitive nanomaterials have been prepared and developed into efficient delivery systems. Stimulus-sensitive nanomaterials, also known as "intelligent," or "environmentally sensitive," or "smart" are more promising than the traditional system of delivery. To enhance the efficacy of anticancer drug delivery, pH-sensitive nanosystems play a crucial role in drug stabilization and controlled release. The newer systems are made to safely keep the drug at physiological pH while promptly releasing it once a definite pH value is reached. Such a mechanism allows for drug accumulation in cells to therapeutically active levels. To achieve these objectives, researchers have explored numerous pH-responsive mechanisms of drug release, each intended to ensure utmost precision and efficacy in cancer treatment. One of the primary approaches in the enhancement of pH-sensitive drug release is through the incorporation of chemically ionizable groups such phosphoric acids, carboxylic acids, and amines into nanomaterials. These functional groups, with distinct pKa values and chemical structures, serve to accept or donate protons, producing pH-dependent variations in their chemical and physical properties. Changes in swelling properties, solubility, or conformational structure variations brought about by pH changes afford the means of controlled release of drugs. Depending on their composition, nanomaterials applied in such systems may be divided broadly into organic, inorganic, or hybrid types, each providing specific merits for targeted therapeutic use.

Polymers represent a crucial class of materials in the formulation of organic drug delivery systems due to their characteristic to encapsulate therapeutic agents, provide protection, and respond to specific physiological stimuli. Among these, pH-sensitive polymers stand out as a subclass of polyelectrolytes, characterized by ionizable groups located within their backbone, side chains, or terminal structures. These "smart" polymers undergo significant conformational changes when exposed to variations in pH and ionic composition, triggering controlled drug release.

pH-sensitive polymers have been synthesized by incorporating acidic functions such as sulfonic and carboxylic acid or basic functions such as ammonium salts. The groups donate or accept protons due to pH changes in the environment, permitting drug delivery at a specific ph. Conformational transition of pH-sensitive polymers typically occurs through three basic mechanisms: (a) partition coefficient changes between drug and delivery system, (b) collapse or swelling through destabilization, and (c) dissociation. These dynamic characteristics render pH-sensitive polymers a valuable tool for precision-controlled drug delivery in biomedical applications [28].

Oshi et al. developed a novel drug delivery system to enhance treatment for ulcerative colitis (UC), a chronic inflammatory condition affecting the colon's inner lining. One of the main limitations of existing oral therapies is their inability to protect the drug from the harsh conditions of the gastrointestinal (GI) tract, leading to reduced effectiveness and increased systemic side effects. To address this, the researchers formulated curcumin-loaded nanocrystals using an antisolvent crystallization technique. Curcumin was first dissolved in ethanol and then quickly added to a chitosan solution under ultrasonication, producing chitosan-coated nanocrystals (CH1@CUNCs) with an approximate diameter of 380 nm and a positively charged surface.

To improve the site-specific release and enhanced stability, a layer-by-layer coating process was applied using alternate layers of alginate and chitosan, followed by an outer layer of cellulose acetate phthalate (CAP), forming CAP1AG4CH5@CUNCs. This multilayered structure protected the drug from premature release in the stomach and small intestine. Microscopy confirmed a well-formed, core-shell architecture with a uniform particle size of about 421 nm and a low polydispersity index. Zeta potential shifts confirmed successful layer deposition. In vitro drug release studies conducted at different pH levels showed minimal curcumin release at pH 1.2 and 4.5 but a significant release at pH 7.2, mimicking colonic conditions. This pH-triggered response was due to the dissolution of the CAP coating in alkaline environments, allowing targeted release in the colon. Overall, this nanocarrier system shows promise for improving curcumin delivery in UC therapy by enhancing colon-specific drug release and minimizing systemic exposure [32].

6.5.2 Enzyme-Triggered Release Systems

Some of the common classes of enzymes that are used as initiators in enzyme responsive controlled drug delivery are described below. Release of drugs occurs directly from different nanocarriers by targeted cleavage using enzymes. Drug incorporation into nanomaterials may be achieved via physical entrapment or covalent attachment, such as caged porous structure or self-assembled system cross-linked matrix. Enzymes may also be utilized to create specific products, such as acidic environment, to release the drug from carriers.

Proteases play a crucial role in physiological processes and are often overexpressed in diseased tissues, making them potential targets for advanced drug delivery. Researchers have developed Peptide-modified polymers that trigger gene expression in reaction to protease activity, enabling targeted transfection in cancer cells. Self-assembled peptide matrices release therapeutic agents when exposed to specific proteases like urokinase-type plasminogen. Matrix metalloproteinases, linked to cancer invasion, have been used to trigger site-specific drug delivery, enhancing tumor specificity. Protease-activated nanoparticles and imaging techniques further enable precise drug release, with promising applications in glioma treatment and tumor-targeted therapies [29, 30]. *Phospholipase enzyme* upregulation is linked to various cancers and diseases, including inflammation and neurodegeneration. Phospholipase a2 (pla2) is highly expressed in tumors, promoting carcinogenesis through arachidonic acid and lysophospholipid release. Researchers have developed pla2-responsive drug delivery systems, such as enzymatically activated liposomes carrying antitumor ether lipids. These systems enable localized drug activation, minimizing side effects. Pla2 hydrolysis facilitates drug release at tumor sites, as demonstrated with capsaicin prodrugs in liposomal vesicles. Studies show efficient drug activation upon pla2 exposure, highlighting its potential for targeted, tumor-specific chemotherapy with reduced systemic toxicity.

Oxidoreductases play a vital role in oxidative environments linked with diseases like diabetes and cancer, making them good triggers for DDS. Glucose oxidase is widely used in glucose-responsive controlled release mechanisms. Gu et al. developed an injectable nano-network utilizing glucose oxidase to enable self-regulated insulin release. In response to high glucose levels, glucose oxidase catalyzes glucose into gluconic acid, breaking down the nano-network and releasing insulin, stabilizing blood sugar for up to 10 days. Similarly, glucose-responsive microgels using glucose oxidase demonstrated effective glucose-triggered insulin release, showcasing promising advancements in long-term diabetes management through enzyme-triggered delivery systems [31].

Liang et al. developed a smart drug delivery system using doxorubicin nanocrystals (DOX NCs) coated with a modified form of chondroitin sulfate (CSOA) to enhance cancer treatment through targeted and responsive release. The DOX NCs were synthesized using a solvent exchange method and then coated with CSOA via electrostatic and hydrophobic interactions. This design allowed for high drug loading (up to 70%) and excellent colloidal stability in physiological conditions. The CSOA coating provided dual advantages: it promoted cancel cell targeting through CD44 receptor-mediated uptake and enables drug release in response to acidic pH and hyaluronidase, and enzyme overexpressed in tumors. In vitro studies showed that these nanocrystals selectively released doxorubicin PEG-PCL micelles, CSOA/NCs demonstrated superior cellular uptake and anticancer activity. The system showed sustained drug release under tumor-like conditions and minimized side effects associated with systemic toxicity. Overall, this strategy illustrates the potential of combining nanocrystal technology with responsive surface modification for precise and efficient cancer therapy [32].

6.5.3 Thermosensitive and Light-Activated Nanocrystal Formulations

Light-controlled drug delivery systems are of three categories: photochemical, photothermal, and photoisomerization-mediated therapies. Light-controlled drug delivery allows for selective temporal modulation and spatial resolution and of the release of many therapeutic agents, such as but not restricted to small molecule drugs, proteins, and genes. Light-controlled drug delivery is triggered by photochemical, photothermal, or photo isomerization-mediated energy. Light is an especially attractive therapeutic intervention agent because it is not invasive, easy to apply, and biologically inert. Its adaptability of intensity and precision control in space and time makes it highly efficient. Light also possesses a number of advantages over other stimuli like temperature and pH, including negligible interaction with the biological system—a factor of significant importance for clinical use. Moreover, it also enables extremely localized drug delivery with unprecedented spatiotemporal resolution in a contactless manner, ultimately improving the efficacy of treatment and patient comfort.

Consequently, light has been utilized very extensively in intelligent targeted delivery systems to deliver a range of therapeutic molecules, such as chemotherapeutic drug molecules, photosensitizers, molecules, proteins, nucleic acids, and enzymes, to the target site. Light responsiveness can be initiated by various wavelengths, including UV (200–400 nm), visible (400–750 nm), and near-IR (750–2000 nm). UV light is a less desirable candidate since it possesses shallow tissue penetration and can be carcinogenic following prolonged exposure. In contrast, near infrared light possesses greater penetration depth in biological tissues, less phototoxicity and lesser background signal and is therefore an improved candidate for biological applications. However, the low energy of near-infrared photons prevents many photochemical processes from being activated, which require high-energy wavelengths. Two-photon absorption will at least partially address this issue. Other methods used in light-activated delivery systems include oxidation-reduction reactions and photo-driven isomerization, photo-uncaging, bond breakage, crosslinking, and electrostatic assembly [33].

Zubik et al. developed injectable hydrogels combining poly (N-isopropylacrylamide) (PNIPAAm) with cellulose nanocrystals (CNCs) for potential use in wound care. These hybrid hydrogels were synthesized without additional crosslinkers, using free-radical polymerization. The resulting materials demonstrated tunable thermal responsiveness and improved mechanical strength, directly influenced by the CNC content. The volume phase transition temperature (VPTT) of the hydrogels ranged from 36 to 39 °C—ideal for activation near body temperature.

Rheological analysis revealed that higher CNC concentrations enhanced both elasticity and injectability. When tested at 37 °C, the hydrogels transitioned from liquid to gel form, suggesting suitability for in situ application. Metronidazole, a common antibiotic for wound infections, was used as a model drug to study release behavior. The hydrogels absorbed the drug efficiently and released it in a two-phase pattern: an initial burst followed by sustained release over 24 h. This release profile aligns with therapeutic needs in wound management, allowing rapid bacterial control and extended drug availability. Overall, the study highlights the potential of PNIPAAm-CNC hydrogels as smart, injectable dressings capable of delivering drugs responsively and maintaining structural integrity at physiological temperatures [34].

6.6 Integration of Nanocrystals with Advanced Technologies

The fusion of nanocrystals with cutting-edge technologies is revolutionizing drug formulation by enhancing solubility, bioavailability, and targeted delivery. Innovations like AI, 3D printing, and lipid-based carriers optimize drug performance, enabling precise dosing, controlled release, and improved therapeutic outcomes for conditions such as cancer and neurological disorders.

6.6.1 Incorporation with 3D Printing for Personalized Medicine

Nanocrystals have been incorporated into 3D-printed solid oral dosage forms, including printlets and tablets, through the melting solidification printing process [35].

For nanocrystals, oral polymer film formulations may be beneficial. Using indomethacin nanocrystals, oral polymeric film formulations were produced. The impact of nanocrystalline drug amount on final formulation features was examined, and the ideal HPMC concentrations for drug-loaded films have been identified. For immediate drug delivery, these 3D-printed nanocrystalline drug-loaded polymer films offer an acceptable substitute [36].

Olanzapine, a model hydrobhobic chemical, has been put into implants created using 3d printing processes. The implants were examined using thermal analysis, X-ray diffraction, SEM, and infrared spectroscopy. In-vitro release studies revealed a maximum percentage of about 60% and continuous drug release for 200 days [37].

6.6.2 AI and Machine Learning in Nanocrystal Design and Optimization

Nanocrystal optimization and design are being revolutionized by artificial intelligent AI and machine learning (ML). They let researchers anticipate features based on structure and composition, analyze large data sets, and identify the best synthesis parameters [38].

In addition to reducing experimental trial and error, this speeds up the development process. Key attributes, including size, shape, surface area, optical characteristics, and reactivity, can be predicted by AI models, and machine learning algorithms can determine the best parameter combinations [39, 40]. Based on anticipated properties, virtual screening aids in identifying promising designs. In addition, complex data from characterization methods like spectroscopy, X-ray diffraction, and electron microscopy can be analyzed using AI. Support vector machines, regression analysis, genetic algorithms, and artificial and neural networks are a few instances of AI and ML techniques. The interpretability of the model, computational complexity, and data quality are challenges, though [41, 42].

6.6.3 Combination with Microfluidic Platforms for Precision Manufacturing

The ability of microfluidic flow reactors to quickly screen through reaction parameters and generate nanocrystals of the appropriate size and monodispersity renders them ideal for nanocrystal synthesis [43, 44]. The synthesis process's outcome is greatly influenced by flow regimes in microfluidic devices, which regulate the shape, size and characteristics of the produced nanocrystals. To create nanocrystals with a variety of sizes, shapes, and narrow size distributions, many types and configurations of microreactors are employed, such as separated flow reactors, droplet flow reactors, and reactors based on microcapillary tubes. Because it allows for fine control over reaction conditions and makes it possible to produce a variety of chemicals, microfluidics has had a considerable impact on the synthesis of nanocrystals [45, 46]. These uses include drug delivery systems, the development of sophisticated materials for CO_2 reduction applications, and the synthesis of hierarchical Cu_2O nanocrystals [47].

6.7 Biological Barriers: Future Strategies

Nanocrystals offer a promising approach to overcoming biological barriers that limit drug delivery, enhancing solubility, stability, and bioavailability. Future strategies focus on surface modifications, targeted delivery, and advanced formulations to improve cellular uptake and therapeutic efficacy. Innovations in nanotechnology will drive more effective treatments across various biomedical applications. Few of these are discussed below:

6.7.1 Managing Protein Corona Formation and Preventing Immune Clearance

Researchers are developing nanocrystals to combat protein corona formation and prevent immune clearance, using surface modifications, biomimetic coatings, and nanocrystal design to enhance therapeutic efficacy [48]. Nanocrystals enter the bloodstream and form a protein corona, affecting their interaction with the immune system and their biological fate. This can lead to rapid uptake and clearance of nanocrystals, limiting their therapeutic potential. To mitigate this, strategies include PEGylation, zwitterionic polymers, biomimetic coatings, surface charge manipulation, nanocrystal size and shape optimization, and targeted delivery [49]. However, the protein corona is dynamic and can change over time, and patient-specific factors can influence its formation. In-vivo characterization is crucial to evaluate the effectiveness of these strategies [50].

6.7.2 Novel Approaches for Enhancing Nanocrystal Penetration in Tissues

Effective penetration without discomfort is a challenge for dermal administration of drugs. Nanocrystals increase a drug's solubility and bioavailability while offering benefits like deeper penetration and a greater drug loading capacity [4]. Several strategies are being investigated, with an emphasis on the tissue environment as well as the nanocrystals themselves. Targeting particular cell types or tissue components, utilizing surface changes such as "stealth" coatings, and optimizing size and shape are all part of nanocrystal engineering. Changes in charge can affect how nanocrystals interact with cell membranes and the extracellular matrix. Target tissues' particular enzymes cause enzyme-responsive coatings to break down, enabling regulated release and better penetration [51]. Enzyme-mediated extracellular matrix breakdown, increased vascular permeability, mechanical disruption, immunomodulation, and targeting cancer-associated fibroblasts are all components of tissue microsurrounding alteration. Multiple strategies are combined in combination techniques to improve nanocrystal penetration. Target specificity, biodegradability and clearance, biocompatibility and toxicity, and real-time monitoring are important factors. These methods are always changing, and resolving the tissue penetration issue is essential to the future of nanomedicine [52, 53].

6.7.3 Strategies for Improved Cellular Uptake and Intracellular Trafficking

Improved intracellular trafficking and cellular uptake are two implications for nanocrystals [54]. Targeting using ligands, altering surface charges, optimizing nanocrystal size and form, and using physical techniques like electroporation and sonoporation are some strategies [55]. Drug delivery can be improved by functionalizing nanocrystals with ligands that attach to certain receptors on target cells. With the goal to balance charge and lessen non-specific interactions, zwitter-ionic coatings are becoming more and more prevalent. Surface charge modulation is essential. Direct translocation across the cell membrane is made possible by the integration of cell-penetrating peptides to the surfaces of nanocrystals. In certain applications, including gene therapy, physical techniques like electroporation and sonoporation are often utilized to improve uptake [56]. Additionally, nanocrystals can be tailored for controlled drug release, intracellular trafficking alteration, intra-cellular escape, and targeting certain organelles [57]. Surface chemistry, stability, biocompatibility, and exact control are important factors for nanocrystals [58].

6.7.4 Mucus-Penetrating Nanocrystal Technologies

Pulmonary inhalation is ideal for treating lung diseases and achieving systemic absorption, but formulations face challenges due to physiological barriers, particle size, and solubility. Inhalable nanocrystals-in-microparticles (INMP) offer an effective inhalation formulation that is dry powder-based with aerodynamic properties. A study aims to develop a novel formulation strategy for curcumin, drug with strong antioxidant, anticancer, and anti-inflammatory properties, by designing inhalable mucus-penetrating nanocrystals-in-microparticles. The study focuses on preparing curcumin suspensions, evaluating their physicochemical characterizations, investigating interactions with mucus, converting nanocrystal into curcumin inhalable nanocrystals-in-microparticles, and studying drug retention and in-vivo bioavailability after pulmonary administration [59].

6.8 Scaling Up and Manufacturing Challenges

Scaling up nanocrystal production poses challenges in maintaining particle uniformity, stability, and efficiency. Issues such as high energy consumption, aggregation, and process reproducibility complicate large-scale manufacturing. Optimizing formulations, refining production techniques, and adhering to regulatory standards are essential for ensuring quality, cost-effectiveness, and successful commercialization of nanocrystal-based drugs. The following are a few more challenges that are faced during scaling up and manufacturing;

1. Batch-to-batch consistency
   Ensuring consistent quality across batches becomes difficult when transitioning from laboratory-scale production to large-scale industrial manufacturing. Variations in particle size or composition may impact the performance of nanocrystals, potentially affecting their safety and effectiveness. Maintaining uniformity across thousands of liters of product requires sophisticated equipment and tightly regulated manufacturing environments [60].

2. Maintaining precision in formulation
   Producing nanocrystals involves a precise balance of components, which must be mixed at specific ratios to achieve the desired size and structure. In small-scale production, this control is feasible, but scaling up complicates maintaining these precise conditions. Minor variations in mixing or formulation can lead to particles that become oversized, unstable, or less effective [61].
3. Adapting production technology
   Lab effective techniques might not easily scale up. For example, microfluidics widely utilized for nanoparticle production need heavy adjustment for large-scale manufacturing [62].

   Investment in new technologies or the remodelling of existing ones is crucial for scalable and effective production. More recent developments in microfluidics systems have been promising to enhance the reproducibility of nanoparticle synthesis [63].
4. Securing raw materials
   Increasing production depends on assured supplies of high-quality raw materials, which are supply chain-limited. Supply chain disruption may result in delays in production or shortages [64].
5. Sterility and contamination control
   Nanocrystals are commonly utilized in applications that require strict sterility, particularly in the pharmaceutical industry. To prevent contamination, large-scale manufacturing facilities must adhere to stringent sterility standards, which can be both expensive and technically challenging [64]. Therapeutic molecules such as RNA are highly sensitive, necessitating meticulously controlled sterile environments due to the increased risk of contamination in large-scale production [65].
6. Complying with regulatory standards
   Mass production has to follow rigorous regulatory guidelines for safety and efficacy. Every step in the manufacturing process has to be documented and qualified in detail to maintain compliance with health authorities such as the FDA or EMA [65, 78]. This makes the process more intricate, especially when transitioning from small-scale research production to full-scale commercial manufacturing [67].

6.8.1 Strategies for Overcoming Challenges

Several strategies are being explored to overcome the challenges in scaling up nanocrystal production: Advanced manufacturing technologies: investing in new technologies, such as scalable microfluidics systems, can improve production efficiency and consistency [61]. Supply chain management: securing reliable raw material supplies through strategic partnerships, and diversified sourcing can mitigate supply chain risks [64]. Regulatory compliance: implementing robust quality control measures and documentation processes ensures compliance with regulatory standards, facilitating the transition from research to commercial production.

6.9 Regulatory Considerations and Clinical Translation of Nanocrystals

The clinical translation of nanocrystals requires stringent regulatory oversight to ensure safety, efficacy, and quality. Key considerations include formulation stability, pharmacokinetics, toxicity, and large-scale production. Addressing regulatory challenges through standardized guidelines and robust preclinical studies will facilitate the approval and integration of nanocrystal-based therapies into mainstream clinical practice. Regulatory considerations have important points which are discussed below:

1. Consumer safety—Foremost crucial thought for any formulation is its effects on human health and yet the inconsistent and insufficient regulation on nanoparticles has resulted in several bad consequences for human health through different unregulated formulations and their processes and that's primary reason for the urgent regulation on various aspects of nanocrystals as well [66].
   Regulatory measures should be such to protect formulation effectiveness of nanomaterial products such as drugs, cosmetics and provide safety to consumers from possible health hazards due to exposure to nanoparticles (NPs). Most nanomedicines achieve their effects either by direct action on genetic material, or through interaction with biomolecules essential to the normal function of the genome and cell division, all which can lead to mutagenicity and genotoxicity [67].
2. Environmental repercussions—Nanotechnology applications are life-altering, but nanoparticle regulations are in the pipeline of processes. Biotechnology regulation failure over the past few years has had numerous alarming environmental consequences [66]. Environmental evaluations under the National Environmental Policy Act (NEPA) are made on a product-by-product basis.
3. Legislation outlook—There is still a lot of excitement around nanomedicine and its impact on the pharmaceutical industry, but regulatory guidance in this area is urgently needed to provide legal certainty to manufacturers, policymakers, healthcare providers, and the general public [67].
4. Process regulation—In addition to discussing the risks associated with the development of nanomedicine, the potential of using a risk-based methodology from the early stage of the research and development phase is emphasized to ensure safety and efficacy when developing novel nano-based dosage forms [68].
   Clinical translation of nanocrystal-based therapeutics involves addressing challenges related to scalability, reproducibility, and long-lasting stability to meet the regulatory requirements and ensure successful transition from laboratory to clinic.
5. Filtering every step—regulatory considerations and clinical translations of nanocrystals present both opportunities and challenges in the field of nanomedicine. Carrier-free nanocrystals, with their high active pharmaceutical

ingredient loading and simple synthetic procedures, are considered promising candidates for disease theranostics [69]. The refine project (refine nanomed regulatory science framework) is an excellent approach that has solved a series of problems, specifically about nanocrystals. The project has outlined deficiencies and regulatory challenges in existing testing methods for nanomedicines [67].

6. Regulatory framework for agencies—the regulatory landscape for nanocrystals and other nanomedicines is complex, involving considerations of physicochemical properties, Pharmacodynamics, pharmacokinetics, nanotoxicity, process control, and biocompatibility [68].
   Both the FDA and EMA have specific regulations for nanomedicines, emphasizing the importance of product quality assessment and safety evaluation. Interestingly, the quality by design (QbD) approach is recommended by regulatory authorities to facilitate the smooth clinical translation of novel nanomedicines.
7. Way forward—To address challenges in clinical translation, researchers need to enhance the design, formulation, drug delivery mechanisms, in vivo effectiveness, and safety of nanocrystals. Additionally, establishing standardized evaluation methods for key quality attributes of nanomedicines is essential [70]. Developing international guidelines and regulatory frameworks, along with incorporating molecular safety assessments early in the development process, will play a vital role in the successful clinical application of nanocrystals and other nanomedicines [71, 72].
8. Product quality and safety assessment—this includes characterization, quality control, and manufacturing processes. Ensuring consistent formulations with low batch-to-batch variability is crucial [73]. Safety assessment is necessary as it involves evaluating biodistribution, clearance, metabolism, and toxicology. The interaction of nanoscale materials with biological systems and the adequacy of testing approaches are significant concerns.

Clinical translation of nanocrystals should have the following approaches:

1. Characterization and standardization: Standardized methods for characterizing nanocrystals and assessing their physicochemical properties are essential for clinical use [67].
2. Safety and efficacy testing: Comprehensive in vitro, ex vivo, and in vivo testing protocols are needed to evaluate nanotoxicity and ensure safety [73].
3. Risk-based approach: Implementing a risk-based methodology from early research and development stages can help ensure safety and efficacy.
4. Quality by design approach: This approach is recommended for developing robust, standardized methods to evaluate critical quality attributes of nanocrystals [74].
5. Interdisciplinary collaboration: Collaboration among experts from academia, industry, and regulatory bodies is necessary for effective clinical translation

6.10 Emerging Applications Beyond Traditional Pharmaceuticals

Applications beyond traditional medications are driving a revolutionary period in the pharmaceutical sector. These developments transform drug development, research, and patient care by utilizing cutting-edge technologies and approaches. An extensive overview of these new applications can be found below:

6.10.1 Nanocrystals in Gene Delivery

Small carriers called nanoparticles aid in the delivery of genetic material into cells, including DNA and RNA. They preserve the genetic material before it reaches the target cells by acting as protective bubbles. By creating tiny nanoparticles with the ability to transport their cargo precisely where it is required, scientists can increase the effectiveness of medicines and decrease adverse effects [75].

6.10.2 Inhalable Nanocrystal-Based Aerosols

An innovative development in pulmonary medication delivery, inhalable nanocrystal-based aerosols provide major advantages for the intervention of respiratory disorders. Budesonide when inhaled was created to treat asthma, it showed better pharmacokinetics than the commercially available inhaled budesonide suspension [16].

Nanocrystals improve the delivery of poorly soluble drugs across the blood-brain barrier, which is a big challenge in treating neurodegenerative illness. Example: paeoniflorin nanocrystals demonstrated significant neuroprotective effects by efficiently crossing the BBB and delivering therapeutic concentrations to the brain [76]

6.10.3 Diagnostic Applications of Nanocrystals

Because fluorescent nanocrystals may produce strong, consistent fluorescence signals, they are frequently employed for brain imaging. It is possible to tailor these NCs to selectively target brain tissues and penetrate the blood-brain barrier, allowing for precise imaging of abnormalities and brain structures.

6.10.4 Drug Monitoring for Target Verification

Drug location and behavior in the brain can be monitored by functionalizing nanocrystals with fluorescent markers. Understanding medication efficacy, maximizing dosage schedules, and confirming therapeutic aims all depend on this application. Fluorescent nanocrystals based on curcumin—curcumin nanocrystal have been used to monitor the distribution of drugs in the brain. Their fluorescent characteristics facilitate the investigation of curcumin's neuroprotective effects by enabling researchers to see how it interacts with neural tissues [77].

6.10.5 Nanocrystals in Scaffolds for Tissue Engineering

Scaffolds offer a structural foundation in tissue engineering that facilitates cell adhesion, growth, and differentiation. The functionality of these scaffolds can be greatly improved by adding nanocrystals.

6.10.6 Nanocrystals of Cellulose

Nanocrystal of cellulose are rod-shaped nanocrystals with remarkable mechanical strength and biocompatibility that come from renewable resources. They are attractive options for bone and cartilage tissue engineering since their integration into scaffolds has been demonstrated to enhance mechanical qualities and promote cellular activity [77].

6.10.7 Nano Neuro Knitting

This technology uses peptide-based nanofiber scaffolds to simulate the extracellular matrix, facilitating axonal development and neural tissue repair. According to research, these scaffolds can promote partial reinnervation in injured areas, resulting in functional recovery in animal models [78].

6.10.8 Nanoparticles as Vaccine Development

Nanoparticles are adaptable platforms for vaccine production, with various advantages over traditional vaccines. Nanoparticles improve antigen stability and immunogenicity by protecting them from degradation, presenting them to immune cells more effectively, and stimulating a powerful immune response [79]. Nanoparticles can be designed to target specific immune cells or tissues, resulting in prolonged antigen exposure and persistent immune responses.

6.11 Summary

The future of nanocrystal-based drug delivery systems holds immense potential for revolutionizing pharmaceutical formulations, particularly in enhancing drug solubility, stability, and bioavailability. Continuous advancements in nanocrystal engineering are expected to enable more precise control over particle size, surface characteristics, and drug release kinetics, making them highly adaptable for diverse therapeutic applications. Hybrid nanocrystal systems integrating polymers, lipids, or inorganic materials are likely to emerge, offering synergistic benefits such as targeted delivery and extended circulation time. The evolution of surface modification techniques, including functionalization with ligands, peptides, or antibodies, will further enhance the specificity and efficacy of nanocrystals for precision medicine, particularly in oncology, neurodegenerative disorders, and infectious diseases.

Furthermore, the scalability of nanocrystal production is expected to improve with the advancement of high-throughput manufacturing technologies, including continuous processing, green synthesis methods, and automation-driven approaches, which will enhance reproducibility and cost-effectiveness. Regulatory challenges remain a critical aspect of the future outlook, as harmonization of guidelines and risk assessment protocols will be essential for ensuring the safety and efficacy of nanocrystal-based therapeutics. The integration of artificial intelligence (AI) and machine learning (ML) into nanocrystal design and optimization is anticipated to accelerate drug development by predicting formulation stability and therapeutic outcomes more efficiently.

Additionally, patient-centric drug development strategies, such as personalized medicine and novel routes of administration, will drive innovations in nanocrystal drug delivery systems. Transdermal, mucosal, and pulmonary delivery approaches are gaining traction, promising enhanced patient compliance and reduced systemic side effects. The expansion of nanocrystal technology into biopharmaceuticals, including peptides and proteins, represents another frontier that could address challenges related to macromolecule stability and absorption. The convergence of nanocrystals with smart drug delivery technologies, including stimuli-responsive and programmable release systems, will open new possibilities for precision therapeutics. Despite these promising developments, challenges such as long-term stability, large-scale reproducibility, and comprehensive safety evaluations need to be addressed. Future research must focus on improving the predictability of in vivo performance and ensuring seamless translation from laboratory to clinical applications. As interdisciplinary collaboration between materials science, pharmaceutical technology, and biomedical engineering intensifies, nanocrystal-based drug delivery systems are poised to play a transformative role in next-generation medicine.

References

1. Junghanns J-UAH, Müller RH, Department of Pharmaceutical Technology, Biotechnology and Quality Management, Free University of Berlin (2008) Nanocrystal technology, drug delivery and clinical applications. Int J Nanomedicine 3:295–309
2. Patra JK, Das G, Fraceto LF et al (2018) Nano based drug delivery systems: recent developments and future prospects. J Nanobiotechnol 16:71. https://doi.org/10.1186/s12951-018-0392-8
3. Lu L, Xu Q, Wang J, Wu S, Luo Z, Lu W (2022) Drug nanocrystals for active tumor-targeted drug delivery. Pharmaceutics 14(4):797. https://doi.org/10.3390/pharmaceutics14040797

4. Alnaim AS (2024) Nanocrystals in dermal drug delivery: a breakthrough for enhanced skin penetration and targeted skin disorder treatments. Pharmaceutics 16(12):1561. https://doi.org/10.3390/pharmaceutics16121561
5. Wu L, Man C, Wang H, Lu X, Ma Q, Cai Y, Ma W (2012) PEGylated multi-walled carbon nanotubes for encapsulation and sustained release of oxaliplatin. Pharm Res 30(2):412–423. https://doi.org/10.1007/s11095-012-0883-5
6. Gigliobianco MR, Casadidio C, Censi R, Di Martino P (2018) Nanocrystals of poorly soluble drugs: drug bioavailability and physicochemical stability. Pharmaceutics 10(3):134. https://doi.org/10.3390/pharmaceutics10030134
7. Shegokar R, Müller RH (2010) Nanocrystals: industrially feasible multifunctional formulation technology for poorly soluble actives. Int J Pharm 399(1–2):129–139. https://doi.org/10.1016/j.ijpharm.2010.07.044
8. Halevas EG, Avgoulas DI, Katsipis G, Pantazaki AA (2022) Flavonoid-liposomes formulations: physico-chemical characteristics, biological activities and therapeutic applications. Eur J Med Chem Rep 5:100059. https://doi.org/10.1016/j.ejmcr.2022.100059
9. Gao L, Liu G, Ma J, Wang X, Zhou L (2013) Application of drug nanocrystal technologies on oral drug delivery of poorly soluble drugs. Pharm Res 30(2):307–324. https://doi.org/10.1007/s11095-012-0883-z
10. Chang T, Zhan H, Liang D, Liang JF (2015) Nanocrystal technology for drug formulation and delivery. Front Chem Sci Eng 9(1):1–14. https://doi.org/10.1007/s11705-015-1509-3
11. Deng J, Huang L, Liu F (2010) Understanding the structure and stability of paclitaxel nanocrystals. Int J Pharm 390(2):242–249. https://doi.org/10.1016/j.ijpharm.2010.02.034
12. Kasai H, Nakanishi H, Oikawa H (2018) Fabrication technique of organic nanocrystals and their optical properties and materialization. In: Nanoparticle technology handbook. Elsevier eBooks, pp 739–744. https://doi.org/10.1016/b978-0-444-64110-6.00070-6
13. Junyaprasert VB, Morakul B (2014) Nanocrystals for enhancement of oral bioavailability of poorly water-soluble drugs. Asian J Pharmaceut Sci 10(1):13–23. https://doi.org/10.1016/j.ajps.2014.08.005
14. Bhaladhare S, Bhattacharjee S (2022) Chemical, physical, and biological stimuli-responsive nanogels for biomedical applications (mechanisms, concepts, and advancements): a review. Int J Biol Macromol226:535–553.https://doi.org/10.1016/j.ijbiomac.2022.12.076
15. Sharma S, Lohan S, Murthy RSR (2013) Formulation and characterization of intranasal mucoadhesive nanoparticulates and thermo-reversible gel of levodopa for brain delivery. Drug Dev Ind Pharm 40(7):869–878.https://doi.org/10.3109/03639045.2013.789051
16. Zingale E, Bonaccorso A, Carbone C, Musumeci T, Pignatello R (2022) Drug nanocrystals: focus on brain delivery from therapeutic to diagnostic applications. Pharmaceutics 14(4):691. https://doi.org/10.3390/pharmaceutics14040691
17. Sharma S, Batra S, Sinha VR (2022) Nanoemulsions: a potential advanced nanocarrier platform for herbal drug delivery. In: Biomedical translational research: drug design and discovery. Springer Nature Singapore, Singapore, pp 351–368
18. Lohan SB, Saeidpour S, Colombo M, Staufenbiel S, Unbehauen M, Wolde-Kidan A, Netz RR, Bodmeier R, Haag R, Teutloff C, Bittl R, Meinke MC (2020) Nanocrystals for improved drug delivery of dexamethasone in skin investigated by EPR spectroscopy. Pharmaceutics 12(5):400. https://doi.org/10.3390/pharmaceutics12050400
19. Kumar L, Verma S, Joshi K, Utreja P, Sharma S (2021) Nanofiber as a novel vehicle for transdermal delivery of therapeutic agents: challenges and opportunities. Future J Pharm Sci 7(1):175
20. Pireddu R, Schlich M, Marceddu S, Valenti D, Pini E, Fadda AM et al (2020) Nanosuspensions and microneedles roller as a combined approach to enhance diclofenac topical bioavailability. Pharmaceutics 12(12):1140
21. Gautam A, Waldrep CJ, Densmore CL (2002) Delivery systems for pulmonary gene therapy. Am J Respir Med 1(1):35–46. https://doi.org/10.1007/bf03257161

22. Geng F, Fan X, Liu Y, Lu W, Wei G (2024) Recent advances in nanocrystal-based technologies applied for ocular drug delivery. Expert Opin Drug Deliv 21(2):211–227. https://doi.org/10.1080/17425247.2024.2311119
23. Torchilin VP (2006) Recent approaches to intracellular delivery of drugs and DNA and organelle targeting. Annu Rev Biomed Eng 8(1):343–375. https://doi.org/10.1146/annurev.bioeng.8.061505.095735
24. Bareford L, Swaan P (2007) Endocytic mechanisms for targeted drug delivery☆. Adv Drug Deliv Rev 59(8):748–758. https://doi.org/10.1016/j.addr.2007.06.008
25. Miersch S, Sidhu SS (2016) Intracellular targeting with engineered proteins. F1000Res 5:1947. https://doi.org/10.12688/f1000research.8915.1
26. Chali SP, Westmeier J, Krebs F, Jiang S, Neesen FP, Uncuer D et al (2024) Albumin nanocapsules and nanocrystals for efficient intracellular drug release. Nanoscale Horizons 9(11):1978–1989
27. Rana B, Sharma S, Batra S (2022) Targeted nano size vesicles to colon: a state of art. ECS Trans 107(1):10259
28. Liu J, Huang Y, Kumar A, Tan A, Jin S, Mozhi A, Liang X (2013) pH-sensitive nano-systems for drug delivery in cancer therapy. Biotechnol Adv 32(4):693–710. https://doi.org/10.1016/j.biotechadv.2013.11.009
29. Oshi MA, Lee J, Naeem M, Hasan N, Kim J, Kim HJ et al (2020) Curcumin nanocrystal/pH-responsive polyelectrolyte multilayer core–shell nanoparticles for inflammation-targeted alleviation of ulcerative colitis. Biomacromolecules 21(9):3571–3581
30. Andresen TL, Thompson DH, Kaasgaard T (2010) Enzyme-triggered nanomedicine: drug release strategies in cancer therapy (invited review). Mol Membr Biol 27(7):353–363. https://doi.org/10.3109/09687688.2010.515950
31. Hu Q, Katti PS, Gu Z (2014) Enzyme-responsive nanomaterials for controlled drug delivery. Nanoscale 6(21):12273–12286. https://doi.org/10.1039/c4nr04249b
32. Liang Y, Fu X, Du C, Xia H, Lai Y, Sun Y (2020) Enzyme/pH-triggered anticancer drug delivery of chondroitin sulfate modified doxorubicin nanocrystal. Artif Cells Nanomed Biotechnol 48(1):1114–1124
33. Tao Y, Chan HF, Shi B, Li M, Leong KW (2020) Light: a magical tool for controlled drug delivery. Adv Funct Mater 30(49):2005029. https://doi.org/10.1002/adfm.202005029
34. Zubik K, Singhsa P, Wang Y, Manuspiya H, Narain R (2017) Thermo-responsive poly(N-isopropylacrylamide)-cellulose nanocrystals hybrid hydrogels for wound dressing. Polymers 9(4):119
35. Lopez-Vidal L, Real JP, Real DA, Camacho N, Kogan MJ, Paredes AJ, Palma SD (2021) Nanocrystal-based 3D-printed tablets: semi-solid extrusion using melting solidification printing process (MESO-PP) for oral administration of poorly soluble drugs. Int J Pharm 611:121311. https://doi.org/10.1016/j.ijpharm.2021.121311
36. Germini G, Peltonen L (2021) 3D printing of drug nanocrystals for film formulations. Molecules 26(13):3941. https://doi.org/10.3390/molecules26133941
37. Picco CJ, Utomo E, McClean A, Domínguez-Robles J, Anjani QK, Volpe-Zanutto F, McKenna PE, Acheson JG, Malinova D, Donnelly RF, Larrañeta E (2022) Development of 3D-printed subcutaneous implants using concentrated polymer/drug solutions. Int J Pharm 631:122477. https://doi.org/10.1016/j.ijpharm.2022.122477
38. Olawade DB, Ige AO, Olaremu AG, Ijiwade J, Adeola AO (2024) The synergy of artificial intelligence and nanotechnology towards advancing innovation and sustainability- a mini-review. Nano Trends 8:100052. https://doi.org/10.1016/j.nwnano.2024.100052
39. Nazar S, Yang J, Amin MN, Khan K, Ashraf M, Aslam F, Javed MF, Eldin SM (2023) Machine learning interpretable-prediction models to evaluate the slump and strength of fly ash-based geopolymer. J Mater Res Technol 24:100–124. https://doi.org/10.1016/j.jmrt.2023.02.180
40. Jena GK, Patra CN, Sruti J, Parhi R, Chand S (2024) Artificial intelligence (AI) and machine learning (ML) implemented drug delivery systems: a paradigm shift in the pharmaceutical industry. J Bio-X Res 7:0016. https://doi.org/10.34133/jbioxresearch.0016

41. Nandipati M, Fatoki O, Desai S (2024) Bridging nanomanufacturing and artificial intelligence—a comprehensive review. Materials 17(7):1621. https://doi.org/10.3390/ma17071621
42. Quazi S (2022) Artificial intelligence and machine learning in precision and genomic medicine. Med Oncol 39(8):120. https://doi.org/10.1007/s12032-022-01711-1
43. Yen BK, Stott NE, Jensen KF, Bawendi MG (2003) A continuous-flow microcapillary reactor for the preparation of a size series of CdSe nanocrystals. Adv Mater 15(21):1858–1862
44. Ejeta F (2021) Recent advances of microfluidic platforms for controlled drug delivery in nanomedicine. Drug Des Devel Ther 15:3881–3891. https://doi.org/10.2147/dddt.s324580
45. Hamdallah SI, Zoqlam R, Erfle P, Blyth M, Alkilany AM, Dietzel A, Qi S (2020) Microfluidics for pharmaceutical nanoparticle fabrication: the truth and the myth. Int J Pharm 584:119408. https://doi.org/10.1016/j.ijpharm.2020.119408
46. Lignos I, Stavrakis S, Nedelcu G, Protesescu L, Demello AJ, Kovalenko MV (2016) Synthesis of cesium lead halide perovskite nanocrystals in a droplet-based microfluidic platform: fast parametric space mapping. Nano Lett 16(3):1869–1877
47. Epps RW, Felton KC, Coley CW, Abolhasani M (2017) Automated microfluidic platform for systematic studies of colloidal perovskite nanocrystals: towards continuous nano-manufacturing. Lab Chip 17(23):4040–4047
48. Miao Y, Li L, Wang Y, Wang J, Zhou Y, Guo L, Zhao Y, Nie D, Zhang Y, Zhang X, Gan Y (2024) Regulating protein corona on nanovesicles by glycosylated polyhydroxy polymer modification for efficient drug delivery. Nat Commun 15(1):1159. https://doi.org/10.1038/s41467-024-45254-7
49. Guo F, Luo S, Wang L, Wang M, Wu F, Wang Y, Jiao Y, Du Y, Yang Q, Yang X, Yang G (2023) Protein corona, influence on drug delivery system and its improvement strategy: a review. Int J Biol Macromol 256:128513. https://doi.org/10.1016/j.ijbiomac.2023.128513
50. Panico S, Capolla S, Bozzer S, Toffoli G, Bo MD, Macor P (2022) Biological features of nanoparticles: protein Corona formation and interaction with the immune system. Pharmaceutics 14(12):2605. https://doi.org/10.3390/pharmaceutics14122605
51. Graván P, Peña-Martín J, De Andrés JL, Pedrosa M, Villegas-Montoya M, Galisteo-González F, Marchal JA, Sánchez-Moreno P (2023) Exploring the impact of nanoparticle stealth coatings in cancer models: from PEGylation to cell membrane-coating nanotechnology. ACS Appl Mater Interfaces 16(2):2058–2074. https://doi.org/10.1021/acsami.3c13948
52. Guo J, Zeng H, Chen Y (2020) Emerging nano drug delivery systems targeting cancer-associated fibroblasts for improved antitumor effect and tumor drug penetration. Mol Pharm 17(4):1028–1048. https://doi.org/10.1021/acs.molpharmaceut.0c00014
53. Patel V, Sharma OP, Mehta T (2018) Nanocrystal: a novel approach to overcome skin barriers for improved topical drug delivery. Expert Opin Drug Deliv 15(4):351–368. https://doi.org/10.1080/17425247.2018.1444025
54. Foroozandeh P, Aziz AA (2018) Insight into cellular uptake and intracellular trafficking of nanoparticles. Nanoscale Res Lett 13(1):339. https://doi.org/10.1186/s11671-018-2728-6
55. Tharkar P, Varanasi R, Wong WSF, Jin CT, Chrzanowski W (2019) Nano-enhanced drug delivery and therapeutic ultrasound for cancer treatment and beyond. Front Bioeng Biotechnol 7:324. https://doi.org/10.3389/fbioe.2019.00324
56. Azevedo C, Macedo MH, Sarmento B (2017) Strategies for the enhanced intracellular delivery of nanomaterials. Drug Discov Today 23(5):944–959. https://doi.org/10.1016/j.drudis.2017.08.011
57. Ran Q, Wang M, Kuang W, Ouyang J, Han D, Gao Z, Gong J (2022) Advances of combinative nanocrystal preparation technology for improving the insoluble drug solubility and bioavailability. Crystals 12(9):1200. https://doi.org/10.3390/cryst12091200
58. Huang G, Shuai S, Zhou W, Chen Y, Shen B, Yue P (2022) To enhance mucus penetration and lung absorption of drug by inhalable nanocrystals-in-microparticles. Pharmaceutics 14(3):538. https://doi.org/10.3390/pharmaceutics14030538

59. Costabile G, Provenzano R, Azzalin A, Scoffone VC, Chiarelli LR, Rondelli V, Grillo I, Zinn T, Lepioshkin A, Savina S, Miro A, Quaglia F, Makarov V, Coenye T, Brocca P, Riccardi G, Buroni S, Ungaro F (2019) PEGylated mucus-penetrating nanocrystals for lung delivery of a new FtsZ inhibitor against Burkholderia cenocepacia infection. Nanomed Nanotechnol Biol Med 23:102113. https://doi.org/10.1016/j.nano.2019.102113
60. Paliwal R, Babu RJ, Palakurthi S (2014) Nanomedicine scale-up technologies: feasibilities and challenges. AAPS PharmSciTech 15(6):1527–1534. https://doi.org/10.1208/s12249-014-0177-9
61. Möschwitzer JP (2012) Drug nanocrystals in the commercial pharmaceutical development process. Int J Pharm 453(1):142–156. https://doi.org/10.1016/j.ijpharm.2012.09.034
62. Srivalli KMR, Mishra B (2014) Drug nanocrystals: a way toward scale-up. Saudi Pharm J 24(4): 386–404. https://doi.org/10.1016/j.jsps.2014.04.007
63. Tenchov R, Hughes KJ, Ganesan M, Iyer KA, Ralhan K, Lotti Diaz LM et al (2025) Transforming medicine: cutting-edge applications of nanoscale materials in drug delivery. ACS Nano 19(4):4011–4038
64. Desai N (2012) Challenges in development of nanoparticle-based therapeutics. AAPS J 14(2): 282–295. https://doi.org/10.1208/s12248-012-9339-4
65. Raval M, Chavda P, Patel P (2023) Regulatory aspects of nanomaterials: current and future perspective. In: Role of nanotechnology in cancer therapy. Bentham Science Publishers, Sharjah, pp 255–277
66. Foulkes R, Man E, Thind J, Yeung S, Joy A, Hoskins C (2020) The regulation of nanomaterials and nanomedicines for clinical application: current and future perspectives. Biomater Sci 8(17): 4653–4664. https://doi.org/10.1039/d0bm00558d
67. Csóka I, Ismail R, Jójárt-Laczkovich O, Pallagi E (2021) Regulatory considerations, challenges and risk-based approach in nanomedicine development. Curr Med Chem 28(36):7461–7476
68. Fang F, Chen X (2024) Carrier-free nanodrugs: from bench to bedside. ACS Nano 18(35): 23827–23841. https://doi.org/10.1021/acsnano.4c09027
69. De Jong WH, Geertsma RE, Borchard G (2022) Regulatory safety evaluation of nanomedical products: key issues to refine. Drug Deliv Transl Res 12(9):2042–2047
70. Cicha I, Chauvierre C, Texier I, Cabella C, Metselaar JM, Szebeni J, Dézsi L, Alexiou C, Rouzet F, Storm G, Stroes E, Bruce D, MacRitchie N, Maffia P, Letourneur D (2018) From design to the clinic: practical guidelines for translating cardiovascular nanomedicine. Cardiovasc Res 114(13):1714–1727. https://doi.org/10.1093/cvr/cvy219
71. Singh AV, Bhardwaj P, Upadhyay AK, Pagani A, Upadhyay J, Bhadra J et al (2024) Navigating regulatory challenges in molecularly tailored nanomedicine. Explor BioMat-X 1:124–134. https://doi.org/10.37349/ebmx.2024.00009
72. Hua S, De Matos MBC, Metselaar JM, Storm G (2018) Current trends and challenges in the clinical translation of nanoparticulate nanomedicines: pathways for translational development and commercialization. Front Pharmacol 9:790. https://doi.org/10.3389/fphar.2018.00790
73. Jarvis M, Krishnan V, Mitragotri S (2018) Nanocrystals: a perspective on translational research and clinical studies. Bioeng Transl Med 4(1):5–16. https://doi.org/10.1002/btm2.10122
74. Duan L, Ouyang K, Xu X, Xu L, Wen C, Zhou X et al (2021) Nanoparticle delivery of CRISPR/Cas9 for genome editing. Front Genet 12:673286
75. Yue P, Zhou W, Huang G, Lei F, Chen Y, Ma Z, Chen L, Yang M (2022) Nanocrystals based pulmonary inhalation delivery system: advance and challenge. Drug Deliv 29(1):637–651. https://doi.org/10.1080/10717544.2022.2039809
76. Domingues RMA, Gomes ME, Reis RL (2014) The potential of cellulose nanocrystals in tissue engineering strategies. Biomacromolecules 15(7):2327–2346. https://doi.org/10.1021/bm500524s
77. Ellis-Behnke RG, Liang YX, You SW, Tay DK, Zhang S, So KF, Schneider GE (2006) Nano neuro knitting: peptide nanofiber scaffold for brain repair and axon regeneration with functional return of vision. Proc Natl Acad Sci USA 103(13):5054–5059. https://doi.org/10.1073/pnas.0600559103

78. Kheirollahpour M, Mehrabi M, Dounighi NM, Mohammadi M, Masoudi A (2019) Nanoparticles and vaccine development. Pharmaceutical. Nanotechnology 8(1):6–21. https://doi.org/10.2174/2211738507666191024162042
79. Liang J, Yao L, Liu Z, Chen Y, Lin Y, Tian T (2024) Nanoparticles in subunit vaccines: immunological foundations, categories, and applications. Small 12:e2407649. https://doi.org/10.1002/smll.202407649

Particokinetics and Biofate of Nanocrystals 7

Xiaoyun Du, Wei Wu, and Yi Lu

Abstract

Nanocrystals offer a promising strategy for enhancing the bioavailability of poorly water-soluble drugs and enabling versatile administration routes. However, their clinical translation is hindered by an incomplete understanding of their *in vivo* fate due to the inability to differentiate intact nanocrystals from dissolved drug molecules. This chapter introduces advances in strategies for probing nanocrystal biofate, including transmission electron microscopy, physiologically based pharmacokinetic modeling, self-fluorescent drugs, and hybrid nanocrystal technology. Hybrid nanocrystals embedded with environment-responsive probes enable real-time tracking of intact particles, revealing that nanocrystals dissolve gradually and remain intact across the epithelial membrane. Key findings highlight prolonged gastrointestinal retention after oral administration, RES-organ accumulation post-intravenous injection, follicular deposition in transdermal delivery, and slow dissolution kinetics. Quantitative particokinetic data derived from fluorescence signals demonstrate that intact nanocrystals contribute significantly to bioavailability. The integration of *in vivo* dissolution profiling further informs *in vitro–in vivo* correlation development. These insights underscore the dual biofate pathways of nanocrystals (dissolution vs. particle absorption) and emphasize the need for particle-specific tracking to optimize formulation design and clinical outcomes.

X. Du · W. Wu · Y. Lu (✉)
School of Pharmaceutical Sciences, Fudan University, Key Laboratory of Smart Drug Delivery of MOE and National Key Laboratory of Advanced Drug Formulations for Overcoming Delivery Barriers, Shanghai, China
e-mail: 24211030010@fudan.edu.cn; wuwei@shmu.edu.cn; fd_luyi@fudan.edu.cn

N. K. Mehra, K. K. Bansal (eds.), *Nanocrystal Technologies in Biomedical Sciences*,
https://doi.org/10.1007/978-981-95-2968-1_7

Keywords

Biofate · Particokinetics · Hybrid nanocrystals · ACQ fluorophores · In vitro–in vivo correlation

7.1 Introduction

The dissolution and membrane transport properties are essential determinants of the therapeutic potential and absorption efficiency of active pharmaceutical ingredients (APIs) [1, 2], as more than 90% of API candidates and around 40% of approved drugs exhibit poorly aqueous solubility [3, 4]. It is full of challenges to develop liquid formulations of these APIs due to the lack of biologically acceptable excipients that can effectively enhance solubility without inducing adverse effects. Although developing solid/semisolid formulations is out of the question, the poor solubility may limit the bioavailability [5, 6]. The dissolution of APIs at the delivery site is a fundamental requirement for their uptake. Furthermore, the solubilized molecules must pass through physiological barriers, including the mucosa, stratum corneum, and cornea, to achieve therapeutic efficacy. The permeability of the molecules governs the process. Therefore, enhancing solubility and permeability is critical for formulation improvement and new drug development [7, 8].

As per the Noyes-Whitney equation, decreasing particle size is among the most effective approaches for increasing dissolution rates [9, 10]. Consequently, micronization has been extensively utilized to enhance the bioavailability of APIs with low solubility. Despite the success, micronization only achieves limited improvement and may fail to meet dosage requirements in some cases [11, 12]. The insufficiency is due to the limited enhancement in specific surface area by micronization. Conversely, particle surface area experiences exponential growth as the size is reduced to the nanometer range. This concept forms the basis for nanocrystal development, where drug crystals exhibit dimensions from a few dozen to several hundred nanometers [13–15]. In addition to enhancing dissolution, nanocrystals enable the liquid formulation of poorly soluble APIs in the form of nanosuspension [16]. Unlike conventional nanoparticles, nanocrystals contain no carrier materials, theoretically leading to a drug loading of up to 100%, which is particularly advantageous for achieving effective therapeutic outcomes with low-dose treatments. For instance, a single administration of Ryanodex® (dantrolene sodium injectable nanosuspension, manufactured by Eagle Pharmaceuticals, Inc., New Jersey, United States) is adequate for an adult patient, while more than ten vials are required in the case of conventional Dantrium® (20 mg/vial) [17]. Therefore, nanocrystals have found broad application in intravenous, transdermal, ocular, and pulmonary drug delivery systems [18].

Although nanocrystals hold great promise in drug delivery, significant challenges remain, particularly in understanding pharmacokinetics and *in vivo* fate. This gap often results in a disconnect between preclinical results and clinical efficacy, presenting a major obstacle to the clinical translation of nanocrystals. Conventional approaches to investigating nanocrystals' biofate depend on assessing overall drug concentrations, without differentiating between dissolved forms and intact nanocrystals. However, only dissolved drugs perform therapeutic effects. The results obtained from the total drug level may lead to a misunderstanding of the pharmacokinetics and biofate of nanocrystals as well as an inaccurate assessment of their efficacy. The discontinuation of Anjeso® (meloxicam nanocrystals, Baudax Bio, Pennsylvania, United States) serves as a notable example. It was intended for managing moderate to severe pain in adults. Although the company attributed the discontinuation to the reduction in commercial force, the limited clinical value may be the fundamental reason. Anjeso's onset time is significantly delayed than expected for intravenous injection, and there is inadequate late-stage analgesia [19]. This is exactly the consequence of insufficient pharmacokinetic studies. Therefore, investigating the *in vivo* fate of nanocrystals is essential—not only for building a theoretical framework for their behavior in the body but also for guiding future clinical applications.

Due to the coexistence of intact nanocrystals and free drugs, the pharmacokinetics of nanocrystals may be subdivided into two parts, the kinetics of the intact nanocrystals (i.e., the particokinetics) and the free drugs. Separating dissolved drug molecules from intact nanocrystals is the crux of the pharmacokinetic study. Separation techniques like equilibrium dialysis, ultracentrifugation, ultrafiltration, and solid-phase extraction have been widely applied to carrier-based nanomedicines [20–23]. These techniques are not suitable for nanocrystals, as the process inadvertently accelerates the dissolution of nanocrystals. Recent innovations, such as transmission electron microscopy (TEM) and hybrid nanocrystal technology, have introduced new tools for analyzing nanoparticle behavior. TEM offers direct visual evidence of nanocrystal transport within the body, while hybrid nanocrystals combine therapeutic and imaging functions by embedding imaging agents within the crystal structure. Despite these advancements, fluorescence signals from dissolved nanocrystals still interfere with precise tracking. The latest generation of hybrid nanocrystals addresses this challenge by incorporating environment-sensitive probes, enabling accurate monitoring of nanocrystal behavior and even the measurement of particokinetics.

This chapter focused on the research progress of particokinetics and biofate of nanocrystals. Therefore, the chapter started with a detailed summary of the strategies for exploring the biofate of nanocrystals. The advantages and limitations of each strategy were discussed. Then advances in the particokinetics of hybrid nanocrystals via intravenous and oral routes were introduced. Based on the particokinetics, the contribution of intact nanocrystals to the overall oral bioavailability could be evaluated. Subsequently, the biofate of nanocrystals following oral administration, intravenous, ocular, and transdermal routes was updated via the latest findings based on the hybrid nanocrystal technique. Finally, the measurement of *in vivo* dissolution of nanocrystals was introduced.

7.2 Strategies for Exploring the *In Vivo* Fate of Nanocrystals

This section introduces the strategies used to explore the biofate of nanocrystals. The transmission electron microscopy (TEM) allows observation of nanocrystal morphology in organ tissues; the PBPK models are used to predict the pharmacokinetic parameters and tissue distribution characteristics of nanocrystals; the autofluorescent drugs enable direct visualization of distribution in biological tissues; the hybrid nanocrystal technique endows non-fluorescent drug nanocrystals with fluorescence, enabling qualitative and quantitative investigation of biofate of nanocrystals. This section discusses the advantages and limitations of these strategies in detail.

7.2.1 Transmission Electron Microscopy (TEM)

TEM is a microscopic technique that forms high-resolution images by transmitting an electron beam through a sample. Unlike optical microscopes that use photons, TEM employs a high-speed electron beam; the short wavelength of high-energy electrons allows TEM to achieve resolution far superior to that of optical microscopy. TEM enables the observation of nanocrystal morphology within biological tissue sections, as the crystalline structure is histologically visible. As shown in Fig. 7.1a, after intravenous administration of itraconazole nanocrystals, crystallinity was observed in the macrophages of the spleen in rats using TEM [24]. Similarly, Fig. 7.1b shows that crystals were detected in the mesenteric lymph fluid after the delivery of nimodipine nanocrystals through duodenal injection [25]. Analysis of crystalline morphology in these two biological tissue samples suggests that nanocrystals *in vivo* dissolve at a slow rate, enabling their intact transport. This finding contradicts the earlier belief that nanocrystals dissolve rapidly *in vivo*. However, despite its high resolution, TEM has several limitations. First, it cannot

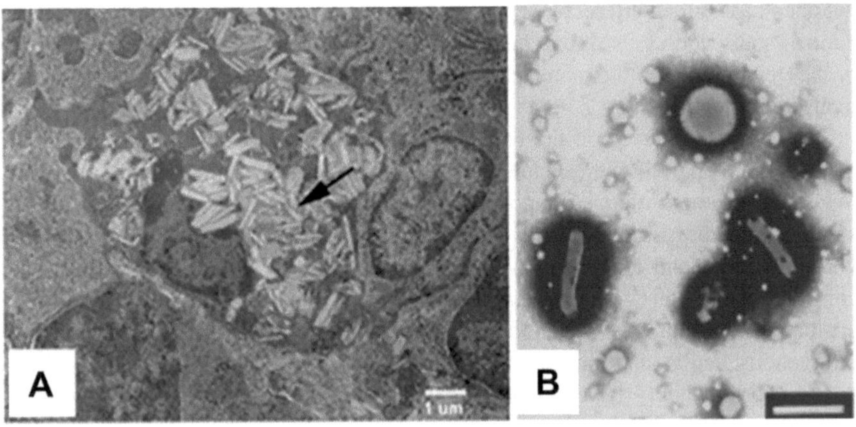

Fig. 7.1 TEM images reveal itraconazole nanocrystals in the spleen and nimodipine nanocrystals in the mesenteric lymphatic fluid. (Reproduced with copyright permission from Refs. [24, 25])

conduct whole-body monitoring and is limited to post-section observation. Second, TEM provides static data, making it impossible to dynamically monitor the *in vivo* kinetics of nanocrystals. Lastly, dehydration in the sample preparation process may lead to the re-crystallization of drug molecules, complicating the accurate assessment of their original state.

7.2.2 PBPK Model

Physiologically Based Pharmacokinetic (PBPK) models offer a numerical framework for describing the absorption, distribution, metabolism, and excretion (ADME) processes of drugs. PBPK models simulate the dynamic behavior of drugs in the body by dividing it into multiple physiologically relevant organ or tissue compartments, thereby predicting the distribution of drug concentrations across different organs over time. Compared to traditional pharmacokinetic models (such as one-compartment or two-compartment models), PBPK models are more detailed and can better reflect individual physiological differences in drug pharmacokinetics.

Creating PBPK models relies on thoroughly comprehending the ADME mechanisms, the interactions between nanocrystals and biological tissues, and the kinetic parameters that govern their transfer across biological barriers. These models can predict the pharmacokinetics and *in vivo* fate of various nanocarriers. For example, a PBPK model was built to quantitatively predict the *in vivo* behavior of intravenously injected SNX-2112 nanocrystals [26]. Figure 7.2a illustrates the non-particulate PBPK model, while Fig. 7.2b presents the version that integrates nanoparticles. A non-particulate PBPK model was initially developed to predict drug distribution *in vivo*, serving as a basis for subsequent development of particulate PBPK model. The model involved compartmentalizing the major physiological organs while grouping other tissues into a "remainder" compartment. Pharmacokinetic and tissue distribution experiments were conducted for the free drug, and the tissue-to-plasma partition coefficient was derived from the AUC ratio of tissue to plasma, representing the drug's distribution across various organs. Each organ's drug concentration was assessed by integrating the plasma flow rate and organ volume. The hepatic apparent clearance rate was determined by adjusting the PBPK model to fit experimental data, as the liver is the primary organ responsible for drug elimination. This rate characterized the elimination process. Using mass balance equations, the drug concentration in each organ over time was modeled, completing the construction of the non-particulate PBPK model. Then, particulate-form drugs were added to the plasma, liver, and spleen compartments to construct the particulate PBPK model. The model incorporated a first-order rate constant to describe nanocrystal dissolution, with the liver and spleen uptake rates for particulate-form drugs denoted as Upli and Upsp, respectively. Drug distribution, dissolution, and organ uptake were described using mass balance equations, finalizing the particle-based PBPK model. The pharmacokinetic and tissue distribution curves predicted by the PBPK model show a high degree of overlap with experimental data, demonstrating the model's ability to predict the pharmacokinetics

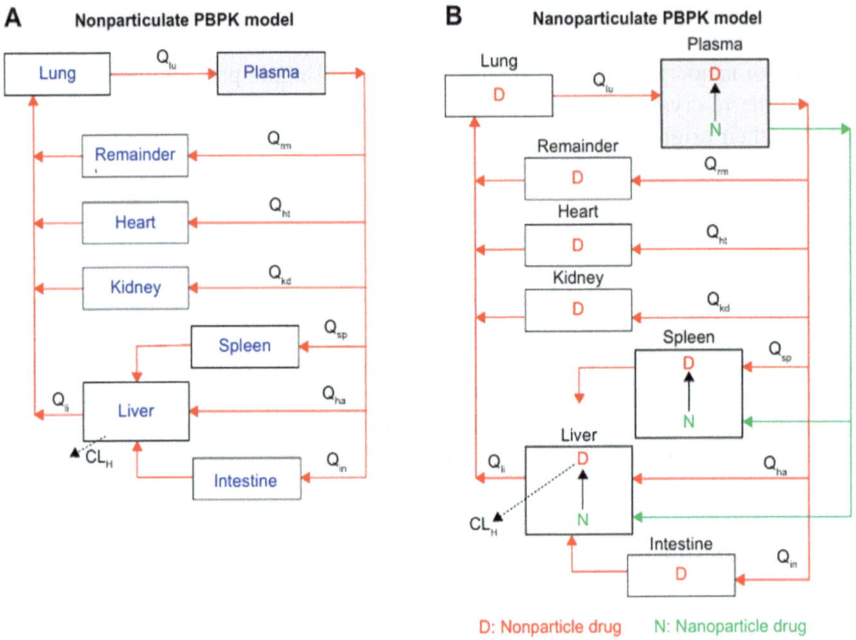

Fig. 7.2 The nonparticulate PBPK model (**a**) and nanoparticulate PBPK model (**b**) of SNX-2112 nanocrystals. (Reproduced with copyright permission from Ref. [26])

and tissue distribution of SNX-2112 nanocrystals effectively. This close alignment indicates that the PBPK model provides a viable approach for gaining deeper insights into nanocrystals' biofate, helping to optimize their use in clinical settings.

Although PBPK models sometimes show good agreement between predicted *in vivo* fate and experimental results, the numerous factors influencing *in vivo* fate such as the formation of protein coronas, uncertainties in cellular uptake mechanisms, and the varying accumulation and clearance in different organs—make PBPK model development highly dependent on experimental data. Relying solely on computational simulations of physiological conditions to develop PBPK models for predicting the *in vivo* fate of nanocrystals is still unrealistic.

7.2.3 Self-Fluorescent Drugs

Fluorescent label is commonly used to achieve dynamic tracking of *in vivo* transfer of nanocarriers. Traditional labeling methods via chemical conjugation do not apply to nanocrystals due to the lack of carrier materials. However, self-fluorescent drugs provide a solution. Curcumin (CUR), a self-fluorescent drug, was adopted to observe the skin penetration of nanocrystals [27]. CUR nanocrystals exhibited better permeability compared to the raw drug. Through comparison of images from optical microscopy and fluorescence microscopy, CUR nanocrystals accumulated more in skin appendages. Similarly, coumarin-6 was used to study the uptake and transport mechanisms of nanocrystals following oral administration. Since zebrafish embryos and larvae were adopted as the model, the *in vivo* transport and distribution of nanocrystals could be observed [28]. A degradable fluorescein diacetate has likewise been employed to investigate the ocular permeability of nanocrystals [29]. Regrettably, there are only a few drugs that exhibit self-fluorescence. Since different drugs exhibit distinct physicochemical characteristics and *in vivo* behaviors, the findings from the self-fluorescent drugs cannot be extrapolated to other nanocrystals. Furthermore, intact nanocrystals and their dissolved counterparts emit identical fluorescence, making it impossible to achieve self-discrimination of the nanocrystals.

7.2.4 Hybrid Nanocrystals

Inspired by guest-host encapsulation in solid-state chemistry, hybrid nanocrystals were developed by doping fluorophores into crystal lattices for biological imaging [30–34]. Figure 7.3 illustrates the structure of a hybrid nanocrystal [35]. In this setup, the drug functions as the host matrix, while the imaging compounds act as guests. Imaging agents may include fluorescent molecules, radioactive nuclides, or contrast metals, enabling precise monitoring [34]. In addition, hydrophilic polymers and ligands can be incorporated to extend circulation time and bind to specific receptors on target cells, respectively. The process doesn't change the physicochemical and crystallographic properties of the APIs, allowing the hybrid system to mimic the *in vivo* behavior of the original nanocrystals.

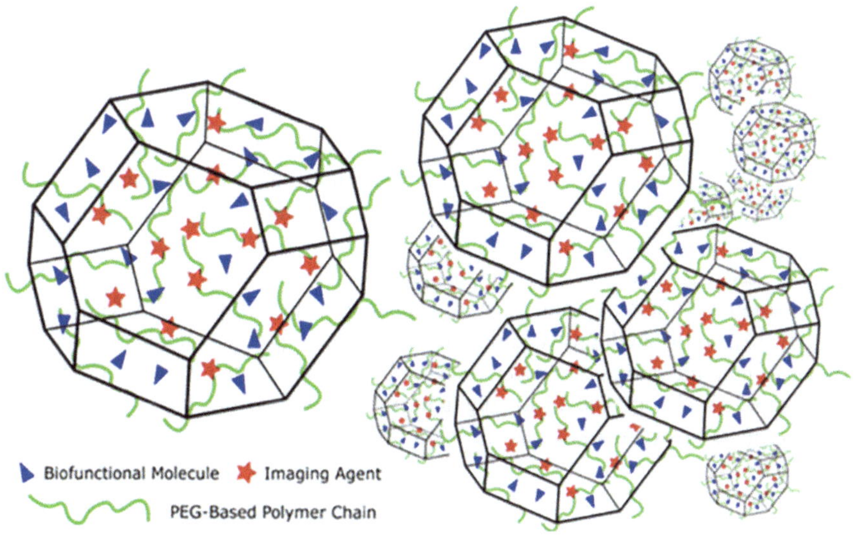

Fig. 7.3 The structural schematic of hybrid nanocrystals with imaging agents embedded in the lattice of the crystals. (Reproduced with copyright permission from Ref. [35])

Since their introduction in 2011, two distinct generations of hybrid nanocrystals have been created. The first generation utilized traditional fluorophores, either the hydrophilic fluorescein, rhodamine B, or the hydrophobic DiD. However, both embedded and released fluorophores contributed to the overall fluorescence signal, complicating the interpretation of the nanocrystals' real-time distribution. In addition, along with the dissolution of hybrid nanocrystals, an increasing number of fluorophores were released as free molecules, leading to fluorescence signals that primarily reflected these free probes rather than the actual distribution of the nanocrystals. This limitation spurred the advancement to a second generation of hybrid nanocrystals [36, 37], which incorporated environment-responsive fluorescent probes such as Förster Resonance Energy Transfer (FRET) pairs [38–40], aggregation-induced emission (AIE) [41, 42], and aggregation-caused quenching (ACQ) fluorophores [37, 43–48]. AIE probes dissipate energy through molecular stretching and rotation when in a dispersed state, resulting in low fluorescence. However, upon aggregation, restricted molecular motion leads to enhanced fluorescent signals. In contrast, ACQ probes can emit fluorescence when molecularly dispersed; while aggregation triggers intermolecular interactions like π-π stacking, causing energy transfer and fluorescence quenching. FRET probes consist of a donor and an acceptor, which have overlapping excitation and emission wavelengths. When the two molecules are in close proximity, the emitted light from the excited donor can be absorbed by the acceptor, illuminating the fluorescence of the acceptor. Conversely, when the FRET pairs separate far apart, they emit distinct fluorescence signals. These sophisticated probes provide fluorescence quenching or signal switching upon dissolution of the hybrid nanocrystals, facilitating the precise identification of intact nanocrystals and allowing for real-time tracking of their transport and distribution [49, 50].

7.3 Particokinetics of Nanocrystals

Pharmacokinetics is vital for understanding the ADME processes of APIs. However, nanocrystals fundamentally differ in composition from conventional drugs, thus presenting more complex *in vivo* processes. Following administration, the dissolved drug molecules and intact nanocrystals coexist in the body. A complete pharmacokinetics study for nanocrystals includes the kinetics of drug molecules and the particokinetics, which is essential to unravel nanocrystals' biofate fully. The requirement was incorporated into a guideline concerning preclinical pharmacokinetic studies of nanomedicines that was issued by the Chinese Center for Drug Evaluation on August 27, 2021. The following part will introduce the latest findings and advancements in particokinetics of nanocrystals based on hybrid nanocrystal technology.

7.3.1 Particokinetics Following *I.V.* Injection

ACQ-based hybrid nanocrystals were first applied to evaluate the particokinetics of CUR nanocrystals following *i.v.* injection [49]. The embedded ACQ probes enabled self-discrimination properties from the free fluorophores. Therefore, the fluorescent intensity correlated well with the quantity of the intact hybrid nanocrystals. The particokinetics were illustrated by the variation in the percentage of blood fluorescence normalized with the initial intensity as 100%. This approach resulted in a semi-quantitative curve that represents the time-dependent changes of the complete nanocrystal form (Fig. 7.4). The particokinetics curve of the CUR hybrid nanocrystals exhibited a typical "L"-shaped curve. Instead of being quickly dissolved, the CUR hybrid nanocrystals presented a long-circulation pattern as the ACQ signals could be observed till 48 h post-administration. The reason was attributed to the hydrophilic stabilizers adsorbed on the surfaces. Quercetin nanocrystals hybridized with ACQ probes exhibited similar behavior after intravenous injection, i.e., the "L"-shaped particokinetic curve [51]. Although the fluorescence intensity in the blood dropped more rapidly, 25% of the fluorescence was still present at 4 h, indicating the slow dissolution characteristics of the nanocrystals.

Comparison between particokinetics and pharmacokinetics may give a clearer understanding of the biofate of nanocrystals. The particokinetic curve of herpetrione hybrid nanocrystals was obtained through a similar method based on ACQ probes [52]. Compared to the pharmacokinetic curves of herpetrione, the particle kinetics exhibited a slower decline. The mean residence times (MRT) of particles were 11.61 ± 0.15 h and 12.87 ± 0.63 h for nanocrystals of 200 nm and 450 nm, respectively, while the corresponding MRTs of herpetrione were 2.53 ± 1.21 h and 3.18 ± 2.13 h, respectively. The residence time of the particles was much longer than that of the drug. The results also indicate that the nanocrystals do not dissolve quickly after injection; instead, a substantial portion remains intact. Interestingly, fluorescence intensity is still detectable at 48 h, which may be due to the return of nanocrystals from tissues into the bloodstream.

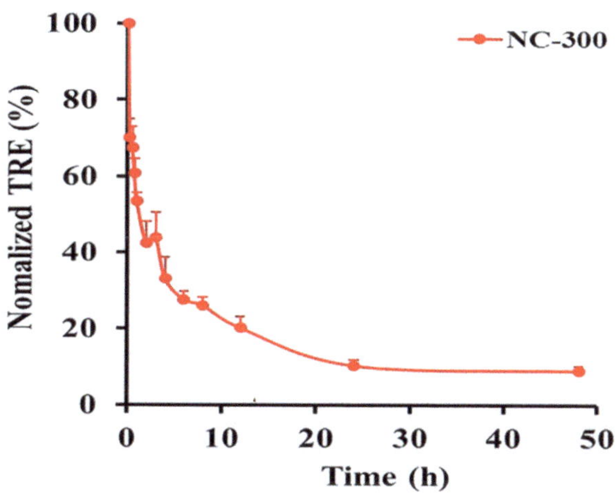

Fig. 7.4 The particokinetic curve of curcumin hybrid nanocrystals. (Reproduced with copyright permission from Ref. [49])

7.3.2 Particokinetics Following Oral Administration

Particokinetics of quercetin nanocrystals were measured via the ACQ-based hybrid nanocrystal technique, which was used to evaluate the contribution of nanocrystals to the overall oral bioavailability [51]. However, the absorbed hybrid nanocrystals predominantly accumulated in the liver after oral administration, leading to low fluorescent intensity in the blood. The variation of the fluorescent intensities in the liver was adopted to represent the particokinetics. The authors attempted to calculate the contribution of particles to bioavailability by using the absolute exposure of particles and the total drug in the liver. They obtained the fluorescence vs. time curves in the liver following oral administration and intravenous injection, respectively. The ratio of the AUC of the two curves represents the absolute exposure of particles in the liver. Similarly, by comparing the AUC of the total drug in the liver after oral administration and intravenous injection, the absolute exposure of the total drug can be determined. Comparing the absolute exposure of particles to that of the total drug gives the contribution of particles to bioavailability. The contribution of particles to oral bioavailability ranged from 40.27% to 50.65%. The findings of this research offer important perspectives on the roles of intact and dissolved forms of nanocrystals in oral bioavailability, enhancing our understanding of the significance of intact nanocrystals in improving overall bioavailability.

7.4 Biofate of Nanocrystals

The *in vitro* dissolution test always leads to the conclusion that nanocrystals dissolve quickly. It is thus estimated that nanocrystals get rapid dissolution at the administration site. The resulting high concentration gradient drives drug absorption. However, the *in vivo* conditions are quite different from that used in the *in vitro* dissolution. The *in vitro* dissolution provides sink conditions and sufficient agitation. The actual *in vivo* behavior of nanocrystals may deviate from the assumptions based on the *in vitro* study. The following section will introduce the latest findings on the biofate of nanocrystals from the hybrid nanocrystal technique.

7.4.1 Oral Route

ACQ-based hybrid nanocrystal technique was adopted to study the biofate of quercetin and cyclosporin A nanocrystals following oral administration [37, 53, 54]. The signals of both quercetin and cyclosporin A hybrid nanocrystals were observed in the gastrointestinal tract for 12–16 h post-administration. The results imply that orally administered nanocrystals may stay undissolved in the gastrointestinal tract for a prolonged time instead of rapid dissolution. Figure 7.5 presents fluorescence intensity from various gastrointestinal segments following oral administration of quercetin hybrid nanocrystals with different sizes (280, 550, and 1100 nm). The signal in the stomach remained consistent during the initial hour

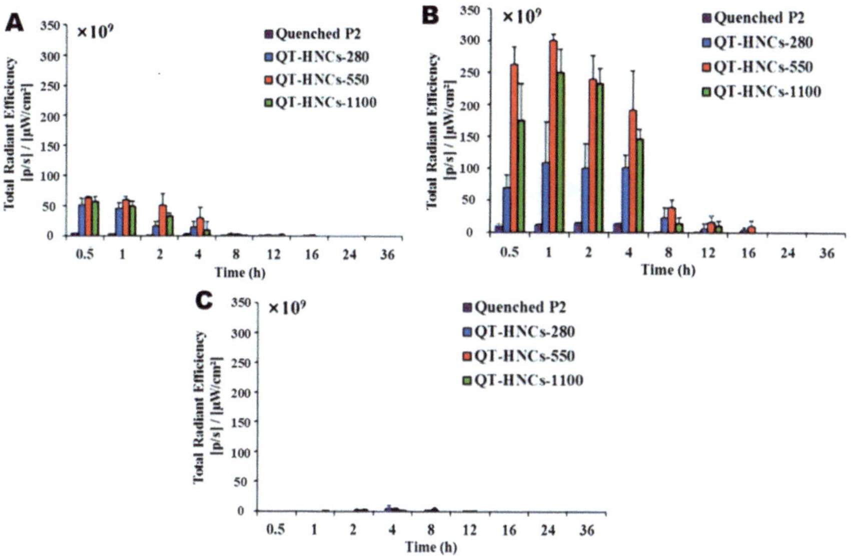

Fig. 7.5 Quantitative data of total fluorescence from ex vivo stomach (**a**), small intestine (**b**), and colon (**c**) of rats treated with quercetin hybrid nanocrystals and quenched P2 (the ACQ probe) solution, respectively. QT-HNC denotes quercetin hybrid nanocrystals, while the numbers in the bracket indicate particle size in nanometers. (Reproduced with copyright permission from Ref. [53])

following administration but vanished at 4 h (Fig. 7.5a). This pattern reflects the process of gastric emptying and nanocrystal dissolution, offering valuable insight into the transit and dissolution of nanocrystals through the gastrointestinal tract and their *in vivo* behavior. The fluorescence signal peaked within 1–2 h in the small intestine, being consistent with gastric emptying characteristics (Fig. 7.5b). However, the fluorescent intensity from 280 nm hybrid nanocrystals was significantly weaker than that from 550 nm and 1100 nm ones, indicating that smaller nanocrystals dissolve more quickly than the bigger ones. Signals in the colon peaked at around 4 h, while the intensities were significantly lower than that in the upper gastrointestinal tract due to the dissolution and absorption of the nanocrystals along transportation (Fig. 7.5c).

The prolonged retention in the gastrointestinal tract allows for extended interaction between the hybrid nanocrystals and the intestinal epithelia. ACQ signals have been detected in key organs including the liver, lung, spleen, and kidney following oral administration of both quercetin and cyclosporine A hybrid nanocrystals (Fig. 7.6) [37, 53, 54], suggesting that intact nanocrystals may be absorbed through the intestinal epithelia and distribute throughout the body. CLSM images of cryostat sections of the jejunum after in situ perfusion with the hybrid nanocrystals to rats confirmed the trans-epithelial transport of the intact hybrid nanocrystals [37, 53, 54]. However, the contribution of the absorbed nanocrystals to the overall

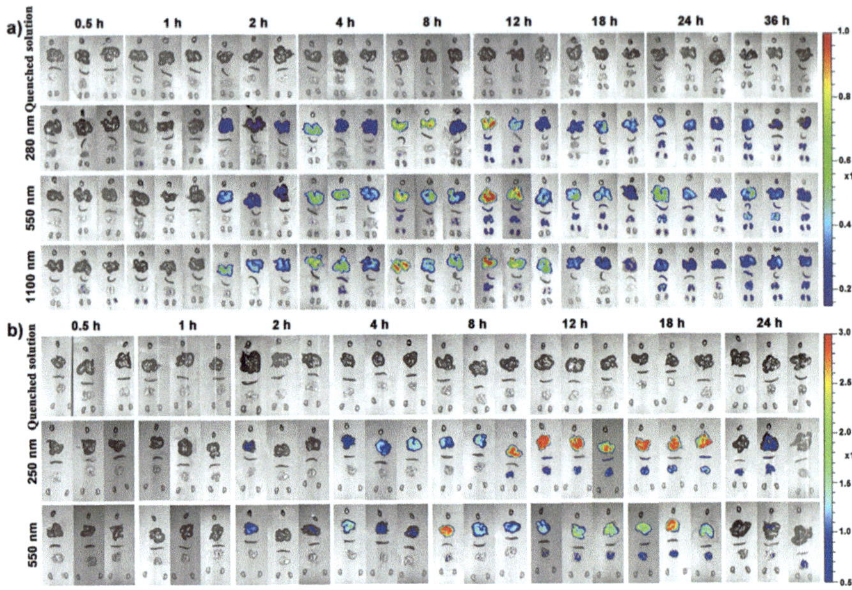

Fig. 7.6 Ex vivo imaging for rats after gavage administration of (**a**) quercetin and (**b**) cyclosporine A hybrid nanocrystals of different sizes. (Reproduced with copyright permission from Refs. [37, 53, 54])

bioavailability may be limited. It is estimated that approximately 3% of quercetin hybrid nanocrystals were absorbed in their intact form, while the majority dissolved in the gastrointestinal tract and were absorbed as free molecules.

MDCK (Madin-Darby canine kidney cells) and Caco-2 cell models were adopted to study the mechanisms of transcellular transport of nanocrystals. MDCK II cells are epithelial cells derived from canine kidney tissue. Due to their good permeability, polarization properties similar to intestinal epithelial cells, and the ease with which they form monolayers, they have the potential to be used in studies of nanocrystal permeability through monolayer cell systems. Studies on the transcellular transport of coumarin six nanocrystals across MDCK II monolayers have identified caveolae-mediated endocytosis as the primary uptake mechanism. The release of nanocrystals from cells occurs through the pathways involving the endoplasmic reticulum to Golgi apparatus and from the Golgi apparatus to the plasma membrane [28]. Likewise, saquinavir nanocrystals were found to accumulate in the cytoplasm of Caco-2 cells following a 2-h co-cultivation duration. The smaller nanocrystals exhibited higher intracellular concentrations than the larger ones. This suggests that nanocrystals are absorbed by cells in their intact form, while the smaller size enhances cellular uptake [55].

There is clear evidence that the biofate of orally administered nanocrystals displays dual characteristics: part of the nanocrystals dissolve in the gastrointestinal tract, while another part is absorbed in its intact form through the intestinal epithelium. This revised understanding holds significant implications for drug delivery, including extended-release, lymphatic transport, cell targeting, and enhanced bioavailability. Upcoming studies need to emphasize quantifying the relative contributions of dissolution and intact absorption on overall drug bioavailability. Additionally, it is important to investigate the elements that affect nanocrystals' behavior in the gastrointestinal tract, including particle size, surface characteristics, and physiological conditions, to optimize oral nanocrystal formulations.

7.4.2 Intravenous Injection

Herpetrione hybrid nanocrystals showed significant accumulation in the liver and lungs, with a small amount of fluorescence in the spleen, following intravenous administration [52]. The total fluorescence in the liver and lungs accounted for 85% of the overall fluorescence. As indicated in the particokinetics, the herpetrione hybrid nanocrystals presented a long-circulating pattern like other carrier-based nanoparticles. Therefore, they were taken up and accumulated in reticuloendothelial (RES) organs like the liver, the lung, and the spleen. Even at 36 h post-administration, fluorescence remained in the lungs, suggesting that the dissolution of the nanocrystals was not as rapid as expected. The ex vivo fluorescence quantification data of quercetin hybrid nanocrystals showed a similar distribution pattern, with cumulative fluorescence accumulation in the liver and lungs remaining the highest [51]. Besides, the fluorescence intensities in the RES organs exhibited first increasing and then decreasing trend for the two hybrid nanocrystals [51, 52], being due to the accumulation and dissolution of the hybrid nanocrystals.

7.4.3 Ocular Administration

Traditional ocular drug delivery faces significant challenges due to multiple barriers in the eye and the flushing effect of tear fluid, often resulting in a bioavailability of less than 5%. Nanocrystals, due to their extremely small size, may adhere to the ocular surface, extending their retention time. Meanwhile, the dissolution of nanocrystals may provide a remarkably high concentration gradient, allowing more drugs to cross the cornea by diffusion. Experiments with fluorescein diacetate nanocrystals demonstrated intense fluorescence signals in the corneal stroma, suggesting that the nanocrystals can be absorbed intact by corneal epithelial tissue, thereby enhancing their absorption in the eye [29]. ACQ probes embedded coumarin 6 nanocrystals were prepared to study the ocular fate of nanocrystals [56]. Observed from fluorescence confocal microscopy, the ACQ signals that represent the intact hybrid nanocrystals were observed in the cornea but absent in the lens. The results indicate that the intact coumarin 6 nanocrystals could permeate into but not across the cornea. Conversely, the fluorescence of coumarin 6 was observed in both the cornea and the lens, indicating the dissolution of the nanocrystals in the cornea and the diffusion of coumarin 6 molecules to the lens. Therefore, although intact nanocrystals can be absorbed by the cornea, they often do not penetrate the barrier effectively.

7.4.4 Transdermal Administration

CUR nanocrystals hybridized with ACQ probes were used to explore the transdermal delivery mechanism of nanocrystals [57]. The traditional view suggests that nanocrystals enhance skin penetration by dissolving rapidly on the skin surface to create a high concentration gradient. However, this study found that even the propylene glycol solution of CUR could not effectively permeate into the skin due to poor permeability. Conversely, the nanocrystals enhanced the skin permeation of CUR. Based on the co-localization analysis of the CUR and the ACQ signals by confocal microscopy, the hybrid nanocrystals could permeate into the stratum corneum and accumulate in the hair follicles, where they dissolved and the CUR molecules diffused into the skin. Moreover, the thickness of the stratum corneum in hair follicles decreases with the depth of the skin, making it easier for CUR in deeper follicles to diffuse into the perifollicular skins.

7.5 *In Vivo* Dissolution

The *in vitro* dissolution is crucial for preparation development by providing information assessing the preparation's performance, particularly for developing generic drugs [49]. However, current methods for setting *in vitro* dissolution conditions are often based on empirical data and experimental observations, which may not truly represent the real *in vivo* dissolution conditions. Only the conditions verified with IVIVC (in vitro-in vivo correlation) get the predictability and reliability. However, the establishment of IVIVC calls for *in vivo* dissolution data [58]. Due to the challenges in measuring the *in vivo* dissolution, an alternative approach is based on the relationship between *in vitro* dissolution and the absorption fractions that are deconvoluted from plasma concentrations. The absorption fractions-based IVIVC always led to bias, especially for Class III and IV drugs under Biopharmaceutics Classification System [59]. The *in vivo* dissolution is thus of great significance for drug formulation development, quality control, and clinical application, thereby accelerating the research and commercialization of new drugs.

The rationality of ACQ probe-based hybrid crystals for the measurement of *in vivo* dissolution was proved with fenofibrate as a model [60]. The hybrid crystals got a good linear correlation between the fluorescence and the residual crystal amount during dissolution, laying a foundation for quantifying crystals with fluorescence intensity. The *in vitro* dissolution profiles acquired by fluorescent quenching correlated well with the routine one from measurements of fenofibrate. The *in vivo* dissolution of the hybrid crystals was obtained by normalizing the quenched fluorescence intensities that were recorded through live imaging. The *in vivo* dissolution curve matched the *in vitro* one with a level A point-to-point correlation, validating the accuracy of the ACQ probe-based quantification. Besides, a good correlation was also obtained between the absorption fraction and the *in vivo* dissolution, further confirming the accuracy of this method. The *in vivo* dissolution could guide the establishment of *in vitro* dissolution methods. The *in vivo* dissolution profile of aripiprazole was obtained via the ACQ probe-based hybrid crystals following the process established on fenofibrate hybrid crystals [50]. The accuracy of the *in vivo* dissolution profile was validated by a level A point-to-point correlation with the absorption fraction. Then, the *in vitro* dissolution protocol was set up via a similarity factor larger than 50 in comparison with the *in vivo* dissolution profile.

7.6 Conclusion

Nanocrystal formulations present an effective approach for improving the druggability and bioavailability of poorly soluble drugs. Besides oral and intravenous administration, nanocrystals show considerable promise in transdermal, ocular, and pulmonary drug delivery. With the emergence of poorly soluble APIs, the need for nanocrystals will continue to rise.

The clinical translation of nanocrystals is hindered by a lack of comprehensive understanding of their fate, especially the pharmacokinetics and the *in vivo* transport. Traditional methodologies are deficiency in differentiating intact nanocrystals from the dissolved molecules, leading to inaccurate assessments. Techniques like TEM, hybrid nanocrystals, and PBPK models offer novel tools for studying biofate of nanocrystals, and the results obtained are renewing people's understanding of nanocrystals. Each technique has its limitations, necessitating further refinement for more precise analysis. The environment-sensitive probe-based hybrid nanocrystals have shown promise in real-time monitoring and quantification of *in vivo* dissolution and distribution of nanocrystals, particularly the ACQ probes. Instead of rapid dissolution at the administration site, nanocrystals may remain intact for a prolonged time *in vivo*. Consequently, nanocrystals get chance to transport across the intestinal epithelia, circulate in the bloodstream and accumulate in RES organs, and accumulate in hair follicles depending on the administration routes. The fluorescence of the ACQ-based hybrid nanocrystals correlates linearly with their amount, enabling semi-quantification of the particokinetics, the *in vivo* distribution, and the *in vivo* dissolution of nanocrystals via fluorescence intensity. The quantitative data provides deeper insights into the biofate of nanocrystals. Besides the intact nanocrystals, the pharmacokinetics and biodistribution of free drugs are also critical to unravel the biofate of nanocrystals. Due to the coexistence of intact nanocrystals and free drugs *in vivo*, analyzing the dynamic changes between these two forms can fully reveal the biofate of nanocrystals. Furthermore, the factors influencing the biofate of nanocrystals can be disclosed, which optimizes the formulation design.

Conflicts of Interest The authors have no conflicts of interest to declare.

References

1. Patel VF, Liu F, Brown MB (2011) Advances in oral transmucosal drug delivery. J Control Release 153(2):106–116. https://doi.org/10.1016/j.jconrel.2011.01.027
2. Kumar R, Islam T, Nurunnabi M (2022) Mucoadhesive carriers for oral drug delivery. J Control Release 351:504–559. https://doi.org/10.1016/j.jconrel.2022.09.024
3. Vithani K, Jannin V, Pouton CW, Boyd BJ (2019) Colloidal aspects of dispersion and digestion of self-dispersing lipid-based formulations for poorly water-soluble drugs. Adv Drug Deliv Rev 142:16–34. https://doi.org/10.1016/j.addr.2019.01.008
4. Davis M, Walker G (2018) Recent strategies in spray drying for the enhanced bioavailability of poorly water-soluble drugs. J Control Release 269:110–127. https://doi.org/10.1016/j.jconrel.2017.11.005
5. Salunke S, O'Brien F, Cheng Thiam Tan D, Harris D, Math MC, Ariën T et al (2022) Oral drug delivery strategies for development of poorly water soluble drugs in paediatric patient population. Adv Drug Deliv Rev 190:114507. https://doi.org/10.1016/j.addr.2022.114507
6. Salim M, Eason T, Boyd BJ (2022) Opportunities for milk and milk-related systems as 'new' low-cost excipient drug delivery materials. Adv Drug Deliv Rev 183:114139. https://doi.org/10.1016/j.addr.2022.114139
7. Luo Z, Klein Cerrejon D, Römer S, Zoratto N, Leroux JC (2023) Boosting systemic absorption of peptides with a bioinspired buccal-stretching patch. Sci Transl Med 15(715):eabq1887. https://doi.org/10.1126/scitranslmed.abq1887
8. Han W, Liu F, Li Y, Liu G, Li H, Xu Y et al (2023) Advances in natural polymer-based transdermal drug delivery systems for tumor therapy. Small 19(35):e2301670. https://doi.org/10.1002/smll.202301670
9. Okeyo PO, Rajendran ST, Zór K, Boisen A (2021) Sensing technologies and experimental platforms for the characterization of advanced oral drug delivery systems. Adv Drug Deliv Rev 176:113850. https://doi.org/10.1016/j.addr.2021.113850
10. Thompson SA, Williams RO 3rd. (2021) Specific mechanical energy – an essential parameter in the processing of amorphous solid dispersions. Adv Drug Deliv Rev 173:374–393. https://doi.org/10.1016/j.addr.2021.03.006
11. Nobili S, Micheli L, Lucarini E, Toti A, Ghelardini C, Di Cesare ML (2024) Ultramicronized N-palmitoylethanolamine associated with analgesics: effects against persistent pain. Pharmacol Ther 258:108649. https://doi.org/10.1016/j.pharmthera.2024.108649
12. Kankala RK, Xu PY, Chen BQ, Wang SB, Chen AZ (2021) Supercritical fluid (SCF)-assisted fabrication of carrier-free drugs: an eco-friendly welcome to active pharmaceutical ingredients (APIs). Adv Drug Deliv Rev 176:113846. https://doi.org/10.1016/j.addr.2021.113846
13. Lu Y, Li Y, Wu W (2016) Injected nanocrystals for targeted drug delivery. Acta Pharm Sin B 6(2):106–113. https://doi.org/10.1016/j.apsb.2015.11.005
14. DuBose JT, Kamat PV (2022) Energy versus electron transfer: managing excited-state interactions in perovskite nanocrystal-molecular hybrids. Chem Rev 122(15):12475–12494. https://doi.org/10.1021/acs.chemrev.2c00172
15. Qi Y, Zhang T, Jing C, Liu S, Zhang C, Alvarez PJJ et al (2020) Nanocrystal facet modulation to enhance transferrin binding and cellular delivery. Nat Commun 11(1):1262. https://doi.org/10.1038/s41467-020-14972-z
16. Junghanns JU, Müller RH (2008) Nanocrystal technology, drug delivery and clinical applications. Int J Nanomedicine 3(3):295–309. https://doi.org/10.2147/ijn.s595
17. (2015) Ryanodex—a new dantrolene formulation for malignant hyperthermia. Med Lett Drugs Ther 57(1472):100
18. Patravale VB, Date AA, Kulkarni RM (2004) Nanosuspensions: a promising drug delivery strategy. J Pharm Pharmacol 56(7):827–840. https://doi.org/10.1211/0022357023691
19. FDA (2020) New Drug Application (NDA): 210583. https://www.accessdata.fda.gov/scripts/cder/daf/index.cfm?event=overview.process&ApplNo=210583. Accessed 10 Oct 2024

20. Wang L, Zhang W, Shao Y, Zhang D, Guo G, Wang X (2022) Analytical methods for obtaining binding parameters of drug-protein interactions: a review. Anal Chim Acta 1219:340012. https://doi.org/10.1016/j.aca.2022.340012
21. Xie Y, Shao N, Jin Y, Zhang L, Jiang H, Xiong N et al (2018) Determination of non-liposomal and liposomal doxorubicin in plasma by LC-MS/MS coupled with an effective solid phase extraction: in comparison with ultrafiltration technique and application to a pharmacokinetic study. J Chromatogr B Analyt Technol Biomed Life Sci 1072:149–160. https://doi.org/10.1016/j.jchromb.2017.11.020
22. Wang H, Zheng M, Gao J, Wang J, Zhang Q, Fawcett JP et al (2020) Uptake and release profiles of PEGylated liposomal doxorubicin nanoparticles: a comprehensive picture based on separate determination of encapsulated and total drug concentrations in tissues of tumor-bearing mice. Talanta 208:120358. https://doi.org/10.1016/j.talanta.2019.120358
23. Song W, Tweed JA, Visswanathan R, Saunders JP, Gu Z, Holliman CL (2019) Bioanalysis of targeted nanoparticles in monkey plasma via LC-MS/MS. Anal Chem 91(21):13874–13882. https://doi.org/10.1021/acs.analchem.9b03367
24. Rabinow B, Kipp J, Papadopoulos P, Wong J, Glosson J, Gass J et al (2007) Itraconazole IV nanosuspension enhances efficacy through altered pharmacokinetics in the rat. Int J Pharm 339(1–2):251–260. https://doi.org/10.1016/j.ijpharm.2007.02.030
25. Fu Q, Sun J, Ai X, Zhang P, Li M, Wang Y et al (2013) Nimodipine nanocrystals for oral bioavailability improvement: role of mesenteric lymph transport in the oral absorption. Int J Pharm 448(1):290–297. https://doi.org/10.1016/j.ijpharm.2013.01.065
26. Dong D, Wang X, Wang H, Zhang X, Wang Y, Wu B (2015) Elucidating the in vivo fate of nanocrystals using a physiologically based pharmacokinetic model: a case study with the anticancer agent SNX-2112. Int J Nanomedicine 10:2521–2535. https://doi.org/10.2147/ijn.S79734
27. Vidlářová L, Romero GB, Hanuš J, Štěpánek F, Müller RH (2016) Nanocrystals for dermal penetration enhancement – effect of concentration and underlying mechanisms using curcumin as model. Eur J Pharm Biopharm 104:216–225. https://doi.org/10.1016/j.ejpb.2016.05.004
28. Miao X, Li Y, Wang X, Lee SM, Zheng Y (2016) Transport mechanism of coumarin 6 nanocrystals with two particle sizes in MDCKII monolayer and larval zebrafish. ACS Appl Mater Interfaces 8(20):12620–12630. https://doi.org/10.1021/acsami.6b01680
29. Chan CP, Bruemmel Y, Seydack M, Sin KK, Wong LW, Merisko-Liversidge E et al (2004) Nanocrystal biolabels with releasable fluorophores for immunoassays. Anal Chem 76(13): 3638–3645. https://doi.org/10.1021/ac0353740
30. Zhao R, Hollis CP, Zhang H, Sun L, Gemeinhart RA, Li T (2011) Hybrid nanocrystals: achieving concurrent therapeutic and bioimaging functionalities toward solid tumors. Mol Pharm 8(5):1985–1991. https://doi.org/10.1021/mp200154k
31. Hollis CP, Weiss HL, Leggas M, Evers BM, Gemeinhart RA, Li T (2013) Biodistribution and bioimaging studies of hybrid paclitaxel nanocrystals: lessons learned of the EPR effect and image-guided drug delivery. J Control Release 172(1):12–21. https://doi.org/10.1016/j.jconrel.2013.06.039
32. Hollis CP, Weiss HL, Evers BM, Gemeinhart RA, Li T (2014) In vivo investigation of hybrid paclitaxel nanocrystals with dual fluorescent probes for cancer theranostics. Pharm Res 31(6): 1450–1459. https://doi.org/10.1007/s11095-013-1048-x
33. Chen Y, Li T (2015) Cellular uptake mechanism of paclitaxel nanocrystals determined by confocal imaging and kinetic measurement. AAPS J 17(5):1126–1134. https://doi.org/10.1208/s12248-015-9774-0
34. Hollis CP, Dozier AK, Knutson BL, Li T (2019) Preparation and characterization of multimodal hybrid organic and inorganic nanocrystals of camptothecin and gold. Acta Pharm Sin B 9(1): 128–134. https://doi.org/10.1016/j.apsb.2018.03.005
35. Lu Y, Lv Y, Li T (2019) Hybrid drug nanocrystals. Adv Drug Deliv Rev 143:115–133. https://doi.org/10.1016/j.addr.2019.06.006

36. Gao W, Lee D, Meng Z, Li T (2017) Exploring intracellular fate of drug nanocrystals with crystal-integrated and environment-sensitive fluorophores. J Control Release 267:214–222. https://doi.org/10.1016/j.jconrel.2017.08.031
37. Xie Y, Shi B, Xia F, Qi J, Dong X, Zhao W et al (2018) Epithelia transmembrane transport of orally administered ultrafine drug particles evidenced by environment sensitive fluorophores in cellular and animal studies. J Control Release 270:65–75. https://doi.org/10.1016/j.jconrel.2017.11.046
38. Algar WR, Hildebrandt N, Vogel SS, Medintz IL (2019) FRET as a biomolecular research tool – understanding its potential while avoiding pitfalls. Nat Methods 16(9):815–829. https://doi.org/10.1038/s41592-019-0530-8
39. Wu L, Huang C, Emery BP, Sedgwick AC, Bull SD, He XP et al (2020) Förster resonance energy transfer (FRET)-based small-molecule sensors and imaging agents. Chem Soc Rev 49(15):5110–5139. https://doi.org/10.1039/c9cs00318e
40. Fang B, Shen Y, Peng B, Bai H, Wang L, Zhang J et al (2022) Small-molecule quenchers for Förster resonance energy transfer: structure, mechanism, and applications. Angew Chem Int Ed Engl 61(41):e202207188. https://doi.org/10.1002/anie.202207188
41. Liu Y, Chen X, Liu X, Guan W, Lu C (2023) Aggregation-induced emission-active micelles: synthesis, characterization, and applications. Chem Soc Rev 52(4):1456–1490. https://doi.org/10.1039/d2cs01021f
42. Zhou Y, Hua J, Ding D, Tang Y (2022) Interrogating amyloid aggregation with aggregation-induced emission fluorescence probes. Biomaterials 286:121605. https://doi.org/10.1016/j.biomaterials.2022.121605
43. Chen T, Li C, Li Y, Yi X, Lee SM, Zheng Y (2016) Oral delivery of a nanocrystal formulation of Schisantherin a with improved bioavailability and brain delivery for the treatment of Parkinson's disease. Mol Pharm 13(11):3864–3875. https://doi.org/10.1021/acs.molpharmaceut.6b00644
44. Hu X, Zhang J, Yu Z, Xie Y, He H, Qi J et al (2015) Environment-responsive aza-BODIPY dyes quenching in water as potential probes to visualize the in vivo fate of lipid-based nanocarriers. Nanomedicine 11(8):1939–1948. https://doi.org/10.1016/j.nano.2015.06.013
45. Qi J, Hu X, Dong X, Lu Y, Lu H, Zhao W et al (2019) Towards more accurate bioimaging of drug nanocarriers: turning aggregation-caused quenching into a useful tool. Adv Drug Deliv Rev 143:206–225. https://doi.org/10.1016/j.addr.2019.05.009
46. He H, Wang L, Ma Y, Yang Y, Lv Y, Zhang Z et al (2020) The biological fate of orally administered mPEG-PDLLA polymeric micelles. J Control Release 327:725–736. https://doi.org/10.1016/j.jconrel.2020.09.024
47. He HS, Liu C, Ming J, Lv YJ, Qi JP, Lu Y et al (2022) Accurate and sensitive probing of onset of micellization based on absolute aggregation-caused quenching effect. Aggregate 3(5):e163. https://doi.org/10.1002/agt2.163
48. Cai YF, Ji X, Zhang YS, Liu C, Zhang ZC, Lv YJ et al (2023) Near-infrared fluorophores with absolute aggregation-caused quenching and negligible fluorescence re-illumination for in vivo bioimaging of nanocarriers. Aggregate 4(2):e277. https://doi.org/10.1002/agt2.277
49. Wang T, Qi J, Ding N, Dong X, Zhao W, Lu Y et al (2018) Tracking translocation of self-discriminating curcumin hybrid nanocrystals following intravenous delivery. Int J Pharm 546(1–2):10–19. https://doi.org/10.1016/j.ijpharm.2018.05.020
50. Yang Y, Lv Y, Shen C, Shi T, He H, Qi J et al (2021) In vivo dissolution of poorly water-soluble drugs: proof of concept based on fluorescence bioimaging. Acta Pharm Sin B 11(4):1056–1068. https://doi.org/10.1016/j.apsb.2020.08.002
51. Shen B, Shen C, Zhu W, Yuan H (2021) The contribution of absorption of integral nanocrystals to enhancement of oral bioavailability of quercetin. Acta Pharm Sin B 11(4):978–988. https://doi.org/10.1016/j.apsb.2021.02.015
52. Hang L, Shen C, Shen B, Yuan H (2022) Insight into the in vivo fate of intravenous herpetrione amorphous nanosuspensions by aggregation-caused quenching probes. Chin Chem Lett 33(11):4948–4951. https://doi.org/10.1016/j.cclet.2022.03.108

53. Shen C, Yang Y, Shen B, Xie Y, Qi J, Dong X et al (2017) Self-discriminating fluorescent hybrid nanocrystals: efficient and accurate tracking of translocation via oral delivery. Nanoscale 10(1):436–450. https://doi.org/10.1039/c7nr06052a
54. Wang K, Qi J, Weng T, Tian Z, Lu Y, Hu K et al (2014) Enhancement of oral bioavailability of cyclosporine a: comparison of various nanoscale drug-delivery systems. Int J Nanomedicine 9: 4991–4999. https://doi.org/10.2147/IJN.S72560
55. He Y, Xia DN, Li QX, Tao JS, Gan Y, Wang C (2015) Enhancement of cellular uptake, transport and oral absorption of protease inhibitor saquinavir by nanocrystal formulation. Acta Pharmacol Sin 36(9):1151–1160. https://doi.org/10.1038/aps.2015.53
56. Liu D, Wan B, Qi J, Dong X, Zhao W, Wu W et al (2018) Permeation into but not across the cornea: bioimaging of intact nanoemulsions and nanosuspensions using aggregation-caused quenching probes. Chin Chem Lett 29(12):1834–1838
57. Shi T, Lv Y, Huang W, Fang Z, Qi J, Chen Z et al (2020) Enhanced transdermal delivery of curcumin nanosuspensions: a mechanistic study based on co-localization of particle and drug signals. Int J Pharm 588:119737. https://doi.org/10.1016/j.ijpharm.2020.119737
58. FDA (1997) Guidance for industry extended release oral dosage forms: development, evaluation, and application of in vitro/in vivo correlations
59. Wang Y, Nedelman J (2002) Bias in the Wagner-Nelson estimate of the fraction of drug absorbed. Pharm Res 19(4):470–476
60. Man F, Yang Y, He H, Qi J, Wu W, Lu Y (2023) Establishment of in vitro dissolution based on similarity with in vivo dissolution: a case study on aripiprazole. Mol Pharm 20(5):2579–2588. https://doi.org/10.1021/acs.molpharmaceut.3c00014

Recent Breakthroughs in Nanocrystal Development

8

Ekta R. Pardhi, Ruchita Patil, Kuldeep Kumar Bansal, and Neelesh Kumar Mehra

Abstract

Nanocrystals are defined as materials comprising atoms arranged in single crystals or polycrystalline structures with dimensions below 100 nm. These nanoscale systems offer unique benefits compared to conventional therapeutic preparation, including high bioavailability, improved targeting selectivity, and controlled drug release, making them highly valuable in drug delivery applications. Their distinct physical and chemical characteristics enable the efficient delivery of diverse therapeutic agents, including genetic materials, anticancer drugs, and poorly water-soluble compounds, offering a broad spectrum of potential applications. Advancements in nanocrystal technology have continuously refined their development, leading to several promising formulations reaching clinical evaluation. Despite these advancements, challenges such as safety concerns and stability issues persist, emphasizing the need for further research in this area. This book chapter outlines a detailed exploration of recent progress in the field of nanocrystals, focusing on novel preparation strategies Intended to enhance the bioavailability of poorly soluble drugs, critical physicochemical properties, advanced characterization

E. R. Pardhi · R. Patil
Pharmaceutical Nanotechnology Research Laboratory, Department of Pharmaceutics, National Institute of Pharmaceutical Education and Research (NIPER), Hyderabad, Telangana, India

K. K. Bansal
Pharmaceutical Sciences Laboratory, Faculty of Science and Engineering Åbo Akademi University, Biocity, Turku, Finland
e-mail: kuldeep.bansal@abo.fi

N. K. Mehra (✉)
Department of Pharmaceutics, National Institute of Pharmaceutical Education and Research (NIPER), Hyderabad, Telangana, India
e-mail: neelesh@niperhyd.ac.in

N. K. Mehra, K. K. Bansal (eds.), *Nanocrystal Technologies in Biomedical Sciences*,
https://doi.org/10.1007/978-981-95-2968-1_8

methodologies, and innovations in drug administration routes. Additionally, the chapter explores the therapeutic potential of nanocrystals in various pharmaceutical applications, aiming to highlight their transformative role in modern drug delivery systems.

Keywords

Nanocrystals · Hybrid nanocrystals · Nano co-crystals · Route of administration · Microneedles · In-vivo imaging

8.1 Introduction

Nanocrystals (NCs) represent nanoparticle technology for drug delivery, offering distinct properties that link conventional formulations and nanomedicine-based products. This evolving field holds significant potential, with current market applications and emerging clinical prospects that could reshape the healthcare industry [1]. Look into cutting-edge drug delivery strategies to boost the solubility of drugs with limited solubility. The latest innovations and trends in this rapidly advancing area of research and development. Notable developments have been achieved in recent years in developing drug delivery systems. Recent findings from advanced bioimaging tools have broadened our understanding, suggesting that intact NCs may provide enhanced long-term stability in-vivo while offering drug delivery potential comparable to other nanocarriers [2]. Moreover, NCs are formulated without the use of carrier excipients, allowing them to reach a 100% theoretical drug loading capacity [3, 4].

It provides NCs with exceptional drug delivery efficiency in contrast to systems based on nanocarriers, minimizing the risk of carrier-related toxicities. Numerous NCs formulations are available commercially for oral and parenteral administration [5]. In the continued finding of poorly soluble drugs, NCs are expected to remain a key area of industrial and academic research interest [6]. In drug delivery, the progress of NCs technology is paving the way for more effective drug targeting. These tailored NCs act as key building blocks for flexible and multifunctional theranostic platforms, making it easier to develop precise and adaptable nanomedicines [7, 8]. This overview highlights where NCs technology stands today, including the market solutions currently in use and the exciting clinical opportunities that could reshape the field of pharmaceuticals. We also take a look at innovative tactics to boost the solubility of poorly soluble drugs while keeping up with the latest trends and advancements in this rapidly developing field.

There is a growing focus on combining therapeutic and diagnostic functions into a single vehicle, driving the progress of advanced nanomedicines. These cutting-edge formulations, called "theranostic," offer regulated drug release and active targeting [9]. Multifunctional NCs, or hybrid NCs, are an evolving area and show great promise in cancer therapy. Recently, a study showed that the intensity of fluorescence in hybrid NCs can be used for accurate analysis by correlating in-vitro dissolution data with fluorescence quenching, inspired by guest inclusion phenomena in solid-state chemistry [10]. NCs engineering is mainly considered a strategy to improve the dissolution of drug crystals; however, information about the in-vivo behavior of NCs remains restricted. Tracking NCs within the body is challenging because of their ongoing dissolution and the lack of effective detection methods. Hybrid NCs, which combine active pharmaceutical ingredients (APIs) with fluorescent dyes or diagnostic probes, offer the potential for tracking after deposition [11]. Similarly, pharmaceutical cocrystals (CCs), widely studied for improving solubility, represent a multicomponent solid form that integrates two or more molecular components, such as an API and a coformer. Within this category, nano-cocrystals (NCCs), a category of particles with sizes smaller than 1 nm, have gained significant attention for their enhanced performance and versatility [12]. This unique feature eliminates the complications associated with additional excipients, enabling simpler formulations. Furthermore, NCs maintain a neutral pH, exhibit long-term stability, and are readily scalable, making them a versatile and practical approach for delivering hydrophobic pharmacologically active compounds through diverse administration routes [13].

An innovative application of NCs involves their incorporation into microneedle-based drug delivery systems, representing a breakthrough in transdermal administration. By embedding NCs into dissolvable or biodegradable microneedles, this approach enables localized and controlled drug release while effectively bypassing the stratum corneum, which is the skin's primary protective barrier. The microneedles create microchannels that facilitate the administration of poorly soluble drugs in their NC form, significantly improving their dissolution and bioavailability. This hybrid strategy synergistically combines the advantages of nanotechnology with microneedle technology, offering benefits such as high drug loading, targeted delivery, and minimal invasiveness. As a result, NCs-incorporated microneedles are emerging as a promising platform for non-invasive and efficient drug administration [14, 15].

To address stability challenges and prevent uncontrolled drug release often observed with standalone NCs, novel formulation strategies have been explored. One such advancement, as highlighted by Liang et al., involves encapsulating NCs within liposomes, creating a "nano-in-nano" system. This approach leverages the protective and controlled release properties of liposomal carriers while retaining the benefits of NCs, thereby enhancing stability, preventing premature drug release, and optimizing therapeutic outcomes. This dual-system strategy exemplifies the potential of advanced nano-formulations in addressing the challenges of traditional NCs systems and opens new pathways for the development of robust and efficient drug delivery modalities [16]. This chapter offers an in-depth exploration of the latest progress in NCs formulations, with a detailed exploration of their design, preparation, and applications as innovative tools in contemporary drug delivery and diagnostics.

8.2 Diverse Formulation Approaches of NCs

8.2.1 Hybrid NCs

Hybrid NCs represent an integrated system with multifunctional capabilities in pharmaceuticals, depending on the components incorporated within the system. Hybrid NCs may consist of drugs, imaging agents, ligands, or biocompatible polymers [10] and are widely employed for enhancing stability, absorption, targeting, *in vivo* drug tracking [17], co-delivery [11], and other applications. Depending on their intended purpose, hybrid NCs may function as nanocrystal stabilizers [18], nanocrystal-bio imaging agents [19], or as vehicles for the concurrent delivery of other therapeutic agents [20]. These versatile systems offer significant advancements in both therapeutic and diagnostic fields.

The preparation of NCs typically follows three main approaches: top-down, bottom-up, or a combination of the two. The top-down approach involves size reduction techniques, such as media milling (Nanocrystal® technology) or high-pressure homogenization (HPH). The HPH method incorporates technologies like piston-gap homogenization in water (Dissocubes® technology), microfluidizer technology (IDD-P® technology), or in water blends and non-aqueous fluids (Nanopure® technology). In contrast, the bottom-up approach involves techniques such as Supercritical fluid technology, evaporative precipitation, high-gravity anti-solvent precipitation, sonoprecipitation, high-gravity controlled precipitation, antisolvent precipitation, solvent evaporation, and spray drying. Among these, the bottom-up approach is particularly suited for the preparation of HNCs as it facilitates the incorporation of fluorescent probes [21]. Importantly, the morphology, size, and crystallinity of NCs are minimally affected by the inclusion of fluorophores [10]. The combined approach leverages technologies like Nanoedge®, Smart Crystals®, and Nano Pure®, along with methods such as spray drying (H42), antisolvent precipitation (H69), and lyophilization (H96), followed by milling to produce hybrid NCs [21]. Hybrid NCs are developed for theranostic, which integrates therapeutic and diagnostic effects, and this system shows the tremendous potential to revolutionize cancer treatment. It offers instantaneous, non-intrusive tracking of the drug's tissue distribution. and its accumulation in tumors by incorporating a fluorescent probe in the crystal lattice.

For instance, innovative hybrid NCs of Paclitaxel have been prepared for theranostic applications, aimed at improving tumor targeting while minimizing toxicity. These hybrid NCs prepared using the antisolvent method, incorporate fluorescent probes such as MMPSense 750 and NIR FPR-648, which allow for non-invasive tracking of biological distribution and metastatic markers. *In vivo* studies demonstrated that MMPSense dyes were released in response to metalloproteinases (MMP-2, MMP-9), and a 20 mg/kg dose of the hybrid NCs showed comparable efficacy to Taxol in inhibiting MCF-7 cells, with reduced toxicity compared to solubilized formulations [22].

Aligned with this approach, Zhao et al. developed novel Paclitaxel hybrid NCs using an antisolvent method that integrates fluorescent molecules like FPR −749 fluorophore and dyes fluorescein and rhodamine B. The resultant regular, elongated

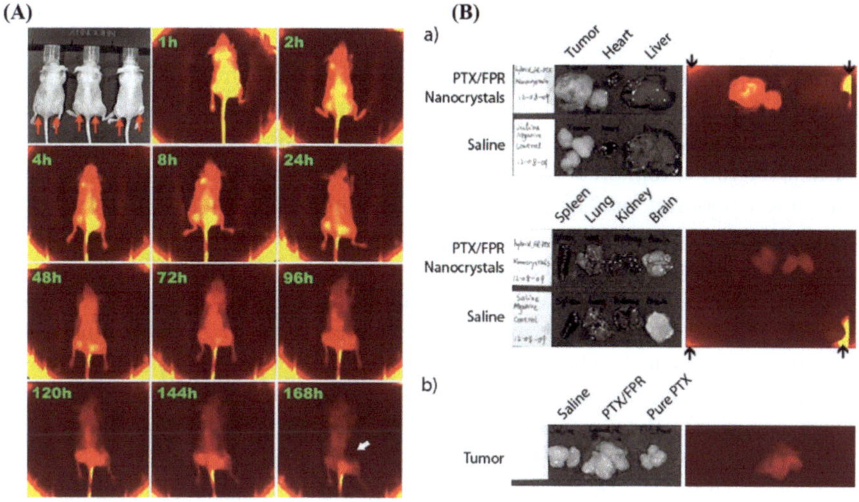

Fig. 8.1 (**a**) MCF-7 tumor-bearing mice were imaged using IVIS imaging. At various intervals following injections, the fluorescence pictures were captured at 780 nm using the same exposure length, F-stop, and binning. Pure drug NCs were administered to the left mouse, hybrid paclitaxel/FPR-749 NCs to the center mouse, and paclitaxel/Rhodamine B NCs to the right mouse. Excitation and emission filters caused optical interference, which resulted in the high-intensity rim. (**b**) Postmortem tumor and tissue samples from mice with tumors treated with pure and dye HNCs of paclitaxel (PTX) and saline as a negative control. (**a**) Evaluating the influence of saline and HNCs on key organs and malignancies. (**b**) Comparing tumors with subjected to various treatment. (Reprint with the permission from Shen et al. [24])

prismatic HNCs were studied on the MCF-7 cell line and female nude mice (20 mg/kg dose of paclitaxel). Specifically, the FPR-749 fluorophore was excited at 740 nm and emitted at 780 nm. Rhodamine B-injected mice displayed very little background fluorescence at these wavelengths. For visualizing Rhodamine B, excitation and emission wavelengths of 535 nm and 580 nm, respectively, were used. The pure and hybrid NCs were used to study their antitumor effects. Results concluded that high fluorescence intensity was observed at the tumor site within 4 h after IV administration, indicating the accumulation of hybrid NCs at the tumor site. Figure 8.1 explains that the harvested lungs and kidneys showed weak fluorescence, possibly because of the reticuloendothelial system (RES) immune cells' filtration and removal of the NCs from the circulation. These results suggest that hybrid NCs not only improve therapeutic outcomes for solid tumors but also enable real-time monitoring of drug particles during treatment [23].

Hybrid NCs can be classified into first-generation and second-generation systems based on the types of fluorophores employed. First-generation hybrid NCs utilize water-soluble fluorophores and less soluble ones, including ethyl rhodamine, DiD, DiR, and Cy5. In contrast, second-generation hybrid NCs incorporate advanced fluorophores such as tetraphenylethene (TPE), which is a typical aggregation-induced emission (AIE) fluorophore, and self-discriminating aggregation-caused quenching (ACQ) probes like P2/P4. Environment-responsive fluorescent probes, often featuring Aza-BODIPY or BODIPY structures, have excellent quantum yield, chemical and photostability, hydrophobicity, and near-infrared emission. These probes enable the creation of self-discriminating hybrid NCs, leveraging ACQ properties for enhanced diagnostic accuracy. This integration of innovative fluorophores with hybrid NCs underscores their potential to advance theranostic applications. Through real-time imaging, precise targeting, and effective treatment of diseases like cancer, hybrid NCs represent a significant leap forward in pharmaceutical and medical technology.

Their responsiveness to environmental changes is evident in their water-quenching sensitivity. This means the fluorescence is instantaneously and completely quenched when the probes aggregate upon transferring from a favourable solvent (like acetonitrile or dichloromethane) to an unfavourable solvent (such as water) [10]. Other than fluorescent probes, gold, copper, zinc, Manganese, and iron, like different materials, are used as contrast agents. Hollis et al. prepared distinctive multifunctional hybrid NCs composed of organic camptothecin and inorganic gold, which showed promising application in cancer therapy and bioimaging, prepared by the antisolvent method. The gold atoms were incorporated into the crystal lattice as a contrast agent as NCs defects. No chemical conjugation was needed. Due to the high attenuation coefficient of gold, just a small amount is needed to achieve a high noise-to-contrast ratio in CT imaging. NCs can reach and concentrate at the tumor site as a consequence of the enhanced permeation and retention (EPR) effect [25]. Another study by Shen et al. reported developing self-differentiating HNCs to seek out the biological consequences, biodistribution, and pharmacokinetics of quercetin (QT) HNCs. The QT-HNCs were fabricated using the antisolvent method with slight modification to achieve higher fluorescence intensity with ACQ probe P2. Oral and IV routes administered QT-HNCs prepared with 280 nm and 550 nm diameters. Ex-vivo and liver imaging concluded that QT-HNCs stay as solid NCs *in vivo* until 48 h following oral and IV administration, as shown in (Fig. 8.2a–c). An increased concentration was detected in the liver and lungs for both routes. The oral delivery of QT-HNCs-280 offers a higher distribution of QT but less accumulation in the liver. Meanwhile, for the IV. QT-HNCs −550 shows slower dissolution, enhanced RES recognition, and higher liver distribution. There was a notable difference in the absorption and distribution of QT-HNCs following oral and IV routes shown in (Fig. 8.2d, e) [26].

In another application of the HNCs, Xu et al. developed innovative versatile PVP-NiS2/FeS2 NCs that demonstrate chemo dynamic/photodynamic/photothermal effects in response to a just one wavelength NIR radiation and have the ability to perform both magnetic resonance (MR) and photoacoustic imaging (PA). Polyvinyl pyrrolidone (PVP) is used to modify the NiS2/FeS2 NCs in order to improve their biocompatibility. This novel drug delivery suppressed the breast tumor developed in BABL/c mice by providing synergistic therapy; firstly, PVP-NiS2/FeS2 NPs impede the EMT pathway, which suppresses tumor metastasis. Secondly, in response to NIR, the NPs might trigger several pathways leading to programmed cell death. An

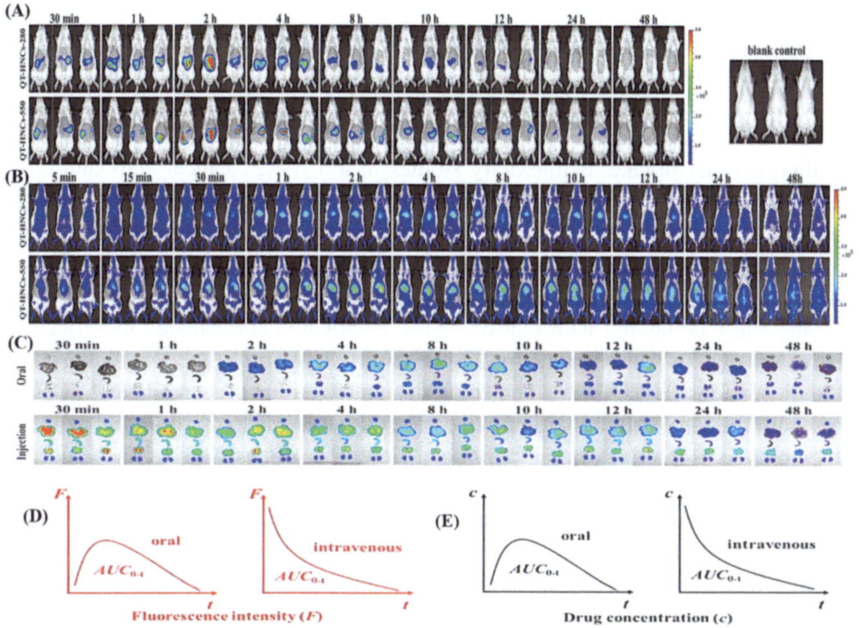

Fig. 8.2 SD rat *in vivo* live imaging following treatment of both QT-HNCs-280 and 550, (**a**) oral and (**b**) intravenous. Untreated rats were included as blank controls for reference. (**c**) Ex vivo analysis of key organs post-oral and IV treatment of QT-HNCs; The organs in every group are arranged from top to bottom in the following sequence: heart, liver, spleen, lungs, kidneys, and brain, for a comprehensive comparison of organ distribution. (**d**) Fluorescence quantification of oral and IV-administered QT-HNCs (**e**) Plasma drug level Vs time curve of oral and IV-administered QT-HNCs. (Reprint with the permission from Shen et al. [26])

808 nm laser wavelength was used for the *in vitro* PA imaging. They show the promising ability to eliminate tumors without apparent toxicity to normal organs [27].

Hybrid NCs technology was also utilized to develop the multi-drug co-delivery of ciprofloxacin-curcumin-N-acetylcysteine for dry power inhalations (DPI). Curcumin nanosuspension and ciprofloxacin N-acetylcysteine NCs were formulated using an antisolvent precipitation method with different stabilizers. Whereby Poloxamer 188 stabilized ciprofloxacin-N-acetylcysteine (NAC) HNCs and PVP (K-30) stabilized curcumin nanosuspension demonstrated satisfactory dispersion and a narrow particle size distribution. L-leucine was utilized to enhance DPIs' powder flow and aerodynamic characteristics. When spray-dried with mannitol, ciprofloxacin, and curcumin (CPFX/NAC/Cur) showed the amorphous condition, and the spray-dried DPIs performed better in terms of fluidity, homogeneity, and dispersion than the freeze-dried ones. This treatment shows potential for improving clinical outcomes and quality of life in patients with pulmonary infections [28]. According to the latest study, the disulfiram-copper complex, also referred to as copper-diethyldithiocarbamate (CuET), is pivotal in suppressing tumor growth [29]. Xiao et al. prepared CuET@HES NCs, which are stabilized by hydroxyethyl starch, have the ability to inhibit cancer stem cells (CSC) both *in vitro* and *in vivo*. HES acts as a stabilizer, preventing NCs aggregation and enhancing colloidal stability. Its effectiveness in suppressing CSC in newly removed tumor tissues from patients with hepatocellular carcinoma is also increased. It is also characterized by colloidal stability, cellular uptake, cytotoxicity studies in 4 T1, H22 cancer cells and CSC, and CSC apoptosis [18]. Compared to amorphous solid dispersions (ASDs), drug nanocomposites exhibit a significant drawback in their inability to reach high levels of supersaturation in dissolution media. In this study, drug hybrid nanocrystal–amorphous solid dispersions (HyNASDs) were prepared, and their efficacy was systematically compared with that of ASDs. Rahman et.al evaluated the effectiveness of amorphous solid dispersions and griseofulvin hybrid nanocrystal-amorphous solid dispersions (HyNASDs). The spray drying technique prepares HyNASDs and ASD with the same drug: polymer ratio. Soluplus (sol) and hydroxypropyl cellulose are employed as matrix-forming polymers. Herein, a sol-based one showed notable supersaturation of up to 180% for HyNASDs and 360% for ASDs. Compared to pure griseofulvin nanocomposites, the HyNASDs showed improved drug supersaturation [30].

Ting Wang et al. developed the self-discriminating hybrid NCs of curcumin (CUR) by incorporating environment-responsive fluorophore into the crystal lattice of the CUR. The HNCs glow due to the ACQ effect. In Fig. 8.3a. fluorescent hybrid CUR NCs' water-quenching sensitivity indicates their self-discriminating property. The CUR-HNCs were administered to the rats through an IV route. The central portion of NCs was accumulated mainly in the liver and lungs, and only some

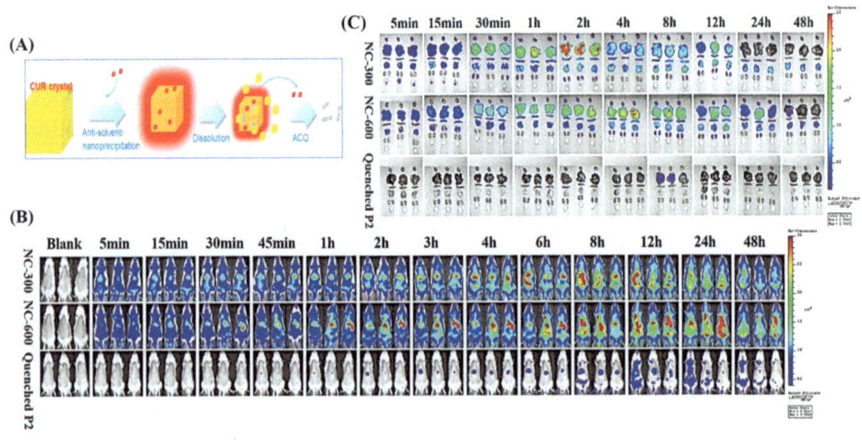

Fig. 8.3 (**a**) Concept of formulation. (**b**) IV-administered CUR NCs with sizes of NC-300 and NC-600, respectively, were imaged *in vivo* using quenched P2 dispersion as a reference. (**c**) CUR NCs, with sizes of NC-300 and NC-600, were intravenously injected, and *in vivo* imaging of major organs was carried out using a quenched P2 solution as a reference. Heart, liver, spleen, lung, kidney, and brain from top to bottom are included in each group. (Reprint with the permission from Hollis et al. [17])

portions were distributed to other RES organs. The circulation of a small fraction of CUR NCs in blood for at least 48 h due to the surface coating of NCs with stabilizer poloxamer 188 provides a sustained and prolonged effect (Fig. 8.3b, c). However, local toxicity of the active ingredient in vital organs should be considered for highly toxic API [19].

The novel fenofibrate hybrid NCs were prepared as a prototype to prove the in-vivo dissolution of poorly water-soluble drugs to monitor real-time residual and undissolved drug crystals in GIT based on live fluorescence imaging in rats. The FNB-HC was prepared using the antisolvent method by incorporating an ACQ probe suited to live, non-invasive imaging. The linearity established between fluorescence and *in vitro* dissolution of FNB was an accurate quantification of the determination of dissolution of FNB crystals. A level A point-to-point correlation between the *in vitro* dissolution profiles and live imaging of the *in vivo* dissolution of FNB HC validates the precision of the ACQ probe-based measurement. Further accuracy of the method is supported by the higher correlation coefficient of 0.9691 between the in-vivo absorption fraction (Fa) and in-vivo dissolution [31].

NCs were prepared for effective bioimaging to track the drug delivery and pharmacokinetics in context with hybrid paclitaxel. Here, the antisolvent technique was used to create tritium-labeled PTX NCs. Through the survival of the human HT-29 colon cancer transgenic mouse model, the anticancer effectiveness and toxicity were investigated. Using scintillation counting and tritium labeling, the drug's tissue distribution in mice was examined immediately after IV administration up to 48 h. The antitumor activity of NCs treatments was assessed, and the traditional Taxol® formulation was compared. The results from a survival study showed that the NCs treatments caused less toxicity. The study states that NCs and solubilized formulation (Taxol) had comparable and negligible EPR effects on drug accumulation. While biodistribution was qualitatively correlated, quantitative divergence exists. These varying outcomes offer potential methods for enhancing the development of hybrid NCs used in cancer treatment and detection [17].

8.2.2 Nano-cocrystals

Nano-cocrystals (NCCs), a subset of pharmaceutical cocrystals with diameters less than 1 nm, have been thoroughly investigated for their ability to enhance solubility and are gaining significant attention. Cocrystals are multi-component solid structures comprising two or more molecules, such as an active pharmaceutical ingredient (API) and a co-former, held together by interactions like van der Waals forces, π–π stacking, and hydrogen bonding [32]. Combining the methods of crystallization and nano-scaling, nano-crystallization has shown promise in improving solubility and bioavailability in pharmaceutical applications, surpassing the capabilities of conventional cocrystals. Since NCCs and NCs do not depend on ionic interactions like salts do, non-ionic materials can be used in their synthesis. Particle aggregation is one of the difficulties that nano-scale formulations may encounter, whereas crystalline structures are thermodynamically stable. NCs may be prepared using a single component and several solvent-antisolvent combinations, making it a relatively simple process. In the case of NCCs, two molecular components are integrated into a single crystal lattice, offering unique advantages in enhancing pharmaceutical performance [33]. Although NCCs offer potential benefits in solubility, especially in water, systematic optimization studies are still required. These nano-systems play a critical role in improving the solubility of poorly soluble drugs, a biggest challenge in the pharmaceutical sector. Key hurdles include finding the proper solvent to co-solubilize the drug and co-former in pharmaceutical cocrystals and creating the proper nano-sized dosage form [34]. Baicalein (BE), a flavonoid known for its potent anticancer, anti-inflammatory, and cardiovascular benefits, is impeded by low solubility and limited oral bioavailability.

In order to boost the bioavailability and dissolution rate of baicalein (BE), BE-NCT NCCs were generated through high-pressure homogenization (HPH). The resultant NCCs had an average particle size of 251.53 nm. The obtained NCCs showed a dissolution rate increase of 2.54 times. Pharmacokinetic studies following oral administration demonstrated that BE-NCT NCCs achieved a 6.02-fold increase in $AUC_{0\text{–}t}$ compared to BE NCs, a 2.87-fold increase compared to BE-NCT cocrystals (CCs), and significantly outperformed BE coarse powder. These results demonstrate the possibility of BE-NCT NCCs to improve the pharmacokinetic properties of BE [35]. Mohammady et al. utilized the antisolvent technique to prepare carvedilol (CAR) NCCs. Using COSMOquick software, tartaric acid (TA) was identified as the optimal coformer, and acetone as the organic solvent, based on enthalpy changes during cocrystallization and solubilization. The resulting CAR NCCs demonstrated an enormous increase in solubility of 2000 times compared to pure CAR, establishing this method as a safe and industrially scalable approach [36]. Similarly, Uracil (UR) was employed to enhance the biopharmaceutical properties of cytarabine (ARA) for antitumor applications.

The co-crystallization of cytarabine (ARA) with uracil (UR) yielded a 1:1 ARA-UR cocrystal, stabilized by cytosine-UR hydrogen bonding and π-π stacking interactions. These were then converted into NCCs with a particle size of 562.70 nm and a zeta potential of −33.40 mV. The advantages of using NCCs for marine hydrophilic antineoplastic drugs (ARA) by creating a cocrystal nano-situation strategy involving co-crystallization and size reduction by the antisolvent method. The cumulative permeability compared to the co-crystal form was 1.25 times higher. From the in-vitro permeability assay, it was evaluated that NCCs showed higher cumulative permeability compared to a co-crystal form derived. The NCCs had the most potent anti-proliferative effect compared to other crystals, with the lowest IC50 of 8.78 μM on EL-4 cells at 48 h, indicating improved drug penetration and absorption. The enhanced cytotoxicity of the cocrystal may simply be due to its superior transport ability, leading more drugs across the membrane and into cells and thus resulting in higher antitumor efficacy. In line with this, in its in-vivo study, it was observed that NCCs showed superior *in vivo* performance regarding higher plasma concentrations, prolonged presence, and increased efficacy over the unsolidified cocrystal or plain ARA. The C_{max} values for ARA, cocrystals, and NCCs were reported as follows: 3.32 ± 0.63 μg/mL, 5.24 ± 0.29 μg/mL,

6.58 ± 0.28 μg/mL, respectively. It was concluded that the synthesized ARA-UR NCCs form showed superior permeability regulation and improved pharmacokinetic behaviors *in vivo*. By incorporating NCC technology, this innovative ARA dosage form fulfills an unmet need in the development of marine drug formulations [34]. Another innovative method for the preparation of NCCs is sonochemical precipitation. Utilizing this approach, the *in vitro* release profiles of PAP-HCl, PAP–PI cocrystals, and PAP–PI NCCs were systematically evaluated in a pH 6.8 phosphate buffer using the USP dissolution apparatus II. The release studies revealed that the in-vitro dissolution of PAP from PAP-HCl and PAP–PI cocrystals reached 67% and 43% after 180 min, respectively. In contrast, PAP–PI NCCs demonstrated a remarkable improvement, achieving complete release within 90 min. These results underscored the significantly enhanced dissolution rate of NCCs compared to both conventional cocrystals and the hydrochloride salt. To investigate whether this enhanced *in vitro* dissolution translated into superior *in vivo* oral bioavailability, PAP–PI NCCs were further evaluated alongside PAP–PI cocrystals and PAP-HCl as control groups. The pharmacokinetic analysis demonstrated that PAP–PI NCCs achieved a maximum plasma concentration (C_{max}) of 111.17 ± 45.52 ng/mL and an area under the curve (AUC) of 131.56 ± 41.72 ng.h/mL, corresponding to a 1.39-fold and 2.44-fold improvement, respectively, compared to PAP-HCl. These outcomes emphasize the potential of NCCs as a transformative solubilization technique, offering significant advantages in enhancing the bioavailability and development prospects of poorly soluble pharmaceutical compounds [37].

This research compares two techniques for creating NCCs and their effect on dissolution rates. The top-down method, using a Retsch MM200 Mixer Mill, led to aggregation of particles, while the bottom-up method involved dissolving Myricetin (MYR) and Nicotinamide (NIC) in separate solutions, followed by rapid injection and sonication, producing uniform nano-sized particles without agglomeration. The re-precipitation step in the bottom-up process, based on TPD, eliminated the need for a non-solvent, optimizing the process. The bottom-up NCCs showed a higher dissolution rate (42 μg/mL) than cocrystals (33 μg/mL). The samples were filtered through a 178 μm filter, and the dissolution medium was 0.1 mol/L acetate buffer (pH 4.5) with 0.5% Tween-80. Overall, both methods improved dissolution rates, but the bottom-up approach offers a simpler, more efficient process for generating MYR-NIC NCCs [38]. Common antiviral medications used to treat HIV/AIDS infections include lamivudine (3TC) and zidovudine (AZT). These compounds present challenges such as irregular bioavailability, frequent dosage needs, and a side effect profile that includes hematotoxicity and hepatotoxicity. To address these issues, 3TC and AZT NCCs were developed, produced, and optimized using a unique pseudo-one-solvent bottom-up technique analysed with the Design of Experiments. This approach utilized sodium dodecyl sulfate (SDS) and α-tocopheryl polyethylene glycol succinate 1000 (TPGS 1000) to stabilize the NCCs electrostatically. Particle size, PDI, and zeta potential were determined experimentally to be 27.1 ± 92.0 nm, 0.467 ± 0.073, and − 41.9 ± 3.9 Mv, respectively. The optimization of electrostatic NCCs showed significantly improved cell viability compared to 3TC and AZT raw materials, likely due to the stabilizing effects of surfactants like TPGS 1000. The optimized nano co-crystals significantly decreased the toxic effects of 3TC and AZT on HeLa cells, according to in-vitro results [39].

Paclitaxel-Disulfiram (PTX-DSF NCCs) were synthesized using the antisolvent precipitation method to enhance therapeutic efficacy and address drug resistance. Cellular uptake and intracellular tracking studies conducted on sensitive (A549) and Taxol-resistant (A549/TAX) human lung adenocarcinoma cell lines demonstrated significantly higher intracellular accumulation of FITC-PTX-DSF NCCs in A549/TAX cells. Furthermore, the IC50 values of PTX-DSF in A549 cells were 3.82 and 3.24 μg/mL, whereas in A549/TAX cells they were 4.65 and 3.17 μg/mL, respectively, indicating a notable synergistic effect. These findings highlight the potential of PTX-DSF NCCs in achieving potent cytotoxic effects on Taxol-resistant cells and reducing dosing requirements, offering a promising approach for combating multidrug resistance in cancer therapy [40]. In Table 8.1, the examples illustrating advancements in NCCs formulations have been consolidated to offer a thorough overview.

8.2.3 Versatile Routes of Administration for Nanocrystals

Recent advancements in NCs technology have expanded their versatility across multiple routes of administration, enhancing their potential in modern therapeutics. Oral NCs formulations continue to improve bioavailability by addressing solubility challenges, with innovative coatings and carrier systems further optimizing gastrointestinal absorption [43]. Injectable NCs have seen progress in stabilizing formulations for controlled release, particularly for oncology and cardiovascular treatments. Transdermal NCs applications now incorporate advanced delivery systems, such as microneedles and patches, to achieve efficient skin penetration and sustained drug release. In ocular delivery, recent NCs formulations provide prolonged drug retention and improved therapeutic outcomes for diseases like glaucoma and macular degeneration. Pulmonary NCs delivery systems have also evolved, offering enhanced deposition and absorption for respiratory conditions, aided by dry powder inhalers and aerosol-based formulations. These advancements underscore the growing adaptability of NCs as a platform for precise, patient-centric drug delivery across diverse administration routes [44].

8.2.3.1 Pulmonary Route

Pulmonary drug delivery is a prominent and considered first-line therapy for various diseases like asthma, COPD, and acute respiratory diseases. This delivery offers both local and systemic action due to the presence of thin, highly permeable (0.2–0.7 μm thickness) and highly absorptive areas (100 m^2) [13]. The major barriers for this route are pulmonary retention and absorption, so size reduction and surface modification approaches are used to overcome this. Particles with size <100 nm can transport across the epithelium, reaching the blood circulation, as the mucus mesh is about 300 nm; particles smaller than this can easily escape mucus barriers,

Table 8.1 Advantages of nano-cocrystal formulations for pharmaceutical applications

Drugs-coformer	Route of administration	In-vitro study	In-vivo animal study/*in vivo* cell line study	Conclusions	Reference
Cytarabine-uracil (ARA-UR)	I/m injection	1. Permeability study NCCs have a cumulative permeability of 266.70 nmol/cm, 1.25 times higher than the co-crystals, which have a cumulative permeability of 213.82 nmol.	1. Pharmacokinetic estimations C_{max} for ARA—3.32 ± 0.63 μg/mL C_{max} of NCCs 6.58 ± 0.28 μg/mL AUC_{0-t} for ARA—15.01 ± 3.76 μg·h/mL AUC_{0-t} for NCCs—38.82 ± 14.66 μg·h/mL	The NCCs, in particular, sustains plasma concentration, extends half-life, and increases bioavailability, offering more stable and effective therapy levels.	[34]
PAP-PI	Oral	1. *In-vitro* dissolution study Complete release from PAP-PI NCCs was obtained after 90 min, whereas 67% and 43% release was obtained from PAP-PI and cocrystal form	1. Pharmacokinetic Estimations C_{max} (ng/mL) for PAP-PI HCl and nanosuspension was found to be 79.98 ± 32.06 and 111.17 ± 45.52 AUC_{0-t} (ng·h/mL) for PAP-PI HCl and nanosuspension was found to be 53.92 ± 25.55131.56 ± 41.72.	NCCs is a novel solubilization method that may be used in the pharmaceutical industry to boost the bioavailability and development potential of poorly soluble substances.	[37]
ITRA-BL	Oral	1. Dissolution Study pH-dependent release of ITRA takes place	Capsules containing ITRA-BL1 (F3) formulation showed the highest C_{max} and AUC that is 806.4 ng/mL and 8.8 μg.h/mL. The reference sporanox showed C_{max} and AUC of 424.9 ng/mL and 4.0 μg.h/mL	In contrast to a reference formulation that is sold commercially, the liquid co-crystalline nanosuspension and bead stacking processing produced instant release solid dosage forms.	[41]

(continued)

Table 8.1 (continued)

Drugs-coformer	Route of administration	In-vitro study	In-vivo animal study/*in vivo* cell line study	Conclusions	Reference
Baicalein-Nicotinamide	Oral	1. *In-vitro* dissolution study The dissolution test showed a 2.17-fold improvement in BE-NCT NCCs and a 2.22-fold increase in BE-NCT cocrystals compared to BE coarse powder in both FaSSIF-V2 and FaSSGF media.	The AUC_{0-t} for BE-NCT NCCs displayed an impressive 6.02-fold increase, much more than BE-NCT cocrystals (2.87-fold), BE NCs (3.32-fold), and BE coarse powder (one-fold).	In both *in vitro* and *in vivo* studies, BE-NCT NCCs performed significantly better than BE coarse powder, BE-NCT cocrystals, and BE NCs.	[35]
Indomethacin, furosemide-saccharin, cytosine, and caffeine	–	1. *In-vitro* dissolution study Compared to the nanocrystal formulations of the individual component medications, the NCCs formulations showed higher solubilization effects and faster dissolution rates.	–	The NC formulations showed higher dissolution rates and saturation concentrations than pure and cocrystals, according to dissolution experiments.	[42]
Paclitaxel Disulphiram	–	1. *In vitro* drug release After 12 h, DSF was released at pH 7.4 and 5.4, respectively, in 68% and 83% concentrations.	PTX-DSF nanosuspensions demonstrated a 14-fold increase in A549/TAX cell uptake, a five-fold increase in apoptosis, a seven-fold drop in IC50, and an 8.9-fold decrease in dose in comparison to free PTX.	PTX-DSF nanosuspensions (Ns) ensured sustained DSF release, effectively inactivating the P-glycoprotein (P-gp) pump and promoting apoptosis in cancer cells.	[40]
Lamivudine-Zidovudine	–	–	The NCCs formulation of lamivudine and zidovudine demonstrated significantly higher cell viability of 74.7 ± 2.4% compared to 52.6 ± 5.1% for the physical mixture	Compared to zidovudine and lamivudine given separately as physical mixes, the optimized NCCs showed reduced toxicity.	[39]

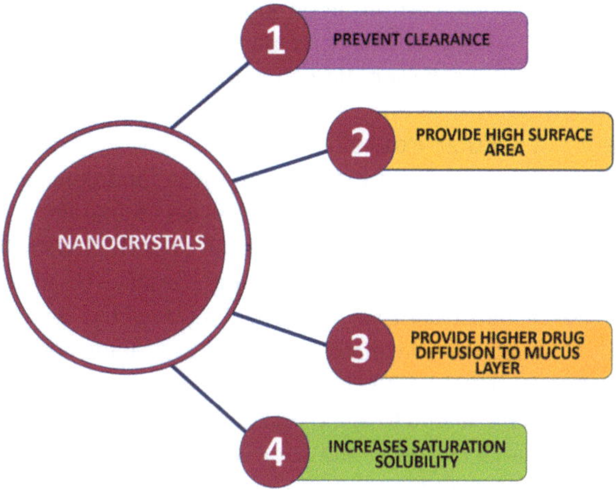

Fig. 8.4 Illustrates the benefits of using NCs in the pulmonary medication delivery system

particles of size <500 nm can evade macrophagic uptake, and mucociliary clearance, thus providing better retention and absorption. The aerodynamic size of the NCs not only helps in enhanced permeation and retention but also decides the site of deposition; >5 μm deposition is probable in the oropharynx, 5–10 μm in the primary bronchi, 1–5 μm in secondary bronchi, 1–3 μm and 0.5–1 μm in bronchioles and alveoli respectively. When administered via the pulmonary route, NCs offer a number of benefits, as shown in Fig. 8.4 [45].

NCs are an attractive drug delivery due to the requirement of a minimum number of excipients or stabilizers and the potential for targeted drug delivery. NCs are targeted by passive (using EPR effect) and active methods (EGFR receptor-based, Transferrin/Folic acid receptor-based, CD44 receptor-based, Sialic acid-based targeting, Integrin adhesion molecule α vβ3 based targeting, Mannose receptors-based targeting) [45]. Rundfeld et al. demonstrated the efficacy of pulmonary drug delivery using a nanosuspension (NS) of itraconazole (ITZ) formulated via wet milling. Nebulized administration for pulmonary aspergillosis resulted in significantly higher drug concentrations in lung tissues compared to oral delivery. Pharmacokinetic studies showed peak pulmonary concentrations of 21,385 ± 1628 ng/g within 5 min of inhalation, which declined to 2961 ± 649 ng/g at 72 h. In contrast, the oral formulation produced much lower serum concentrations, with 216 ± 47 ng/g in fasted and 469 ± 138 ng/g in fed rats. This study underscores the superior efficiency of pulmonary delivery using nanosuspensions over traditional oral treatments for targeted drug delivery [46]. Another piece of work orchestrated by Fasima M. et al. revealed the systemic application of pulmonary drug delivery; they prepared Glibenclamide (Gbn) NS to treat type II DM, Gbn being a BCS class II drug, shows poor solubility and also shows high first-pass metabolism (only 50% oral bioavailability); this NS was delivered using pressurized metered dose inhaler (pMDI); this therapy was compared with the oral GBC therapy. The PDD exhibited a maximum hypoglycemic effect within 180 min, achieving a 61.78 ± 1.1 mean percentage decrease in blood glucose levels. In contrast, oral delivery resulted in only a 23.41 ± 1.0 mean percentage decrease over the same duration, highlighting its potential as a superior alternative to conventional oral treatment [47].

8.2.3.2 Ocular Route

The human eye is an extremely intricate and advanced organ. It is challenging to formulate an ocular drug delivery system due to the presence of various anatomical and physiological barriers, specifically for BCS class II and class III drugs. Ocular Drug Delivery (ODD) systems are particularly advantageous for managing eye diseases like cataracts, glaucoma, retinopathy, and microbial infections because they target the eye directly, avoiding systemic drug administration's limitations and potential side effects. Conventional ocular drug delivery has a bioavailability of up to 1–5% due to anatomical and physiological barriers like tear turnover, high nasolacrimal drainage, and poor corneal penetration [48].

NCs come to the rescue as they help bypass most of the saturated barriers. The NCs penetrate the mucus layer of the tear film, interact with the corneal tissues, and enhance retention in the ocular cul-de-sac. This enhances bioavailability and corneal penetration, likely due to increased retention and saturation solubility of poorly water-soluble drugs. Additionally, the NCs exhibit a dual drug release mechanism, offering both an initial rapid release and a prolonged sustained release [48]. Romero et al. prepared a cationic NCs for ocular delivery, made up of 1% stabilizer (Cetylpyridinium chloride and benzalkonium chloride), 5% dexamethasone acetate, and optionally 1% polymyxin B sulfate (antibiotic) (all w/w). Due to its nanosize and mucoadhesive cationic charge provided by polymyxin B, the formulation enhances drug solubility, increases residence time in the eye, and achieves a − 15 mV zeta potential. This distinct interaction with the mucous membrane results in enhanced bioavailability compared to standard eye drop formulations [49]. Tuomela et al. focus on the development of ocular NCs suspensions for the poorly soluble drug brinzolamide (BZM) to reduce intraocular pressure (IOP). Various stabilizers (hydroxypropyl methylcellulose (HPMC), poloxamer F127 and F68, and polysorbate 80) were screened using a rapid wet milling technique, with HPMC being the most effective. Three formulations were developed: formulation I (BZM nanosuspension in PBS pH 7.4), formulation II (BZM nanosuspension in PBS pH 7.4 with polysorbate 80 as an absorption enhancer), and formulation III (treated with Azopt®, the marketed nanosuspension of BZM). These formulations demonstrated rapid BZM dissolution, achieving 100% dissolution within 1 min at pH 7.4, with enhanced dissolution at pH 4.5. Cytotoxicity testing on human corneal epithelial cells showed no significant difference in cell viability between the nanosuspension formulations and Azopt®. *In vivo*, all formulations significantly reduced elevated IOP in a rat ocular hypertensive model, comparable to Azopt®. Overall, BZM nanosuspensions demonstrated excellent dissolution, absorption, and IOP reduction profiles [50].

8.2.3.3 Injectable Route

Injectables offer critical advantages, such as bypassing the gastrointestinal tract and first-pass metabolism, rapid onset of action, and suitability for hydrophobic and hydrophilic APIs, including proteins and peptides. NCs address the poor water solubility of many anticancer drugs, enhancing saturation solubility and drug targeting efficiency, thereby reducing dosage and minimizing adverse reactions [6, 51]. Intravenous administration is the most common route for NCs formulations in injectables. Yan Gao et al. developed a curcumin nanosuspension (CUR-NS) using high-pressure homogenization to improve the cytotoxicity and intravenous suitability of curcumin. The CUR-NS formulation exhibited significantly greater cytotoxicity, showing IC50 values of 15.52 ± 0.43 μg/mL for Hela cells and 26.45 ± 2.36 μg/mL for MCF-7 cells, in contrast to the free curcumin's IC50 of 35.94 ± 4.5 μg/mL and 36.64 ± 3.16 μg/mL. These findings suggest that CUR-NS could improve therapeutic efficacy and reduce adverse effects, making it a promising intravenous cancer treatment formulation [52].

Sreeharsha et al. prepared Paclitaxel (PT) stable NCs formulation created with the triblock copolymer Pluronic F127 administered by IV route. The NCs exhibited a drug release profile characterized by sustained release over a prolonged duration. O-PT-NC-Folate (Folate receptor (FR)-targeted NCs), was created by conjugating folic acid with Pluronic F127. As the PT concentration rose from 0.08 to 10 mM in FR-positive KB cells, the cell viability of FR-targeted nanocarriers (NCs) containing 10% F127-folate was significantly diminished ($p < 0.001$) in comparison to the non-targeted O-PT-NCs. The PT NCs formulation did not cause hemolysis at pharmacologically significant doses. In an in-vivo anti-tumor study using NCI/ADR-RES cell lines, O-PT-NC-Folate showed longer-lasting anti-cancer effects compared to the non-targeted PT-NCs. These results demonstrate excellent anti-tumor efficacy with low toxicity of conjugated PTNCs formulation [53]. Gautam et al. explored long-acting injectable nanoformulated antiretroviral therapy (nanoART) for improved drug compliance and targeting viral reservoirs. Nanoformulations of atazanavir (ATV) and ritonavir (RTV), prepared via high-pressure homogenization, showed enhanced pharmacokinetics and tissue distribution. ATV and RTV levels were up to 4500 times higher at injection sites and 9 times higher in the liver compared to conventional doses. Multiple dosing further increased plasma and tissue drug levels by up to 270-fold. Sustained release from macrophages created intracellular depot sites, highlighting nanoART's potential to improve antiretroviral therapy outcomes [54].

Sun et al. developed an in situ cross-linkable hydrogel depot with paclitaxel NCs to enhance intraperitoneal chemotherapy for ovarian cancer. The paclitaxel NCs outperformed microparticulate paclitaxel precipitates in suppressing SKOV3 cell proliferation and exhibited a reduced maximum tolerated dose (MTD), indicating better dissolution and cellular uptake. Animals treated with PPT-gel had a median survival of 51 days, similar to Taxol-treated ones. Mice administered with PNC-gel exhibited a notable increase in survival duration (67 days) compared to Taxol ($p < 0.05$, Log rank Mantel-Cox test). Both treatments showed minimal tumor growth until 7 weeks in survivors. Cytotoxicity assays revealed paclitaxel NCs were 3.3 times more potent than microparticulate paclitaxel, highlighting their superior efficacy. These findings demonstrate the potential of paclitaxel NC-based hydrogels as a promising platform for ovarian cancer therapy [55].

8.2.3.4 Oral Route

Convenience and ease of administration have made the oral route of administration the most favorable and opted rote of drug administration; these benefits are behind the figures that around 70% of the drugs sold in the USA are orally prescribed [56–59]. Upon oral administration, the drugs must endure the challenging conditions of the gastrointestinal tract. and then dissolve in the luminal fluid, which for most hydrophobic drugs is the rate-limiting process [60]. Moreover, oral administration methods based on NCs have been reported to reduce variability in medication absorption and food-related effects [61].

Lei et al. developed solid glycyrrhetinic acid (GA) NCs to enhance oral bioavailability. Prepared via anti-solvent precipitation-ultrasonication and freeze-drying, the NCs had a mean size of 220 nm and maintained their crystalline structure. Compared to coarse GA, the NCs showed significantly improved solubility and dissolution rates. In rats, the bioavailability of GA NCs was 4.3 times higher, with C_{max} of 23.00 ± 5.45 μg/mL and AUC_0-24 of 111.99 ± 6.24 μg·h/mL, compared to 4.16 ± 1.26 μg/mL and 25.36 ± 6.92 μg·h/mL for coarse GA powder [62]. Sharma et al. developed paclitaxel NCs (PTX/NCs) to improve oral delivery by overcoming poor solubility and rapid metabolism. Using high-pressure homogenization, stabilizers like Tween 80 and sodium polystyrene sulfonate (PSS) were found most effective in enhancing size reduction and stability. Studies conducted in MCF-7 and MDA-MB breast cancer cell lines showed that PTX/NCs retained their therapeutic potential and were more effective than the PTX solution. Oral pharmacokinetic studies in rats revealed a nine to ten-fold increase in AUC and C_{max} for PTX/NCs compared to plain PTX crystals. PTX/NCs also exhibited 3–4 times greater dissolution rates, suggesting their potential for oral [63].

8.2.3.5 Transdermal Route

For both parenteral and oral routes of drug administration, TDD is more approachable. The system treats both local and systemic diseases by delivering therapeutic moieties through the layers of the skin. No first-pass effect is observed in the transdermal route. This method is satisfactory for drugs having low bioavailability when administered orally or displaying adverse effects due to biotransformation [64]. NC technology enhances the permeation and bioavailability of water-insoluble drugs in transdermal delivery, addressing challenges of low absorption, and has been successfully applied in cosmeceuticals and the treatment of skin diseases [65]. NCs enhance epidermal permeation in transdermal drug delivery by increasing the concentration gradient and dissolving rapidly on the dermal surface, with reduced saturation solubility and diffusion corona layer, further facilitating drug absorption [66].

Over the past 20 years, microneedle arrays have been widely studied to fill the gap between transdermal patches and painful injections. These barely disruptive, painless, tiny projections can pierce the skin's stratum corneum. Cholecalciferol (CL), a hydrophobic chemical, was converted into a nanocrystal and added to hydrophilic polymer-based dissolving microneedles (DMNs) for improved transdermal administration. Two techniques were used to prepare the microneedle filled with NCs. The first method involved nanoprecipitation using poly (vinyl alcohol) and poly (vinyl pyrrolidone) of various molecular weights as stabilizers, using acetone and ethanol as solvents to optimize particle size and zeta potential. The second method involved centrifugation-assisted micro-molding was employed to create DMN arrays. An ex vivo study using porcine skin and Franz diffusion cells was conducted to compare the permeation of CL from CL-NC loaded DMN arrays with MN-free CL-NC patches. Higher skin permeation was found for CL-NC-loaded DMN array (489 ± 89.3 μg) compared to the MN-free CL-NC patches (73.2 ± 26.5 μg) over 24 h. The study concluded that NCs particles were uniformly distributed within the polymeric structure of DMN arrays without aggregation. This innovative NCs-loaded MN system might be beneficial for depot preparation via the transdermal route [67].

In another study, researchers formulated methotrexate (MTX) NCs using the bottom-up technique for the treatment of psoriasis and incorporated this MTX-NCs into microneedle Fig. 8.5. An ex-vivo study with neonatal porcine skin demonstrated that the microneedle shafts dissolve entirely within 20 min of insertion. An *in vivo* study on Sprague Dawley rats assessed various pharmacokinetic

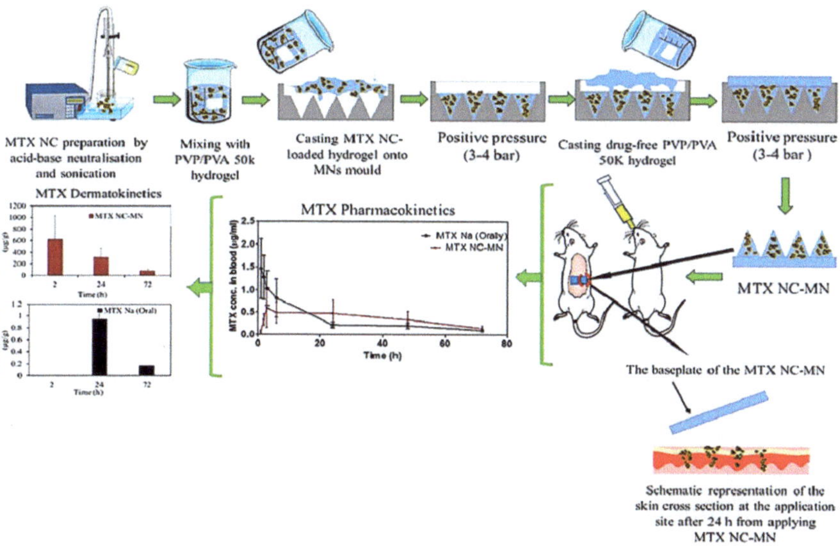

Fig. 8.5 Pictorial representation of the preparation of microneedle delivery system containing NCs and its *in vivo* studies [68]

parameters, concluding that the MTX-NC-MN array effectively delivered MTX-NCs intradermally functioning as a depot that delivered the medication in a sustained manner over 72 h [68].

8.2.3.6 Intranasal Route

Nasal delivery is becoming a promising route for the systemic administration of pharmacologically active molecules, offering a viable alternative to traditional oral and parenteral methods. This approach allows for easy and non-invasive administration, bypasses first-pass metabolism, and many provide direct access to the central nervous system, thus circumventing the BBB [69].

A significant limitation of administering drugs intranasally (IN) involves the small volumes of medication that can be delivered through this route. Therefore, it is essential that the drug be highly potent, even in minimal quantities, and have adequate water solubility to achieve the necessary concentration for administration. This highlights two critical prerequisites: firstly, the effectiveness of the drug at low doses, and secondly, its high solubility in water to support the dosage needed [70]. To treat Parkinson's disease (PD), a formulation of paeoniflorin (PA) nanocrystals (PA-NCs) was created for intranasal administration utilizing an antisolvent precipitation process with TPGS as a stabilizer. The PA-NCs had a particle size of 139.6 ± 1.3 nm, exhibited significantly higher drug release (87.14% ± 5.34% at 24 h) and improved cellular uptake in Calu-3 cells compared to free PA. After intranasal administration, pharmacokinetic investigations revealed increased PA concentration in the brain. FRET imaging confirmed NCs internalization in cells, while molecular dynamics simulations indicated stable PA-TPGS binding. These findings suggest that PA-NCs could be a potentially effective PD therapy through improved brain targeting and bioavailability [71]. In another study, researchers selected curcumin as a model drug to create NCs for potential intranasal administration by combining precipitation and ultrasonication using a probe sonicator. To achieve this, they designed Box Behnken design to develop curcumin NCs. Formulation variables were optimized, and the ideal level for each variable was identified, using a response surface quadratic model. Three levels (low, medium, and high) make up the design, and 51 experiments were carried out to optimize factors such the kind of stabilizer, solvent-to-antisolvent ratio, curcumin concentration, and stabilizer concentration. After the optimized formulation was characterized, olfactory ensheathing cells (OECs) were used to assess the cellular absorption of curcumin and curcumin NCs. After going through all the characterization parameters, Curcumin NCs were shown to have promising formulation properties for transport from the nose to the brain [70].

8.3 Summary and Future Perspective

Despite the considerable progress made in nanocrystal technology, several challenges remain. Issues related to the long-term stability of nanocrystals, particularly in biological environments, need to be addressed. Aggregation and the potential for toxicological concerns are areas of active research, as the size, surface properties, and composition of nanocrystals can influence their safety and efficacy. Furthermore, regulatory approval of nanocrystal-based formulations presents a challenge, as the mechanisms underlying the behavior and interaction of nanocrystals *in vivo* are still being understood.

To overcome these challenges, future research will focus on improving the physical stability of nanocrystals, understanding their long-term therapeutic outcomes, and conducting comprehensive safety assessments. Additionally, advances in formulation strategies and manufacturing processes will be essential to bring nanocrystal-based products from the laboratory to clinical practice. The future of drug nanocrystals lies in their integration with other cutting-edge technologies, such as gene therapy, immunotherapy, and biologics. The combination of nanocrystals with these advanced therapies offers the potential for groundbreaking treatments for complex diseases, including cancer, neurological disorders, and rare diseases. By enhancing the precision of drug delivery and improving the pharmacokinetics of therapeutic agents, nanocrystals can play a critical role in expanding the capabilities of modern medicine. For example, nanocrystals could be employed to deliver gene-editing tools, such as CRISPR, to specific tissues or cells, enabling more targeted and efficient genetic interventions. Similarly, integrating nanocrystals with immunotherapy could enhance the targeted delivery of immune-modulating agents to tumors, improving therapeutic outcomes and minimizing side effects.

The future of drug nanocrystals holds great promise for revolutionizing drug delivery and advancing therapeutic options. With ongoing innovations in nanocrystal formulations, stabilizers, and delivery systems, the pharmaceutical industry is poised to tackle some of the most pressing challenges in drug development. The integration of nanocrystals with advanced drug delivery platforms, along with a deeper understanding of their behavior in biological systems, will enable more targeted, efficient, and personalized treatments. As research continues to evolve, nanocrystal technology is set to become a cornerstone of the next generation of pharmaceutical formulations, offering new hope for patients and addressing unmet medical needs across a range of therapeutic areas.

References

1. Phuna ZX, Panda BP, Shivashekaregowda NKH, Madhavan P (2022) Recent development in nanocrystal based drug delivery for neurodegenerative diseases: scope, challenges, current and future prospects. J Drug Deliv Sci Technol 68:102921. https://doi.org/10.1016/j.jddst.2021.102921
2. Lv Y, Wu W, Corpstein CD, Li T, Lu Y (2022) Biological and intracellular fates of drug nanocrystals through different delivery routes: recent development enabled by bioimaging and PK modeling. Adv Drug Deliv Rev 188:114466. https://doi.org/10.1016/j.addr.2022.114466
3. Kumari NU, Pardhi E, Madan J, Singh PK, Mehra NK (2025) Non-vesicular nanocarriers for breast cancer management: current advances, challenges, and future perspective. Int J Polym Mater Polym Biomater 74(5):424–444. https://doi.org/10.1080/00914037.2024.2338135
4. Lu Y, Chen Y, Gemeinhart RA, Wu W, Li T (2015) Developing nanocrystals for cancer treatment. Nanomedicine 10(16):2537–2552. https://doi.org/10.2217/nnm.15.73

5. Jarvis M, Krishnan V, Mitragotri S (2019) Nanocrystals: a perspective on translational research and clinical studies. Bioeng Transl Med 4(1):5–16. https://doi.org/10.1002/btm2.10122
6. Chen Z, Wu W, Lu Y (2020) What is the future for nanocrystal-based drug-delivery systems? Ther Deliv 11(4):225–229. https://doi.org/10.4155/tde-2020-0016
7. Pardhi E et al (2023) Multifunctional targetable liposomal drug delivery system in the management of leukemia: potential, opportunities, and emerging strategies. Life Sci 325:121771. https://doi.org/10.1016/j.lfs.2023.121771
8. Pardhi E, Vasave R, Srivastava V, Yadav R, Mehra NK (2024) Nanocrystal technologies in biomedical science: from the bench to the clinic. Drug Discov Today 29(3):103913. https://doi.org/10.1016/j.drudis.2024.103913
9. Lu Y, Li Y, Wu W (2016) Injected nanocrystals for targeted drug delivery. Acta Pharm Sin B 6(2):106–113. https://doi.org/10.1016/j.apsb.2015.11.005
10. Lu Y, Lv Y, Li T (2019) Hybrid drug nanocrystals. Adv Drug Deliv Rev 143:115–133. https://doi.org/10.1016/j.addr.2019.06.006
11. Mohammad IS, Teng C, Chaurasiya B, Yin L, Wu C, He W (2019) Drug-delivering-drug approach-based codelivery of paclitaxel and disulfiram for treating multidrug-resistant cancer. Int J Pharm 557:304–313. https://doi.org/10.1016/j.ijpharm.2018.12.067
12. Sander JRG, Bučar D, Henry RF, Zhang GGZ, MacGillivray LR (2010) Pharmaceutical Nano-cocrystals: sonochemical synthesis by solvent selection and use of a surfactant. Angew Chem Int Ed 49(40):7284–7288. https://doi.org/10.1002/anie.201002588
13. McGuckin MB et al (2022) Nanocrystals as a master key to deliver hydrophobic drugs via multiple administration routes. J Control Release 345:334–353. https://doi.org/10.1016/j.jconrel.2022.03.012
14. Xiang M, Yang C, Zhang L, Wang S, Ren Y, Gou M (2024) Dissolving microneedles for transdermal drug delivery in cancer immunotherapy. J Mater Chem B 12(24):5812–5822. https://doi.org/10.1039/D4TB00659C
15. Ruan S, Zhang Y, Feng N (2021) Microneedle-mediated transdermal nanodelivery systems: a review. Biomater Sci 9(24):8065–8089. https://doi.org/10.1039/D1BM01249E
16. Liang H et al (2021) Nanocrystal-loaded liposome for targeted delivery of poorly water-soluble antitumor drugs with high drug loading and stability towards efficient cancer therapy. Int J Pharm 599:120418. https://doi.org/10.1016/j.ijpharm.2021.120418
17. Hollis CP, Weiss HL, Leggas M, Evers BM, Gemeinhart RA, Li T (2013) Biodistribution and bioimaging studies of hybrid paclitaxel nanocrystals: lessons learned of the EPR effect and image-guided drug delivery. J Control Release 172(1):12–21. https://doi.org/10.1016/j.jconrel.2013.06.039
18. Xiao C et al (2023) Hydroxyethyl starch stabilized copper-diethyldithiocarbamate nanocrystals for cancer therapy. J Control Release 356:288–305. https://doi.org/10.1016/j.jconrel.2023.02.043
19. Wang T et al (2018) Tracking translocation of self-discriminating curcumin hybrid nanocrystals following intravenous delivery. Int J Pharm 546(1–2):10–19. https://doi.org/10.1016/j.ijpharm.2018.05.020
20. Rahman M, Radgman K, Tarabokija J, Ahmad S, Bilgili E (2023) Preparation and characterization of spray-dried hybrid nanocrystal–amorphous solid dispersions (HyNASDs) for supersaturation enhancement of a slowly crystallizing drug. Nanomaterials 13(17):2419. https://doi.org/10.3390/nano13172419
21. Fontana F, Figueiredo P, Zhang P, Hirvonen JT, Liu D, Santos HA (2018) Production of pure drug nanocrystals and nano co-crystals by confinement methods. Adv Drug Deliv Rev 131:3–21. https://doi.org/10.1016/j.addr.2018.05.002
22. Hollis CP, Weiss HL, Evers BM, Gemeinhart RA, Li T (2014) In vivo investigation of hybrid paclitaxel nanocrystals with dual fluorescent probes for cancer theranostics. Pharm Res 31(6):1450–1459. https://doi.org/10.1007/s11095-013-1048-x

23. Zhao R, Hollis CP, Zhang H, Sun L, Gemeinhart RA, Li T (2011) Hybrid nanocrystals: achieving concurrent therapeutic and bioimaging functionalities toward solid tumors. Mol Pharm 8(5):1985–1991. https://doi.org/10.1021/mp200154k
24. Zhang L, Zhang Y-X, Qiu J-N, Li J, Chen W, Guan Y-Q (2017) Preparation and characterization of hypoglycemic nanoparticles for oral insulin delivery. Biomacromolecules 18(12): 4281–4291. https://doi.org/10.1021/acs.biomac.7b01322
25. Hollis CP, Dozier AK, Knutson BL, Li T (2019) Preparation and characterization of multimodal hybrid organic and inorganic nanocrystals of camptothecin and gold. Acta Pharm Sin B 9(1): 128–134. https://doi.org/10.1016/j.apsb.2018.03.005
26. Shen B, Shen C, Zhu W, Yuan H (2021) The contribution of absorption of integral nanocrystals to enhancement of oral bioavailability of quercetin. Acta Pharm Sin B 11(4):978–988. https://doi.org/10.1016/j.apsb.2021.02.015
27. Xu S et al (2023) A versatile NiS2/FeS2 hybrid nanocrystal for synergistic cancer therapy by inducing ferroptosis and pyroptosis. Chem Eng J 460:141639. https://doi.org/10.1016/j.cej.2023.141639
28. Ma Y, Cong Z, Wang Y, Gao P (2024) A novel multi-drugs ciprofloxacin-curcumin-N-acetylcysteine co-delivery system based on hybrid nanocrystals for dry powder inhalations. Next Nanotechnol 6:100084. https://doi.org/10.1016/j.nxnano.2024.100084
29. Skrott Z et al (2017) Alcohol-abuse drug disulfiram targets cancer via p97 segregase adaptor NPL4. Nature 552(7684):194–199. https://doi.org/10.1038/nature25016
30. Rahman M, Arevalo F, Coelho A, Bilgili E (2019) Hybrid nanocrystal–amorphous solid dispersions (HyNASDs) as alternative to ASDs for enhanced release of BCS class II drugs. Eur J Pharm Biopharm 145:12–26. https://doi.org/10.1016/j.ejpb.2019.10.002
31. Yang Y et al (2021) In vivo dissolution of poorly water-soluble drugs: proof of concept based on fluorescence bioimaging. Acta Pharm Sin B 11(4):1056–1068. https://doi.org/10.1016/j.apsb.2020.08.002
32. Guo M, Sun X, Chen J, Cai T (2021) Pharmaceutical cocrystals: a review of preparations, physicochemical properties and applications. Acta Pharm Sin B 11(8):2537–2564. https://doi.org/10.1016/j.apsb.2021.03.030
33. Tan J, Liu J, Ran L (2021) A review of pharmaceutical nano-cocrystals: a novel strategy to improve the chemical and physical properties for poorly soluble drugs. Crystals (Basel) 11(5): 463. https://doi.org/10.3390/cryst11050463
34. Yu Y-M, Bu F-Z, Meng S-S, Yan C-W, Wu Z-Y, Li Y-T (2023) The first nano-cocrystal formulation of marine drug cytarabine with uracil based on cocrystal nanonization strategy for long-acting injection exhibiting enhanced antitumor activity. Int J Pharm 644:123300. https://doi.org/10.1016/j.ijpharm.2023.123300
35. Pi J et al (2019) A nano-cocrystal strategy to improve the dissolution rate and oral bioavailability of baicalein. Asian J Pharm Sci 14(2):154–164. https://doi.org/10.1016/j.ajps.2018.04.009
36. Mohammady M, Hadidi M, Iman Ghetmiri S, Yousefi G (2021) Design of ultra-fine carvedilol nanococrystals: development of a safe and stable injectable formulation. Eur J Pharm Biopharm 168:139–151. https://doi.org/10.1016/j.ejpb.2021.08.015
37. Huang Y, Li J-M, Lai Z-H, Wu J, Lu T-B, Chen J-M (2017) Phenazopyridine-phthalimide nano-cocrystal: release rate and oral bioavailability enhancement. Eur J Pharm Sci 109:581–586. https://doi.org/10.1016/j.ejps.2017.09.020
38. Liu M, Hong C, Li G, Ma P, Xie Y (2016) The generation of myricetin–nicotinamide nanococrystals by top down and bottom up technologies. Nanotechnology 27(39):395601. https://doi.org/10.1088/0957-4484/27/39/395601
39. Witika BA, Smith VJ, Walker RB (2020) Top-down synthesis of a lamivudine-zidovudine nano co-crystal. Crystals (Basel) 11(1):33. https://doi.org/10.3390/cryst11010033
40. Mohammad IS, He W, Yin L (2018) A smart paclitaxel-disulfiram nanococrystals for efficient MDR reversal and enhanced apoptosis. Pharm Res 35(4):77. https://doi.org/10.1007/s11095-018-2370-0

41. De Smet L et al (2014) Formulation of itraconazole nanococrystals and evaluation of their bioavailability in dogs. Eur J Pharm Biopharm 87(1):107–113. https://doi.org/10.1016/j.ejpb.2013.12.016
42. Karashima M, Kimoto K, Yamamoto K, Kojima T, Ikeda Y (2016) A novel solubilization technique for poorly soluble drugs through the integration of nanocrystal and cocrystal technologies. Eur J Pharm Biopharm 107:142–150. https://doi.org/10.1016/j.ejpb.2016.07.006
43. Pardhi E, Bhor Y, Singh PK, Mehra NK (2024) An apprise on novel drug delivery systems for management of diabetes mellitus. J Drug Deliv Sci Technol 94:105473. https://doi.org/10.1016/j.jddst.2024.105473
44. Gigliobianco MR, Casadidio C, Censi R, Di Martino P (2018) Nanocrystals of poorly soluble drugs: drug bioavailability and physicochemical stability. Pharmaceutics 10(3):134. https://doi.org/10.3390/pharmaceutics10030134
45. Kumar M, Jha A, Dr M, Mishra B (2020) Targeted drug nanocrystals for pulmonary delivery: a potential strategy for lung cancer therapy. Expert Opin Drug Deliv 17(10):1459–1472. https://doi.org/10.1080/17425247.2020.1798401
46. Rundfeldt C, Steckel H, Scherliess H, Wyska E, Wlaź P (2013) Inhalable highly concentrated itraconazole nanosuspension for the treatment of bronchopulmonary aspergillosis. Eur J Pharm Biopharm 83(1):44–53. https://doi.org/10.1016/j.ejpb.2012.09.018
47. Hashem FM, Abd Allah FI, Abdel-Rashid RS, Hassan AAA (2020) Glibenclamide nanosuspension inhaler: development, *in vitro* and *in vivo* assessment. Drug Dev Ind Pharm 46(5):762–774. https://doi.org/10.1080/03639045.2020.1753062
48. Sharma OP, Patel V, Mehta T (2016) Nanocrystal for ocular drug delivery: hope or hype. Drug Deliv Transl Res 6:399–413. https://doi.org/10.1007/s13346-016-0292-0
49. Romero GB, Keck CM, Müller RH, Bou-Chacra NA (2016) Development of cationic nanocrystals for ocular delivery. Eur J Pharm Biopharm 107:215–222. https://doi.org/10.1016/j.ejpb.2016.07.005
50. Tuomela A et al (2014) Brinzolamide nanocrystal formulations for ophthalmic delivery: reduction of elevated intraocular pressure in vivo. Int J Pharm 467(1–2):34–41. https://doi.org/10.1016/j.ijpharm.2014.03.048
51. Kumari NU, Chary PS, Pardhi E, Mehra NK (2024) Tailoring micellar nanocarriers for pemetrexed in breast cancer: design, fabrication and *in vitro* evaluation. Nanomedicine 19:1145–1166. https://doi.org/10.2217/nnm-2024-0013
52. Gao Y et al (2011) Preparation and characterization of intravenously injectable curcumin nanosuspension. Drug Deliv 18(2):131–142. https://doi.org/10.3109/10717544.2010.520353
53. Sreeharsha N et al (2022) Enhancement of anti-tumoral properties of paclitaxel nano-crystals by conjugation of folic acid to Pluronic F127: formulation optimization, in vitro and in vivo study. Molecules 27(22):7914. https://doi.org/10.3390/molecules27227914
54. Gautam N et al (2013) Preclinical pharmacokinetics and tissue distribution of long-acting nanoformulated antiretroviral therapy. Antimicrob Agents Chemother 57(7):3110–3120. https://doi.org/10.1128/AAC.00267-13
55. Sun B, Taha MS, Ramsey B, Torregrosa-Allen S, Elzey BD, Yeo Y (2016) Intraperitoneal chemotherapy of ovarian cancer by hydrogel depot of paclitaxel nanocrystals. J Control Release 235:91–98. https://doi.org/10.1016/j.jconrel.2016.05.056
56. Pardhi E et al (2025) Enhancing solubility, bioavailability and anti-cancer efficacy of aprepitant through a naringin-based co-amorphous system. J Mol Liq 425:127225. https://doi.org/10.1016/j.molliq.2025.127225
57. Jati S et al (2025) Enhanced bioavailability and therapeutic efficacy of ibrutinib and naringenin-loaded self-nanoemulsifying drug delivery system. J Drug Deliv Sci Technol 109:106963. https://doi.org/10.1016/j.jddst.2025.106963
58. Pardhi E, Tomar DS, Khemchandani R, Samanthula G, Singh PK, Mehra NK (2024) Design, development and characterization of the apremilast and indomethacin coamorphous system. J Mol Struct 1299:137045. https://doi.org/10.1016/j.molstruc.2023.137045

59. Patil SK et al (2025) Development of venetoclax with 2-hydroxypropyl-beta-cyclodextrin inclusion complex for improved bioavailability. J Biomol Struct Dyn 43(9):4847–4864. https://doi.org/10.1080/07391102.2024.2305695
60. Gao L et al (2013) Application of drug nanocrystal technologies on oral drug delivery of poorly soluble drugs. Pharm Res 30(2):307–324. https://doi.org/10.1007/s11095-012-0889-z
61. Chang D et al (2018) Improved oral bioavailability for lutein by nanocrystal technology: formulation development, *in vitro* and *in vivo* evaluation. Artif Cells Nanomed Biotechnol 46(5):1018–1024. https://doi.org/10.1080/21691401.2017.1358732
62. Lei Y, Kong Y, Sui H, Feng J, Zhu R, Wang W (2016) Enhanced oral bioavailability of glycyrrhetinic acid via nanocrystal formulation. Drug Deliv Transl Res 6(5):519–525. https://doi.org/10.1007/s13346-016-0300-4
63. Sharma S, Verma A, Teja BV, Shukla P, Mishra PR (2015) Development of stabilized paclitaxel nanocrystals: in-vitro and in-vivo efficacy studies. Eur J Pharm Sci 69:51–60. https://doi.org/10.1016/j.ejps.2014.11.012
64. Gowda BHJ, Ahmed MG, Sahebkar A, Riadi Y, Shukla R, Kesharwani P (2022) Stimuli-responsive microneedles as a transdermal drug delivery system: a demand-supply strategy. Biomacromolecules 23(4):1519–1544. https://doi.org/10.1021/acs.biomac.1c01691
65. Schafer N, Balwierz R, Biernat P, Ochędzan-Siodłak W, Lipok J (2023) Natural ingredients of transdermal drug delivery systems as permeation enhancers of active substances through the *stratum corneum*. Mol Pharm 20(7):3278–3297. https://doi.org/10.1021/acs.molpharmaceut.3c00126
66. Jayaprakash R, Hameed J, Anupriya A (2017) An overview of transdermal delivery system. Asian J Pharm Clin Res 10(10):36. https://doi.org/10.22159/ajpcr.2017.v10i10.19909
67. Vora LK, Vavia PR, Larrañeta E, Bell SEJ, Donnelly RF (2018) Novel nanosuspension-based dissolving microneedle arrays for transdermal delivery of a hydrophobic drug. J Interdiscip Nanomed 3(2):89–101. https://doi.org/10.1002/jin2.41
68. Tekko IA, Permana AD, Vora L, Hatahet T, McCarthy HO, Donnelly RF (2020) Localised and sustained intradermal delivery of methotrexate using nanocrystal-loaded microneedle arrays: potential for enhanced treatment of psoriasis. Eur J Pharm Sci 152:105469. https://doi.org/10.1016/j.ejps.2020.105469
69. Clementino AR et al (2021) Structure and fate of nanoparticles designed for the nasal delivery of poorly soluble drugs. Mol Pharm 18(8):3132–3146. https://doi.org/10.1021/acs.molpharmaceut.1c00366
70. Bonaccorso A et al (2020) Optimization of curcumin nanocrystals as promising strategy for nose-to-brain delivery application. Pharmaceutics 12(5):476. https://doi.org/10.3390/pharmaceutics12050476
71. Wu C et al (2020) Intranasal delivery of paeoniflorin nanocrystals for brain targeting. Asian J Pharm Sci 15(3):326–335. https://doi.org/10.1016/j.ajps.2019.11.002

Role of Quality by Design (QbD) Approaches in Nanocrystal Fabrication

9

Rati Yadav, Padakanti Sandeep Chary, and Neelesh Kumar Mehra

Abstract

Improving the solubility of pharmaceutical compounds with poor aqueous solubility remains a significant challenge of bioavailability of APIs for product development. Many drugs classified under the Biopharmaceutics Classification System (BCS) Class II experience limited absorption due to slow dissolution rates, necessitating innovative strategies to enhance solubility. These challenges could be ameliorate using the nanocrystal technologies, pure drug nanoparticles stabilized by protective agents, which significantly increase the surface area available for dissolution, thereby improving drug absorption. Nanocrystals can be produced through top-down methods, such as wet milling and high-pressure homogenization, or bottom-up techniques like anti-solvent precipitation. Additionally, hybrid approaches that combine these methods can further refine particle size and improve process efficiency. The QbD framework is integral to the development of nanocrystal formulations that are both stable and effective. By

Rati Yadav and Padakanti Sandeep Chary contributed equally with all other contributors.

R. Yadav
Pharmaceutical Nanotechnology Research Laboratory, Department of Pharmaceutics, National Institute of Pharmaceutical Education and Research (NIPER), Ministry of Chemical and Fertilizer, Hyderabad, Telangana, India

P. S. Chary
College of Pharmacy and Pharmaceutical Sciences, Florida A&M University, Tallahassee, FL, USA

N. K. Mehra (✉)
Department of Pharmaceutics, National Institute of Pharmaceutical Education and Research (NIPER), Hyderabad, Telangana, India
e-mail: neelesh@niperhyd.ac.in

N. K. Mehra, K. K. Bansal (eds.), *Nanocrystal Technologies in Biomedical Sciences*,
https://doi.org/10.1007/978-981-95-2968-1_9

systematically identifying Critical Quality Attributes (CQAs) and optimizing Critical Process Parameters (CPPs), QbD ensures a controlled and reproducible manufacturing process. Important CQAs, such as particle size, polydispersity index, zeta potential, and solid-state characteristics, must be carefully assessed, as they directly impact on drug stability and bioavailability. Critical process parameters (CPPs) such as milling speed, processing time, bead size, stabilizer concentration, and drug load must be meticulously optimized to attain desirable nanocrystal characteristics. This chapter discusses the application of QbD principles in pharmaceutical nano-milling, with a focus on key methodologies, process optimization strategies, and relevant case studies. Establishing a well-defined factorial design and conducting thorough preliminary screening are crucial for determining an optimal design space. As a continuous improvement approach, QbD enhances formulation development, reduces costs, and ensures high-quality drug products that meet regulatory standards.

Keywords

Nanocrystal · QbD · Solubility · Techniques · DoE

9.1 Introduction

One of the major challenges in pharmaceutical sciences is the limited aqueous solubility of APIs [1]. A significant portion of both approved pharmaceutical compounds and new drug candidates exhibit poor water solubility, which poses difficulties in drug formulation and effectiveness [2]. Currently, around 70% APIs are under investigation fall into Biopharmaceutics Classification System (BCS) Class II, where dissolution rate is the primary factor by limiting absorption due to low aqueous solubility. Poor aqueous solubility results in a slower dissolution rate, which in turn affects bioavailability and biodistribution, preventing the achievement of optimal therapeutic concentrations [3]. Addressing these issues is crucial for enhancing drug efficacy and ensuring successful pharmaceutical formulations [4, 5].

Drug nanocrystals offer a promising solution to this challenge. These are pure drug nanoparticles, stabilized by a protective layer, typically measuring a few hundred nanometers in size [6–8]. The primary advantage of drug nanocrystals is their ability to enhance the solubility of poorly aqueous soluble drugs. By reducing particle size, the surface area available for dissolution increases significantly, as described by the Noyes–Whitney equation, which directly correlates interfacial surface area with dissolution rate. In addition to improving solubility, nanocrystals can also be employed for controlled drug release and targeted drug delivery [9–11].

Nanocrystal production methods are categorized into two main approaches: top-down and bottom-up techniques. Top-down methods involve breaking down larger drug particles into nanosized particles using techniques such as wet milling and high-pressure homogenization [2, 12]. In contrast, bottom-up techniques construct nanocrystals at the molecular level through methods like anti-solvent precipitation and liquid atomization [13]. Hybrid approaches combining multiple techniques can enhance process efficiency, help achieve smaller particle sizes, and prevent equipment clogging [14].

Process optimization is crucial in the formulation of nanocrystal. Following preliminary screening, critical process parameters should be systematically assessed to achieve an optimal product profile. The QbD framework facilitates this process by integrating quality into product and process development rather than relying solely on final product testing [15–17]. Regulatory agencies advocate for the QbD approach, emphasizing the selection of stabilizers, definition of Critical Quality Attributes (CQAs), and establishment of a well-defined design space to ensure consistent product quality [18].

QbD is extensively utilized in the biopharmaceutical industry for developing pharmaceuticals, biologics, surgical materials, excipients, and additives. Given their significance in healthcare, these products must undergo rigorous quality evaluations to ensure their safety and efficacy. Regulatory bodies have established stringent guidelines for assessing biopharmaceutical quality. However, challenges remain in effectively monitoring final product quality, leading to prolonged review processes, delayed market approvals, product inconsistencies, decreased patient acceptance, and increased recalls or rejections [19]. This chapter explores the application of the QbD approach in pharmaceutical nano-milling, providing an overview of nano-milling processes, mill designs, and critical process parameters, supported by case studies. Key terminologies related to QbD steps are explained, and practical examples demonstrate the successful implementation of QbD principles in pharmaceutical nano-milling.

9.2 Roadmap for Implementing QbD in Nanocrystals

The process of applying QbD to nanosystems involves several key steps: identifying the Quality Target Product Profile (QTPP), Critical Quality Attributes (CQAs), and Critical Process Parameters (CPPs), as well as developing a design space based on prior research and knowledge as shown in Fig. 9.1. Risk analysis and control strategies are integral throughout the process. Finally, the design space is evaluated and refined to align with manufacturing requirements.

9.2.1 Identifying QTPP

The identification of QTPP is based on prior scientific understanding and its relevance to in vivo outcomes. This encompasses factors like dosage form, route of administration, the release of the therapeutic agent, and the pharmaceutical properties necessary for the intended market product [20].

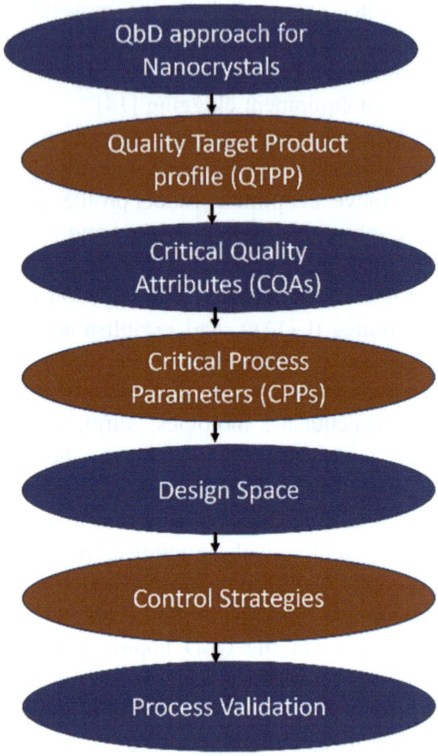

Fig. 9.1 Flow chart for Implementing QbD in Nanoformulations. (Self drawn)

9.2.2 Identifying CQAs

Once the QTPP is defined, the next step is to identify the CQAs. These are typically identified through risk assessment, drawing from previous knowledge and research. Key critical quality attributes (CQAs) for nanosystems typically include parameters such as particle size, polydispersity index, surface charge (zeta potential), drug loading or encapsulation efficiency, and overall formulation stability [18].

9.2.3 Screening for Significant Process Parameters

A Plackett–Burman design is employed to initially identify potential risk factors, which are then subjected to in-depth risk analysis to evaluate their influence on the process. Kuk et al. developed a resveratrol nanosuspension with enhanced oral bioavailability using a novel anti-solvent precipitation method that required no solvent removal or heating, guided by a QbD approach. Key Critical Quality Attributes (CQAs)-particle size, zeta potential, and drug content were identified based on the Quality Target Product Profile (QTPP). Using a Box-Behnken design, the optimal formulation included poly-vinyl-pyrrolidone vinyl acetate (10 mg/mL), PVP K12 (5 mg/mL), sodium lauryl sulfate (1 mg/mL), and diethylene glycol

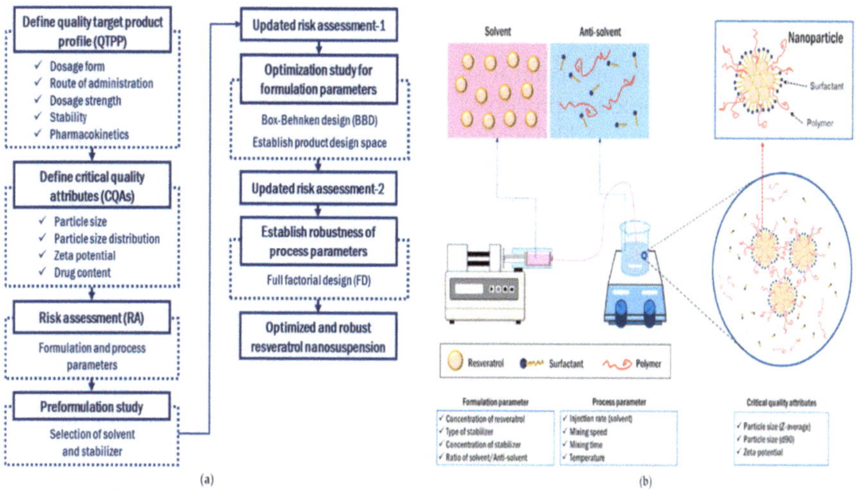

Fig. 9.2 (**a**) Development flow, and manufacturing process (**b**) of the resveratrol suspension using anti-solvent approach employing the QbD approach. (Reproduced with copyright permission from open access Do-Hoon Kuk et al. [21])

monoethyl ether (DEGEE; 5% v/v) at 5 mg/mL resveratrol, yielding a nanosuspension with 46.3 nm particle size and 38.02 mV zeta potential. Robustness was validated via full factorial design [21] (Fig. 9.2).

9.2.4 Defining the Design Space

The design space consists of both the product design space and the process design space. The product design space is established by evaluating various CQAs as the dimensions. After determining the variability of the CQAs, the process design space is developed, which links CQAs to CPPs. Key steps include conducting a Design of Experiments (DoE) to explore the process, performing a risk analysis to evaluate how variables affect CQAs, and determining the operating region based on a predefined confidence level. It's important to note that defining the design space requires multivariate experimentation, considering both the primary effects of factors and their interactions [18, 20].

9.2.5 Defining the Control Strategy

Unlike traditional methods, QbD's control strategy is dynamic and typically involves two levels. Level 1 is real-time, automatic control that monitors CQAs of materials and adjusts process parameters accordingly. Level 2 involves less frequent testing of the end products, with the ability to flexibly adjust CQAs and CPPs within the design space [22].

9.2.6 Validated Process and Filling Steps

After the design space and control strategy have been established, it is crucial to verify whether the product, processed within the defined design space, meets the targeted quality criteria. Additionally, it must be confirmed that the design space, whether at the lab or pilot scale, is suitable for full-scale manufacturing. Finally, regulatory filings must include the design space, specifying the ranges for CQAs and CPPs [18].

9.3 Methods of Nanocrystal Preparation

The synthesis of nanocrystals involves various techniques aimed at controlling their size, shape, and composition. These methods are generally categorized into two main types: bottom-up strategies like chemical precipitation and sol-gel processes, and top-down methods such as ball milling and lithographic techniques [23]. Scientists have devised accurate methods to control the properties of nanocrystals at both atomic and molecular scales, highlighting the importance of comprehending the core principles underlying each synthesis technique [24, 25].

Nanocrystal fabrication methods are designed to present a comprehensive understanding of the different available approaches, highlighting their unique benefits, limitations, and possible areas of application. As nanotechnology advances, a comprehensive understanding of these fabrication techniques is crucial for optimizing nanocrystal properties and unlocking their full potential in scientific and technological innovations [26].

9.3.1 Microfluidization

Microfluidization is a high-shear processing technique that utilizes specifically engineered microchannels to force liquids through at extremely high pressures, often reaching up to 1700 bar (25,000 psi). Powered by compressed air, this system facilitates the high-velocity mixing of lipids and active ingredients within microchannels, producing uniform and stable nanoscale delivery systems generally sized between 50 and 100 nm [27].

In industrial applications, two main types of microfluidizers are employed: the single-step dual-channel and the two-step single-channel systems. The single-step dual-channel microfluidizer provides advantages including lower energy consumption and enhanced efficiency in using expensive lipids and oils during the production of nanoformulations. This technique is widely applied in the development of liposomal formulations designed to encapsulate a variety of therapeutic agents, including nucleic acids, proteins, and small-molecule drugs with diverse physicochemical characteristics [2, 28, 29].

9.3.2 Milling Method

In the 1990s, researchers at Sterling Drug Inc./Eastman Kodak pioneered wet ball milling (WBM), a wet media milling technique. Originally created to decrease the particle size of hydrophobic drugs, this novel approach drew inspiration from milling methods commonly employed in the paint and photography sectors. Today, this technology is implemented in five different devices available commercially, underscoring its broad acceptance and impact in the pharmaceutical field [4]. Low-energy wet ball milling (LE-WBM), the simplest variant of this technique, utilizes a jar containing plain glass beads as the milling medium. In this approach, a coarse drug material—ideally micronized—is dispersed in a medium that contains one or more stabilizing agents. LE-WBM is recognized for generating extremely fine particles with a narrow size distribution, primarily due to size reduction occurring through cleavage and abrasion, particularly when the milling duration is prolonged [30]. To adapt this method for large-scale pharmaceutical manufacturing, more advanced high-energy milling techniques were introduced. High-energy wet ball milling (HE-WBM) serves as the foundation for today's NanoCrystal™ technology. The widespread success of wet ball milling stems from its flexibility, allowing it to process almost any active pharmaceutical ingredient (API). Additionally, it is noteworthy that nanosuspension produced via wet ball milling-whether through low-energy or high-energy methods-typically achieve particle sizes in the 100–300 nm range [31].

9.3.3 Extrusion

Extrusion is a process where material is pushed through a shaped opening, or die, under precise conditions to create products with consistent form and density. One of the most significant innovations in this area is hot-melt extrusion (HME), which has enabled the development of various dosage forms. Basic extrusion equipment typically consists of a motor, an extrusion barrel, an extrusion die, and a rotating screw [32, 33].

When compared to traditional pharmaceutical manufacturing techniques, several advantages of HME technology become evident. These include faster production times, environmental benefits by eliminating the need for solvents during processing, and enhanced drug delivery efficiency that is more patient-focused [34].

9.3.4 Nanoprecipitation

Nanoprecipitation involves adding a solvent phase into an anti-solvent under gentle magnetic stirring, leading to particle formation through rapid precipitation [35]. The formation of nanocrystals (NCs) through nanoprecipitation is mainly driven by the "diffusion-stranding phenomenon." In this process, solvent and polymer chains from the organic phase begin to diffuse into the surrounding aqueous phase. As this diffusion progresses, the drug experiences desolvation, resulting in nanocrystal formation [35]. Following the process, organic solvents are typically eliminated by application of a rotary evaporator operated at a suitable temperature, whereas the aqueous phase is separated through techniques like lyophilization or ultracentrifugation [36]. Nanocrystal (NC) formation via nanoprecipitation occurs through a series of stages: supersaturation, nucleation, growth through condensation, and growth via coagulation. The solvent phase typically contains a film-forming polymer, one or more active pharmaceutical compounds, a lipophilic surfactant, and organic solvents. When the anti-solvent is introduced, the solubility of the solute drops, leading to supersaturation and subsequent precipitation [37].

The anti-solvent coprecipitation technique consists of three core stages: first, the active pharmaceutical ingredient (API) along with excipient(s) is dissolved in a suitable solvent; second, this solution is combined with an anti-solvent; and third, coprecipitation occurs, followed by collection of the resulting product. As solute solubility decreases during mixing, particles begin to aggregate, crystals grow, and nucleation starts within the solution [37, 38].

In a study, nanocrystals of niclosamide through nanoprecipitation, using hyaluronic acid and PVP/PVA as stabilizing agents was developed and evaluated for bioavailability enhancements. The resulting niclosamide nanocrystals (Ncl-NC) had an average size of 173.3 ± 10.2 nm and a PDI of 0.221 ± 0.09. These nanocrystals were then incorporated into Carbopol® 940 hydrogels to create a topical formulation. Using a QbD approach, three formulations were optimized as shown in Fig. 9.3. The nanocrystal-loaded gels exhibited shear-thinning behavior and released the drug following the Higuchi model, with 86.39 ± 2.4%, 77.57 ± 2.36%, and 71.27 ± 3.2% of the drug released within 12 h. By transforming niclosamide into its nanocrystal form, the IC50 value was significantly reduced from 1.25 μM to 0.017 μM and 0.023 μM in the HaCaT cell line, indicating enhanced

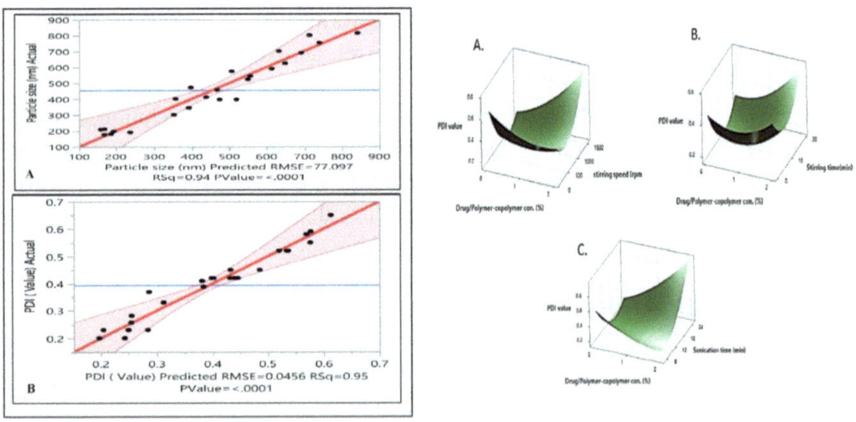

Fig. 9.3 Design of experiment results illustrating the Regression coefficient, *P*-value, and Surface response plots for drug/polymer-copolymer ratio, stirring speed, stirring time, and sonication time. (Reproduced with copyright permission from Kumar et al. [39])

potency. These findings indicate that niclosamide may be repurposed for cancer therapy by addressing its poor solubility and bioavailability challenges. Additionally, the study demonstrated that incorporating Ncl-NCs into a gel matrix significantly improves niclosamide's solubility and permeability. This approach holds promise as a treatment for psoriasis, although further validation through in-vivo studies is necessary [39].

Proposed a structured methodology integrating Quality-by-Design (QbD) principles and Design of Experiments (DoE) for the semi-automated nanocrystallization of active pharmaceutical ingredients (APIs) through the solvent–anti-solvent precipitation technique. Figures 9.4 and 9.5 present the Design of Experiments (DoE) plots, which demonstrate the impact of different CMAs and CPPs on the quality of CUR nanocrystals synthesized using the semi-automated solvent–anti-solvent precipitation technique, offering valuable insights into their influence. Automating the injection step contributed to a stable, dependable, and repeatable process, making it well-suited for industrial-scale manufacturing practices. Curcumin (CUR), selected as a model drug due to its low water solubility (0.6 μg/mL at 25 °C) and log*P* value of 3.29, was used in the study. Based on a risk assessment, the average particle size and PDI in 24 h after preparation were identified as critical quality attributes (CQAs). Critical material attributes (CMAs) and critical process parameters (CPPs) were evaluated using a fractional factorial 24-1 screening design, and significant factors were further investigated with a Central Composite Design (CCD). The optimization process enabled the determination of the design space and the establishment of a control strategy, which is essential for regulatory applications. To achieve more accurate characterization of particle size and PDI in the optimized formulation, an orthogonal method was utilized [40].

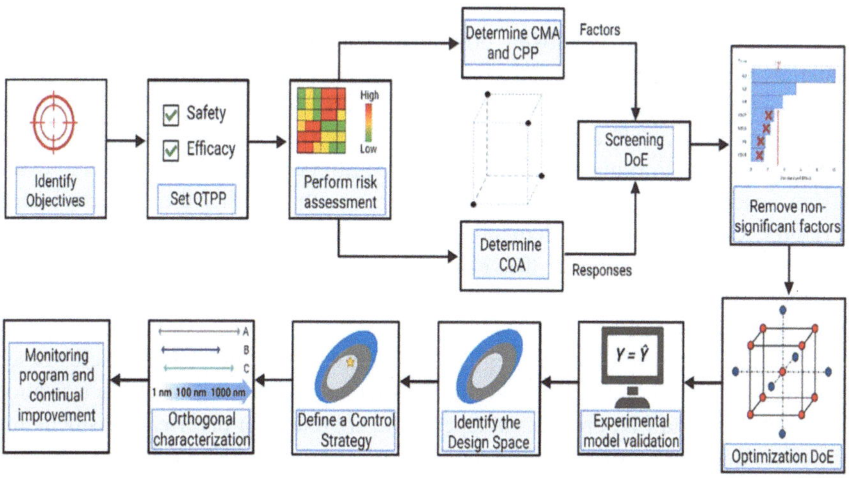

Fig. 9.4 A combined workflow incorporating Quality-by-Design (QbD) principles and Design of Experiments (DoE) methodology. (Reproduced with copyright permission from open access Henríquez et al. [40])

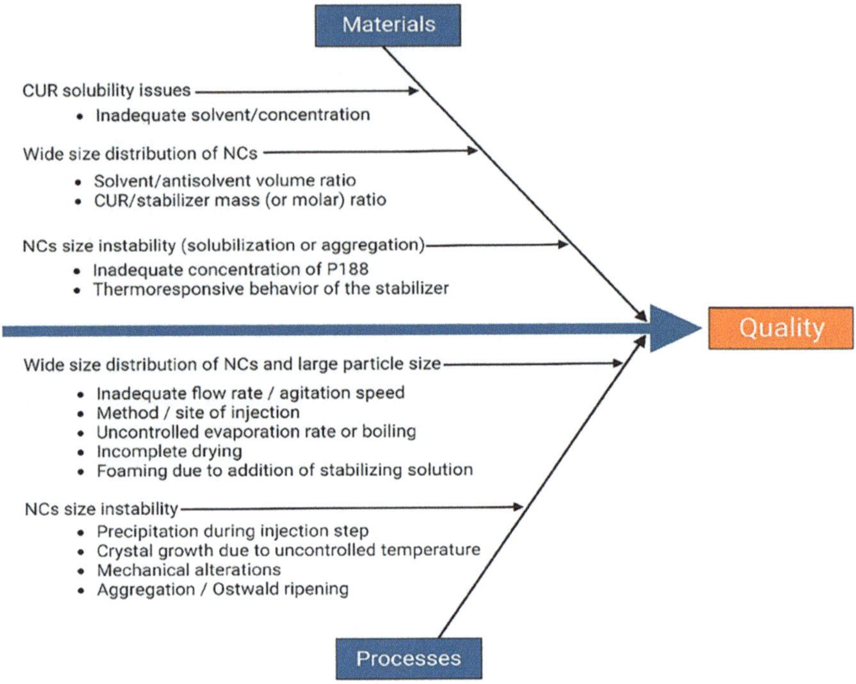

Fig. 9.5 Ishikawa diagram for identifying potential causes and effects on CUR nanocrystals prepared by the solvent-anti-solvent semi-automated nanoprecipitation method, focusing on materials and processes branches. (Reproduced with copyright permission from open access Henríquez et al. [40])

The QbD approach employed provides comprehensive insights into both the materials and the semi-automated synthesis process, facilitating a rational evaluation of the factors involved. When combined with DoE for screening and optimization, QbD proves to be an effective strategy for identifying and exploring the design space in NC synthesis. Working within the optimal combination of factors predicted in the design space is anticipated to produce a product that fulfills the desired quality attributes. Consequently, implementing an appropriate control strategy helps maintain conditions that support the formation of nanocrystals (NCs) meeting the specified criteria for particle size, PDI, and stability, as defined in the Quality Target Product Profile (QTPP).

Although a well-defined regulatory framework for nanoparticles-based formulations is still lacking, employing orthogonal techniques for characterization of particle size and PDI provides a dependable approach for precise assessment and objective data interpretation. The integrating systematic data collection for the standardization of the nano crystallization process is helpful for the successful translation into commercial/practical application. The detailed protocol presented is beneficial to aid in the early-stage development of drug-based nanocrystalline formulations, especially those incorporating automation to enhance product quality. Future investigations should aim to extend this methodology to the biological assessment of nanocrystals and explore its applicability in scale-up procedures [40].

9.3.5 Spray Drying

Spray drying is a continuous, cost-effective, and scalable process used to convert liquid feed into dry powders. This is achieved by atomizing the liquid into a high-temperature drying gas medium through an atomizer. It is a versatile technique widely used in various industries, including food, cosmetics, and pharmaceuticals. Over the years, spray drying has seen continuous improvements, leading to the advanced equipment and methods available today [41, 42].

Kumar et al. applied QbD principles to optimize the spray drying process for converting liquid nanosuspension into solid nano-crystalline dry powders, using indomethacin as a model drug. A full factorial design was used to investigate the impact of critical process variables—namely inlet temperature, flow rate, and aspiration rate-on key critical quality attributes (CQAs) such as particle size, moisture content, percent yield, and crystallinity. Additionally, a central composite design was employed to construct response surfaces for particle size and percent yield. Statistical tools including multiple linear regression analysis and ANOVA were used to identify significant process parameters, define their relationship with CQAs, model the spray drying process, and establish a design space. The results revealed that inlet temperature was the only significant factor influencing particle size ($p < 0.05$), with higher temperatures causing surface melting and aggregation of nano-crystalline powders [43].

One of the key advantages of spray drying is that it often eliminates the need for surfactants and stabilizers, which are typically required in other bottom-up methods to prevent particle agglomeration, thereby improving yield. In addition, because spray drying is a fast process with limited heat exposure, it is especially well-suited for drying heat-sensitive substances like heat-labile drugs. These features make spray drying an ideal and efficient technology for a range of pharmaceutical applications [44–46].

9.3.6 Supercritical Fluid Technology (SCF)

SCFs are recognized for their ability to preserve uniformity even at temperatures exceeding their critical points. Supercritical carbon dioxide (SC-CO_2) is distinguished by its moderate critical parameters, affordability, non-flammable nature, and advantageous safety profile. The use of SCF technology presents multiple benefits, such as accelerated precipitation and the capacity to generate pure drug powders through a single-step procedure. Moreover, SCFs have become increasingly popular as an eco-friendly option for manufacturing biodegradable micro- and nanoparticles [4].

9.4 Case Studies Illustrating the Implementation of Quality by Design in Nanocrystal Formulation

QbD is a scientific and structured method aimed at attractive quality by systematically applying risk-based tactics. It concentrates on accomplishing predefined objectives through a comprehensive understanding of product characteristics and effective process control to ensure the desired quality attributes in newly developed formulations. Real-world applications of QbD in nanocrystal fabrication offer valuable insights into its effectiveness. The case studies below demonstrate the critical steps taken from formulation design to process control.

Ahmad Ainurofiq and his team reported resveratrol (RSV) nanocrystals using a QbD technique, embedding them into a soluble mesoporous carrier to elevate bioavailability and formulation stability. Although RSV possesses potent antioxidant activity, its clinical application is imperfect by poor solubility and limited bioavailability. Traditional nanosuspension-based nanocrystal formulations often face challenges with bulkiness, physical instability, and scalability. To overcome these limitations, the study employed a scientifically justified QbD framework, utilizing a Box-Behnken design (BBD) to optimize critical parameters such as particle size, polydispersity index, zeta potential, drug loading, and supersaturation. The optimized formulation, with a desirability value of 68%, yielded nanocrystals of 133.3 ± 1.2 nm and a zeta potential of 24.6 mV. Moreover, the formulation revealed nearly a two-fold increase in bioavailability compared to a conventional RSV suspension. To optimize the formulation, a BBD was employed, focusing on three independent variables: PEG 400 (factor *A*), Kolliphor EL (factor *B*), and RSV loading (factor *C*), all evaluated at three levels (Fig. 9.6). The experimental data were organized and analyzed based on specific response parameters. A second-order polynomial equation was used to model the influence of these variables and their interactions on the responses, expressed as:

$$Y = \beta + \alpha_1 A + \alpha_2 B + \alpha_3 C + \alpha_4 AB + \alpha_5 AC + \alpha_6 BC + \alpha_7 A2 + \alpha_8 B2 + \alpha_9 C2$$

where Y represents the predicted response, and the α terms represent the coefficients for linear, interaction, and quadratic effects.

The statistical analysis of the BBD was performed using multiple linear regression analysis (MLRA), where β represents the intercept, A–C are the coded variables, and α_1–α_9 denote the coefficients for main, interaction, and quadratic effects. Model evaluation was conducted using observed vs. predicted plots and key statistical parameters, including R^2, adjusted R^2, predicted R^2, adequate precision (AP), and PRESS, all computed via Design Expert v22 (Stat-Ease Inc.). The significance of each variable's impact on responses was assessed through ANOVA with a 95% confidence level ($p < 0.05$). Contour and superimposed contour plots were constructed to identify the optimal formulation, guided by the target product profile of RSV nanocrystals (Fig. 9.6). Optimization criteria included particle size, PDI, zeta potential, drug loading, supersaturation time (ST), peak level (SC), area under the supersaturation curve (AUC), and degree of supersaturation (SD) [47].

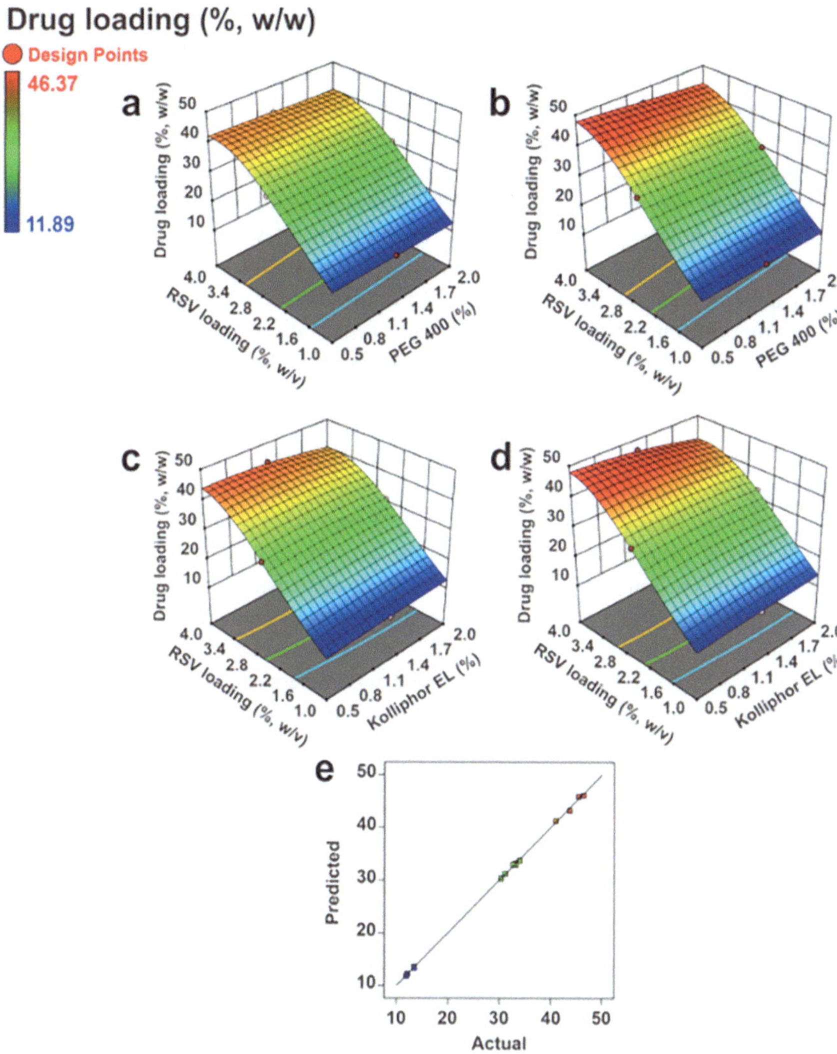

Fig. 9.6 Contour plot of drug loading parameters at a minimum concentration of Kolliphor EL (**a**), a minimum concentration of PEG 400 (**b**), a maximum concentration of PEG 400 (**c**), a maximum concentration of Kolliphor EL (**d**), and predicted vs. actual graph (**e**). (Reproduced with copyright permission from Ainurofiq et al. [47])

Iurian et al. developed oral lyophilizes of meloxicam (MEL) nanocrystals using a Quality QbD approach to enhance drug dissolution. Nanosuspension was prepared via high-pressure homogenization with PVP, poloxamer, or PEG as stabilizers, and mannitol as a cryoprotectant during freeze-drying. The effects of formulation variables on particle size and polydispersity index were analyzed, leading to an optimized formulation. This was used to create oral lyophilizes, with sodium alginate and cross-carmellose sodium tested as matrix agents under three freezing conditions. The formulation was optimized by selecting excipients that provided both strong mechanical properties and rapid meloxicam (MEL) dissolution. Among the tested polymers, poloxamer achieved the greatest particle size reduction down to 477.6 ± 7.5 nm, approximately 10.27% of the original size-though a slight increase was observed post-freeze-drying. PEG was less effective during homogenization but showed further size reduction after lyophilization. Given poloxamer's superior size reduction and ability to maintain MEL crystallinity, it was selected for the final formulation at 1% concentration, along with 5% mannitol as a cryoprotectant. In matrix development, sodium alginate (SA) outperformed cross-carmellose sodium (CCS) by offering better structure and faster drug release, even at low concentrations. The optimal formulation included 1.157% SA and underwent thermal treatment during freeze-drying. It showed a rapid disintegration time of 3.33 s and released 77.14% of MEL within 2 min. The use of a QbD strategy ensured robust product quality and a deep understanding of the formulation process. Drawing from initial experiments and existing literature, the study identified key variables: type of stabilizers (X1), their concentration ratios (X2), and the cryoprotectant-to-drug ratio (X3) as critical parameters. A 3^3 factorial design was implemented to formulate the MEL nanosuspension. To assess how the nanosuspension responded throughout both the preparation and freeze-drying stages, particle size and PDI were measured post-high-pressure homogenization (HPH) (Y1, Y2), after freeze-drying and reconstitution (Y3, Y4), and the changes in these parameters due to freeze-drying (Y5, Y6). Multiple linear regression and ANOVA were applied to construct predictive models for each response. Post-HPH, particle sizes varied between 477.6 nm and 1232.0 nm, with PDI values ranging from 0.29 to 0.88 [48].

The smallest particles were obtained using PVP and poloxamer as stabilizers, whereas PEG resulted in significantly larger sizes, reaching up to 1.2 μm. The broad variability in the results highlights the substantial impact of the choice of stabilizer on the formulation's particle characteristics. Particle size and PDI were evaluated after freeze-drying to determine the impact of the independent variables on the overall process. Meloxicam was fully solubilized in the nanosuspension following high-pressure homogenization, which was reflected in the statistical analysis. Post freeze-drying, the average particle sizes ranged from 489.5 to 836.1 nm, indicating a reduction in the upper size limit compared to the initial measurements [48].

In an another research investigation reported by Panwar et al. [49], formulation development of Bcl-2 inhibitor (venetoclax) loaded hyaluronic stabilized nanocrystal for targeted delivery to breast cancer cells. The HA-nanocrystals were synthesized using the anti-solvent precipitation method and optimized via central composite design (CCD). Both hyaluronic acid-coated (HA-VEN-NCs) and uncoated (VEN-NCs) formulations were assessed for in-vitro drug release, cytotoxic effects on cancer cell lines, binding to CD-44 overexpressing breast cancer cells, and overall anticancer potential. The CCD design was employed to assess both the qualitative and quantitative influence of various formulation variables on the nanocrystals, including their interactive effects. The factors were evaluated using analysis of variance (ANOVA) at a 95% confidence interval, while minimizing the number of experimental runs to reduce error and overall cost. The experiment investigated the effects of four independent variables-HA/VEN ratio (X1), acetone/water ratio (X2), stirring speed (X3), and sonication time (X4)-on particle size and PDI. Using 30 randomized runs, the design minimized uncontrolled variability. ANOVA, based on Fisher's analysis, confirmed that all factors significantly impacted the model ($F \geq 0.05$, $p < 0.05$). High correlation coefficients ($R^2 = 0.85$ for both) confirmed a strong fit between experimental and predicted outcomes. CCD regression enabled the development of predictive models, with particle sizes ranging from 86.4 to 274.4 nm [49].

The mupirocin (MP) nanocrystals for topical use to improve treatment of skin infections, especially those caused by methicillin-resistant *Staphylococcus aureus* (MRSA) was developed and evaluated. The research aimed to formulate mupirocin nanocrystals (MP-NC) using the nanoprecipitation method, with process optimization carried out through a D-optimal response surface design. The optimized formulation produced uniformly spherical nanocrystals with an average particle size of approximately 70 nm and a polydispersity index of 0.2, indicating narrow size distribution. In this research, the optimization of MP-NC was carried out by applying a D-optimal response surface design implemented in Design-Expert® software (v12.0, Stat-Ease Inc., USA). The CMAs encompassed three continuous variables-homogenization time (X1), drug concentration in the organic phase (X2), and the drug-to-stabilizer ratio (X3)-in addition to one categorical variable, which was the type of stabilizer used (X4). These factors were estimated for their impact on critical quality attributes (CQAs). Factor levels were selected to maximize the design space while sustaining practical circumstances for nanocrystal processing. The D-optimal design enabled efficient selection of factor levels and analysis with a reduced number of experimental trials. Two primary responses-particle size (PS, Y1) and polydispersity index (PDI, Y2)-were evaluated, with the objective of minimizing both to maximize overall desirability. The optimal parameters obtained from this analysis were subsequently utilized to formulate MP-NC for integration into a topical cream [50].

Design-Expert® generated 24 randomized runs, including three replicates, to reduce bias and improve model accuracy. Responses were fitted to linear, 2FI, and quadratic models, and evaluated using ANOVA. Higher-order interactions were excluded for model simplicity. Model validity was assessed using p-values, lack-of-fit, Adjusted R^2, Predicted R^2, and R^2. The model with the highest Adjusted and Predicted R^2 and a non-significant lack-of-fit was selected. The aim was to reduce both the particle size, and PDI the formulated MP-NC. A simultaneous optimization approach was applied to achieve this goal. The formulation with the highest desirability score was selected, with greater emphasis placed on minimizing both PS and PDI. A maximum desirability score of 0.965 was achieved for the MP-NC formulation prepared with PVA 4–88 as the stabilizer, 1% w/v mupirocin in the organic phase, a drug-to-stabilizer ratio of 1:4, and a homogenization duration of 5 min (Fig. 9.7). These results suggest that employing a Quality by Design (QbD) strategy enabled the development of optimal MP-NC formulations characterized by ultra-

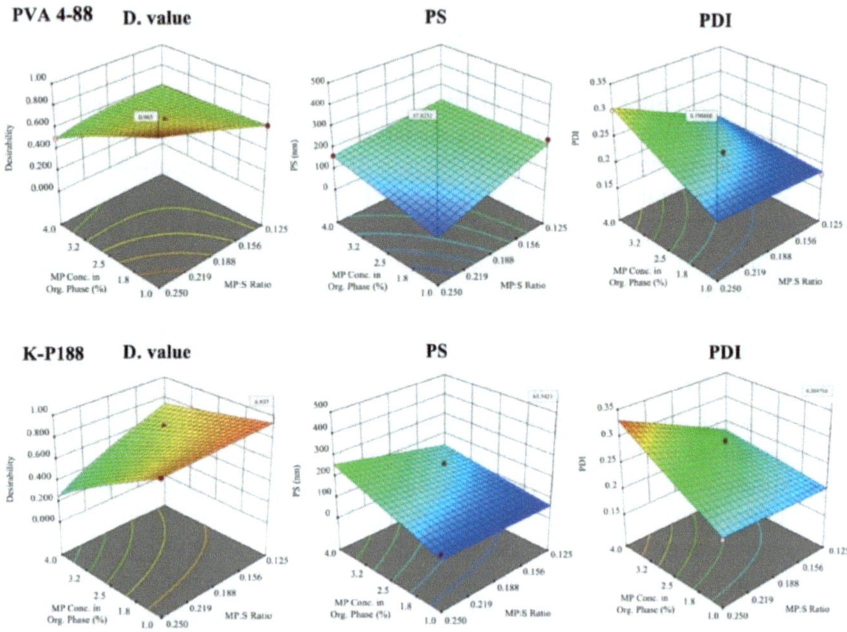

Fig. 9.7 Response surface plots showing the desirability values (*D*. value) and the optimum levels of the studied factors used in the optimization of MP-NC using PVA 4–88 and K-P188 as stabilizers. (Reproduced with copyright permission from Najm et al. [50] (open access))

small particle sizes (less than 100 nm) and acceptable polydispersity index values. As a result, the two best-performing formulations were selected for further investigation [50].

Another interesting study was reported for the development of the ivermectin nanocrystals (IVM-NC) using optimization technique to enhance topical delivery, addressing rising anti-parasitic resistance and limited drug efficacy. Using microfluidization and a D-optimal design, they evaluated key formulation variables. The optimized, lyophilized IVM-NC showed uniform particles (186 nm, PDI 0.4), a 730-fold increase in solubility, and a 24-fold improvement in dissolution. Ex-vivo studies on pig ear skin revealed a three-fold increase in dermal delivery. When formulated into a cream, IVM-NC demonstrated superior drug release compared to a marketed product, supporting its potential for improved topical anti-parasitic therapy [51].

In this investigation, a D-optimal response surface design was applied using Design-Expert® software (Version 13.0, Stat-Ease Inc., Minneapolis, MN, USA) to optimize the formulation of IVM nanocrystals (IVM-NC). Three continuous numerical factors and two categorical variables were selected as CMAs and CPPs to evaluate their influence on key quality attributes (CQAs). These included (1) number of microfluidization cycles (X1), (2) ivermectin concentration (X2), (3) stabilizer-to-drug ratio (X3), (4) stabilizer type (X4), and (5) pre-microfluidization homogenization (X5). The variable levels were chosen to define a suitable design space while maintaining practical feasibility for nanocrystal production. Three key responses were monitored during optimization: (1) particle size (PS, Y1), (2) polydispersity index (PDI, Y2), and (3) zeta potential (ZP, Y3). The goal was to minimize PS and PDI while maximizing ZP to achieve the highest overall desirability. The optimal levels of the independent variables were then applied to prepare the final IVM-NC formulation for subsequent testing and inclusion in the topical cream (Fig. 9.8).

A total of 34 experimental runs, including five replicates, were generated using Design-Expert® software. The responses were simultaneously fitted to linear, two-factor interaction (2FI), and quadratic models. Polynomial equations were validated statistically through analysis of variance (ANOVA), with higher-order interactions (three or more factors) excluded for simplicity. Key statistical metrics such as *p*-value, lack-of-fit *p*-value, adjusted R^2, predicted R^2, and R^2 were evaluated to confirm model significance.

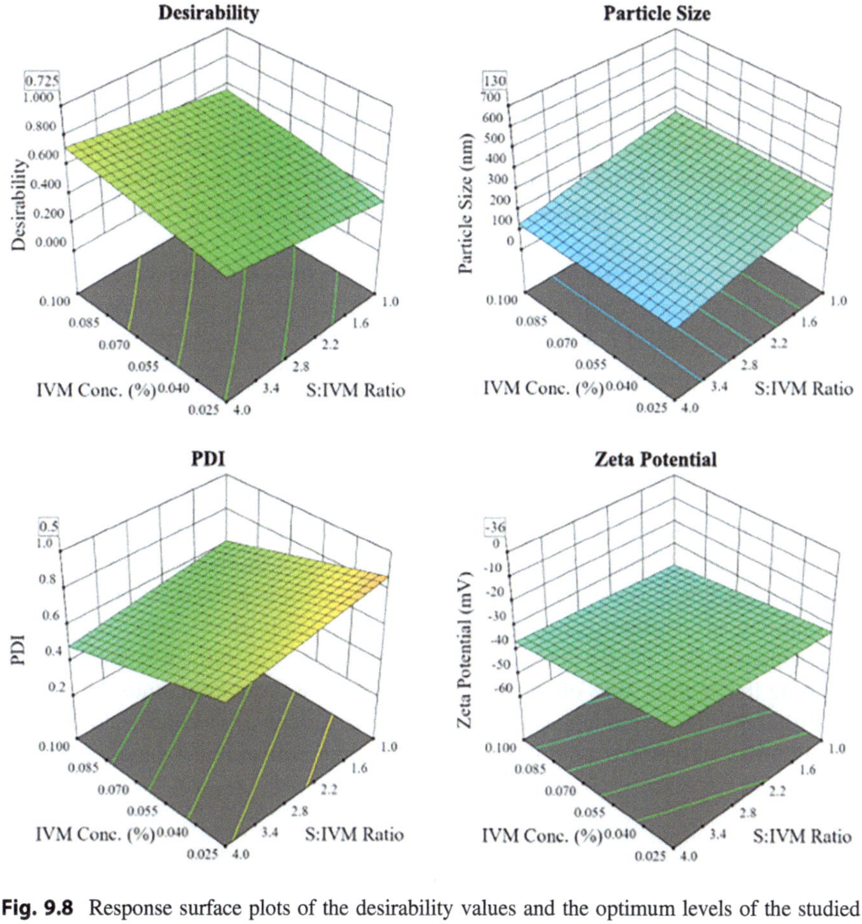

Fig. 9.8 Response surface plots of the desirability values and the optimum levels of the studied factors used in the optimization of IVM-NC using SDS as stabilizer. (Reproduced with copyright permission from Awad et al. [51] (open access))

In this study, a systematic formulation optimization of Ivermectin nanocrystals (IVM-NC) was carried out using response surface analysis based on a D-optimal design. The primary goal was to tailor the formulation to meet predefined critical quality attributes (CQAs), specifically minimizing particle size and polydispersity index while maximizing the absolute zeta potential (ZP). These CQAs are crucial for ensuring nanoparticle stability and uniformity, where a lower PDI indicates a more homogeneous (monodisperse) size distribution. The optimization process identified the ideal formulation parameters: sodium dodecyl sulfate (SDS) as a stabilizer at a 4:1 SDS-to-IVM ratio, with an IVM concentration of 0.1% (*w/v*). The selected formulation underwent a two-step process involving homogenization followed by three cycles of microfluidization. The resulting IVM-NC formulation achieved a desirability score of 0.725, indicating a high level of optimization across all CQAs and confirming the effectiveness of the design approach in developing a robust nanocrystal system (Fig. 9.8) [51].

9.5 Regulatory Approach for Nanocrystals in Drug Delivery

Nanocrystal technology has emerged as a promising approach to enhance the delivery of poorly water-soluble active pharmaceutical ingredients (APIs). This innovative strategy has significantly improved drug solubility and bioavailability, leading to the submission of over 80 drug applications incorporating nanocrystals to the U.S. Food and Drug Administration (FDA). These formulations have been developed for diverse therapeutic applications and are administered via multiple routes [52].

A retrospective analysis of past FDA submissions has been conducted to identify critical developmental challenges associated with nanocrystal-based drug products. Notably, more than 60% of these submissions pertain to orally administered drugs. This pattern is especially significant for BCS Class II drugs, which are characterized by high permeability in the intestines but very poor water solubility. Given these characteristics, nanocrystal formulations have become a preferred strategy to improve drug dissolution and absorption, thereby facilitating regulatory approval by the FDA [25, 53].

The FDA exercises regulatory oversight of nanotechnology-based products within its existing statutory framework, employing a product-specific, science-driven evaluation process. This approach facilitates the integration of advanced pharmaceutical technologies while ensuring compliance with regulatory requirements. The agency has established specific size criteria for nanocrystals, requiring that at least 10% of nanocrystal particles in a formulation must be smaller than 1000 nm. These guidelines aim to standardize quality assessments and enhance the reliability of nanocrystal formulations in pharmaceutical development [4].

9.5.1 Quality Aspects Related to Chemistry, Manufacturing, and Control (CMC) Processes

A comprehensive analysis of regulatory pathways such as New Drug Applications (NDAs), Abbreviated New Drug Applications (ANDAs), and Investigational New Drug (IND) filings highlights the unique hurdles in developing nanocrystal-based drug formulations. Key quality concerns typically center around achieving and preserving the physical and chemical stability of the active pharmaceutical ingredient, ensuring consistent manufacturing of the final product, and adhering to stringent regulatory criteria for quality specifications. These challenges underscore the complexity of bringing nanocrystal formulations to market under current regulatory frameworks.

Chemistry, Manufacturing, and Controls (CMC) considerations are integral to ensuring the quality, safety, and efficacy of pharmaceutical products. CMC activities involve detailed evaluation of the active pharmaceutical ingredient (API), formulation development, process design, analytical methods, and stability data. Regulatory submissions must demonstrate compliance with Good Manufacturing Practices (GMP) and predefined quality standards through comprehensive CMC documentation. Effective implementation of CMC principles facilitates control over critical quality attributes (CQAs) and critical process parameters (CPPs), ensuring consistent product performance. In the case of nanocrystal formulations, quality challenges such as particle size variability, stability concerns, and polymorphic transitions demand meticulous attention. Addressing these factors is essential for regulatory acceptance and therapeutic reliability [4].

9.6 Advancements in Nanocrystal-Based Drug Development

Since the early 2000s, notable progress in nanocrystal technology has enabled the redesign and improvement of micronized active pharmaceutical ingredients (APIs) that were already in state of market approval. The 505(b)(2) regulatory pathway empowers pharmaceutical multinationals to seek FDA approval for reformulated nanocrystal-based drugs by leveraging existing data and published scientific literature, thereby reshuffling the approval process [4].

A notable benefit of nanocrystals lies in their amorphous state, which enhances drug solubility and dissolution rates associated to their crystalline counterparts. Nanocrystalline and amorphous form have ability to remarkably improve the bioavailability by enhancing its saturation solubility. However, maintaining the amorphous state over the product's shelf life is a critical challenge that must be addressed through appropriate formulation tactics [4].

9.7 Conclusion

The application of the QbD approach in nanocrystal formulation offers a systematic and science-based framework for developing robust and effective drug products. By initiating development with a clear definition of the desired critical quality attributes (CQAs) and understanding the influence of critical process parameters (CPPs), QbD enables precise control over the nanomilling process. Although identifying key product characteristics may be straightforward, determining their criticality and establishing appropriate acceptance criteria requires comprehensive process knowledge. The interdependence of CPPs further underscores the need for detailed preliminary studies and well-defined design spaces, particularly during scale-up or scale-down phases. As a lifecycle-oriented methodology, QbD supports continuous improvement by allowing for iterative optimization based on new insights. Incorporating factorial design strategies within the QbD framework not only enhances the efficiency of nanocrystal research but also improves the quality, consistency, and cost-effectiveness of pharmaceutical development.

Conflict of Interest The authors report no conflict of interest related to the manuscript.

References

1. Chen H, Khemtong C, Yang X, Chang X, Gao J (2011) Nanonization strategies for poorly water-soluble drugs. Drug Discov Today 16:354–360. https://doi.org/10.1016/J.DRUDIS.2010.02.009
2. Möschwitzer JP (2013) Drug nanocrystals in the commercial pharmaceutical development process. Int J Pharm 453:142–156. https://doi.org/10.1016/J.IJPHARM.2012.09.034
3. Ige PP, Baria RK, Gattani SG (2013) Fabrication of fenofibrate nanocrystals by probe sonication method for enhancement of dissolution rate and oral bioavailability. Colloids Surf B Biointerfaces 108:366–373. https://doi.org/10.1016/J.COLSURFB.2013.02.043
4. Chary PS, Shaikh S, Bhavana V, Rajana N, Vasave R, Mehra NK (2024) Emerging role of nanocrystals in pharmaceutical applications: a review of regulatory aspects and drug development process. Appl Mater Today 40:102334. https://doi.org/10.1016/J.APMT.2024.102334
5. Petros RA, Desimone JM (2010) Strategies in the design of nanoparticles for therapeutic applications. Nat Rev Drug Discov 9:615–627. https://doi.org/10.1038/nrd2591
6. Müller RH, Gohla S, Keck CM (2011) State of the art of nanocrystals—special features, production, nanotoxicology aspects and intracellular delivery. Eur J Pharm Biopharm 78:1–9. https://doi.org/10.1016/J.EJPB.2011.01.007
7. Pınar SG, Oktay AN, Karaküçük AE, Çelebi N (2023) Formulation strategies of nanosuspensions for various administration routes. Pharmaceutics 15:1520. https://doi.org/10.3390/PHARMACEUTICS15051520
8. Tuomela A, Hirvonen J, Peltonen L (2016) Stabilizing agents for drug nanocrystals: effect on bioavailability. Pharmaceutics 8(2):16. https://doi.org/10.3390/Pharmaceutics8020016
9. Peltonen L, Hirvonen J (2018) Drug nanocrystals—versatile option for formulation of poorly soluble materials. Int J Pharm 537:73–83. https://doi.org/10.1016/J.IJPHARM.2017.12.005
10. Peltonen L, Hirvonen J, Laaksonen T (2014) Drug nanocrystals and nanosuspensions in medicine. In: Handbook of nanobiomedical research. World Scientific Publishing Company, Singapore, pp 169–197. https://doi.org/10.1142/9789814520652_0005

11. Peltonen L (2018) Design space and QbD approach for production of drug nanocrystals by wet media milling techniques. Pharmaceutics 10:104. https://doi.org/10.3390/PHARMACEUTICS10030104
12. He S, Yang H, Zhang R, Li Y, Duan L (2015) Preparation and in vitro–in vivo evaluation of teniposide nanosuspensions. Int J Pharm 478:131–137. https://doi.org/10.1016/J.IJPHARM.2014.11.020
13. Bilgili E, Afolabi A (2012) A combined microhydrodynamics–polymer adsorption analysis for elucidation of the roles of stabilizers in wet stirred media milling. Int J Pharm 439:193–206. https://doi.org/10.1016/J.IJPHARM.2012.09.040
14. Salazar J, Müller RH, Möschwitzer JP (2014) Combinative particle size reduction technologies for the production of drug nanocrystals. J Pharm (Cairo) 2014:265754. https://doi.org/10.1155/2014/265754
15. Bhatia H, Read E, Agarabi C, Brorson K, Lute S, Yoon S (2016) A design space exploration for control of critical quality attributes of mAb. Int J Pharm 512:242–252. https://doi.org/10.1016/J.IJPHARM.2016.08.046
16. Martin-Moe S, Lim FJ, Wong RL, Sreedhara A, Sundaram J, Sane SU (2011) A new roadmap for biopharmaceutical drug product development: integrating development, validation, and quality by design. J Pharm Sci 100:3031–3043. https://doi.org/10.1002/JPS.22545
17. Eon-Duval A, Valax P, Solacroup T, Broly H, Gleixner R, Le Strat C, Sutter J (2012) Application of the quality by design approach to the drug substance manufacturing process of an Fc fusion protein: towards a global multi-step design space. J Pharm Sci 101:3604–3618. https://doi.org/10.1002/JPS.23273
18. Li J, Qiao Y, Wu Z (2017) Nanosystem trends in drug delivery using quality-by-design concept. J Control Release 256:9–18. https://doi.org/10.1016/J.JCONREL.2017.04.019
19. Beg S, Rahman M, Kohli K (2019) Quality-by-design approach as a systematic tool for the development of nanopharmaceutical products. Drug Discov Today 24:717–725. https://doi.org/10.1016/J.DRUDIS.2018.12.002
20. Rathore AS (2009) Roadmap for implementation of quality by design (QbD) for biotechnology products. Trends Biotechnol 27:546–553. https://doi.org/10.1016/j.tibtech.2009.06.006
21. Kuk DH, Ha ES, Ha DH, Sim WY, Lee SK, Jeong JS, Kim JS, Baek IH, Park H, Choi DH, Yoo JW, Jeong SH, Hwang SJ, Kim MS (2019) Development of a resveratrol nanosuspension using the antisolvent precipitation method without solvent removal, based on a Quality by Design (QbD) approach. Pharmaceutics 11(12):688. https://doi.org/10.3390/PHARMACEUTICS11120688. Mdpi.Com. https://www.mdpi.com/1999-4923/11/12/688. Accessed 17 June 2025
22. Yu LX, Amidon G, Khan MA, Hoag SW, Polli J, Raju GK, Woodcock J (2014) Understanding pharmaceutical quality by design. AAPS J 16:771–783. https://doi.org/10.1208/S12248-014-9598-3/TABLES/2
23. Xia D, Gan Y, Cui F (2014) Application of precipitation methods for the production of water-insoluble drug nanocrystals: production techniques and stability of nanocrystals. Curr Pharm Des 20:408–435. https://doi.org/10.2174/13816128113199990397
24. Junghanns J-UAH, Müller RH (2008) Nanocrystal technology, drug delivery and clinical applications. Int J Nanomedicine 3:295. https://doi.org/10.2147/IJN.S595
25. Chen ML, John M, Lee SL, Tyner KM (2017) Development considerations for nanocrystal drug products. AAPS J 19:642–651. https://doi.org/10.1208/S12248-017-0064-X/TABLES/3
26. Mohammad IS, Hu H, Yin L, He W (2019) Drug nanocrystals: fabrication methods and promising therapeutic applications. Int J Pharm 562:187–202. https://doi.org/10.1016/J.IJPHARM.2019.02.045. Elsevier
27. Ganesan P, Karthivashan G, Park SY, Kim J, Choi DK (2018) Microfluidization trends in the development of nanodelivery systems and applications in chronic disease treatments. Int J Nanomedicine 13:6109–6121. https://doi.org/10.2147/IJN.S178077

28. Cheng M, Li Y, Luo X, Chen Z, Wang R, Wang T, Feng W, Zhang H, He J, Li C (2022) Effect of dynamic high-pressure microfluidization on physicochemical, structural, and functional properties of oat protein isolate. Innovative Food Sci Emerg Technol 82:103204. https://doi.org/10.1016/J.IFSET.2022.103204
29. Schuh RS, Poletto É, Fachel FNS, Matte U, Baldo G, Teixeira HF (2018) Physicochemical properties of cationic nanoemulsions and liposomes obtained by microfluidization complexed with a single plasmid or along with an oligonucleotide: implications for CRISPR/Cas technology. J Colloid Interface Sci 530:243–255. https://doi.org/10.1016/J.JCIS.2018.06.058
30. Chang TL, Zhan H, Liang D, Liang JF (2015) Nanocrystal technology for drug formulation and delivery. Front Chem Sci Eng 9:1–14. https://doi.org/10.1007/S11705-015-1509-3/METRICS
31. Parveen N, Abourehab MAS, Thanikachalam PV, Khar RK, Kesharwani P (2023) Nanocrystals as an emerging nanocarrier for the management of dermatological diseases. Colloids Surf B Biointerfaces 225:113231. https://doi.org/10.1016/J.COLSURFB.2023.113231
32. Sarode AL, Sandhu H, Shah N, Malick W, Zia H (2013) Hot melt extrusion (HME) for amorphous solid dispersions: predictive tools for processing and impact of drug–polymer interactions on supersaturation. Eur J Pharm Sci 48:371–384. https://doi.org/10.1016/J.EJPS.2012.12.012
33. Baumgartner R, Eitzlmayr A, Matsko N, Tetyczka C, Khinast J, Roblegg E (2014) Nano-extrusion: a promising tool for continuous manufacturing of solid nano-formulations. Int J Pharm 477:1–11. https://doi.org/10.1016/J.IJPHARM.2014.10.008
34. Lang B, McGinity JW, Williams RO (2014) Hot-melt extrusion—basic principles and pharmaceutical applications. Drug Dev Ind Pharm 40:1133–1155. https://doi.org/10.3109/03639045.2013.838577
35. Legrand P, Lesieur S, Bochot A, Gref R, Raatjes W, Barratt G, Vauthier C (2007) Influence of polymer behaviour in organic solution on the production of polylactide nanoparticles by nanoprecipitation. Int J Pharm 344:33–43. https://doi.org/10.1016/J.IJPHARM.2007.05.054
36. Zeng Z, Dong C, Zhao P, Liu Z, Liu L, Mao HQ, Leong KW, Gao X, Chen Y (2019) Scalable production of therapeutic protein nanoparticles using flash nanoprecipitation. Adv Healthc Mater 8:1801010. https://doi.org/10.1002/ADHM.201801010
37. Martínez Rivas CJ, Tarhini M, Badri W, Miladi K, Greige-Gerges H, Nazari QA, Galindo Rodríguez SA, Román RÁ, Fessi H, Elaissari A (2017) Nanoprecipitation process: from encapsulation to drug delivery. Int J Pharm 532:66–81. https://doi.org/10.1016/J.IJPHARM.2017.08.064
38. Dhondale MR, Nambiar AG, Singh M, Mali AR, Agrawal AK, Shastri NR, Kumar P, Kumar D (2023) Current trends in API co-processing: spherical crystallization and co-precipitation techniques. J Pharm Sci 112:2010–2028. https://doi.org/10.1016/J.XPHS.2023.02.005
39. Kumar A, Valamla B, Thakor P, Chary PS, Rajana N, Mehra NK (2022) Development and evaluation of nanocrystals loaded hydrogel for topical application. J Drug Deliv Sci Technol 74:103503. https://doi.org/10.1016/J.JDDST.2022.103503
40. Henríquez LC, Bahloul B, Alhareth K, Oyoun F, Frejková M, Kostka L, Etrych T, Kalshoven L, Guillaume A, Mignet N, Corvis Y (2024) Step-by-step standardization of the bottom-up semi-automated nanocrystallization of pharmaceuticals: a quality by design and design of experiments joint approach. Small 20:e2306054. https://doi.org/10.1002/SMLL.202306054
41. Vehring R (2008) Pharmaceutical particle engineering via spray drying. Pharm Res 25:999–1022. https://doi.org/10.1007/S11095-007-9475-1/FIGURES/10
42. Kuriakose R, Anandharamakrishnan C (2010) Computational fluid dynamics (CFD) applications in spray drying of food products. Trends Food Sci Technol 21:383–398. https://doi.org/10.1016/J.TIFS.2010.04.009
43. Kumar S, Gokhale R, Burgess D (2014) Quality by design approach to spray drying processing of crystalline nanosuspensions. Int J Pharm 464:1–18. Elsevier. https://www.sciencedirect.com/science/article/pii/S0378517313011150 Accessed 17 June 2025

44. Su CY, Tang HZ, Chu K, Lin CK (2014) Cosmetic properties of TiO2/mica-BN composite powder prepared by spray drying. Ceram Int 40:6903–6911. https://doi.org/10.1016/J.CERAMINT.2013.12.011
45. Ziaee A, Albadarin AB, Padrela L, Femmer T, O'Reilly E, Walker G (2019) Spray drying of pharmaceuticals and biopharmaceuticals: critical parameters and experimental process optimization approaches. Eur J Pharm Sci 127:300–318. https://doi.org/10.1016/J.EJPS.2018.10.026
46. Li X, Anton N, Arpagaus C, Belleteix F, Vandamme TF (2010) Nanoparticles by spray drying using innovative new technology: the Büchi nano spray dryer B-90. J Control Release 147:304–310. https://doi.org/10.1016/J.JCONREL.2010.07.113
47. Ainurofiq A, Rahayu BG, Murtadla FA, Kuncahyo I, Windarsih A, Choiri S (2024) QbD-based formulation development of resveratrol nanocrystal incorporated into soluble mesoporous material: pharmacokinetic proof of concept study. Int J Pharm 661:124471. https://doi.org/10.1016/J.IJPHARM.2024.124459
48. Iurian S, Bogdan C, Tomuţă I, Szabó-Révész P, Chvatal A, Leucuţa SE, Moldovan M, Ambrus R (2017) Development of oral lyophilisates containing meloxicam nanocrystals using QbD approach. Eur J Pharm Sci 104:356–365. https://doi.org/10.1016/J.EJPS.2017.04.011
49. Panwar D, Thakor P, Sharma M, Bakshi AK, Bhavana V, Srivastava V, Mishra PR, Singh SB, Mehra NK (2023) Hyaluronic acid-engineered Bcl-2 inhibitor nanocrystals for site-specific delivery to breast tumor cells. Nanomedicine 18:1005–1023. https://doi.org/10.2217/NNM-2023-0132
50. Najm M, Rawas-Qalaji M, Assar N, Yahia R, El Hosary R, Ahmed IS (2022) Optimization, characterization and in vivo evaluation of mupirocin nanocrystals for topical administration. Eur J Pharm Sci 176:106251. Elsevier. https://www.sciencedirect.com/science/article/pii/S0928098722001361. Accessed 17 June 2025
51. Awad H, Rawas-Qalaji M, El Hosary R, Jagal J, Ahmed IS (2023) Formulation and optimization of ivermectin nanocrystals for enhanced topical delivery. Int J Pharm X 6:100210. https://doi.org/10.1016/j.ijpx.2023.100210
52. Peltonen L, Strachan C, Rades T, Grohganz H, Löbmann K (2015) Understanding critical quality attributes for nanocrystals from preparation to delivery. Molecules 20:22286–22300. https://doi.org/10.3390/MOLECULES201219851
53. Pawar VK, Singh Y, Meher JG, Gupta S, Chourasia MK (2014) Engineered nanocrystal technology: in-vivo fate, targeting and applications in drug delivery. J Control Release 183:51–66. https://doi.org/10.1016/J.JCONREL.2014.03.030

Translating Nanocrystals from Lab Scale to Clinical Applications

10

Niveditha Manne, Vishal Labhade, Vaibhav Dhumal, Pooja Khade, Sourabh Tapekar, Bhavana Gajare, Anitha Sriram, Anish Dhuri, Ankur Sharma, and Pankaj Kumar Singh

Abstract

The journey of nanocrystals from laboratory innovation to clinical application marks a groundbreaking advancement in drug delivery and therapeutic development. By enhancing drug solubility, bioavailability, and targeted delivery, nanocrystals provide effective solutions to persistent pharmacological challenges. However, the journey from lab-scale synthesis to clinical utilization is complex, requiring rigorous optimization and adaptation at every stage. This chapter explores the essential methodologies and technological innovations enabling this translation, including scalable synthesis, surface engineering, and advanced formulation strategies designed to retain nanocrystal stability and bioactivity. Critical regulatory and industrial challenges, such as compliance with Good Manufacturing Practices (GMP), quality control, and formulation reproducibility, are thoroughly examined, along with innovative strategies that address these barriers. The text also highlights successful case studies where nanocrystal-based drugs have achieved clinical success, providing insights into the translational hurdles they overcame. By detailing both the scientific and regulatory milestones necessary for clinical integration, this chapter offers a roadmap for researchers aiming to advance nanocrystal-based therapies from bench to bedside, illuminating the pathways for future innovations that bridge nanotechnology and medicine.

N. Manne · V. Labhade · V. Dhumal · P. Khade · S. Tapekar · B. Gajare · A. Sriram · A. Dhuri
P. K. Singh (✉)
Department of Pharmaceutics, National Institute of Pharmaceutical Education and Research (NIPER), Hyderabad, Telangana, India

A. Sharma
Hyalo Technologies, Somerest, NJ, USA

N. K. Mehra, K. K. Bansal (eds.), *Nanocrystal Technologies in Biomedical Sciences*,
https://doi.org/10.1007/978-981-95-2968-1_10

Keywords

Nanocrystals · Functionalization · Preclinical studies · Clinical translation · Regulatory aspect

Abbreviations

ADME	Absorption, Distribution, Metabolism and Excretion
ATP	Adenosine Triphosphate
CQAs	Critical Quality Attributes
ELPI	Electrical Low Pressure Impactor
EMA	European Medical Agency
ENMs	Engineered Nanomaterials
FDA	Food and Drug Administration
GAS	Gas Antisolvent
GMP	Good Manufacturing Practices
HGAP	High Gravity Antisolvent Precipitation
HPH	High Pressure Homogenization
LDH	Lactate Dehydrogenase Assay
MDA	Malondialdehyde Assay
NCD	Nanocrystal Colloidal Dispersion
NCE	New Chemical Entity
NCs	Nanocrystals
NMs	Nanomedicines
PDI	Polydispersity Index
QbD	Quality by Design
RESS	Rapid Expansion of Supercritical Solutions
ROS	Reactive Oxygen Species
SAS	Supercritical Antisolvent
SEDS	Solution Enhanced Dispersion by Supercritical Fluids

10.1 Introduction

10.1.1 Overview of Nanocrystals

The poorly soluble and least permeable development has been possible using nanocrystals. It is the most significant attempt, showing improved dissolution, bioavailability, and permeability due to its greater surface area/volume ratio. There are three distinct strategies to improve the solubility of pharmaceutical compounds. The primary approach focuses on molecular modifications, including the use of cosolvent systems, salt formation, prodrugs, and cyclodextrins. The secondary strategy targets the particulate level, emphasizing the development of amorphous systems, metastable forms, cocrystals, and extensive research on particle size reduction. The tertiary approach involves colloidal systems, such as emulsions and other lipid-based formulations, which enhance surface energy and solubility. Notably, nanocrystals exhibit unique properties that position them as an effective bridge between conventional formulations and advanced nanomedicine products [1].

Compared with nanoparticles, these can be scalable and applicable to all manufacturing processes and dosage forms. The low oral bioavailability and unpredictable drug absorption are primarily due to insufficient solubility and limited dissolution, resulting in a very small concentration of the drug between the gut and blood vessels [2]. Nano-sizing is a significant method for enhancing the delivery of poorly soluble drugs. The two primary techniques used for particle size reduction, both in industrial and academic settings, are wet ball milling and high-pressure homogenization. Nano-sizing has emerged as a pivotal strategy for enhancing the delivery of poorly soluble drugs. The primary techniques for particle size reduction, widely employed in both industrial and academic settings, are wet ball milling and high-pressure homogenization. Initial challenges have primarily been addressed, such as material abrasion, prolonged milling times, and downstream processing complexities. Improved understanding of the biopharmaceutical aspects of poorly soluble drugs has significantly increased the success rate of nanocrystals entering the market. The leading pharmaceutical companies have recognized the potential of drug nanocrystals and integrated them into drug development pipelines. Nanocrystals are a prominent topic in drug delivery technology, representing carrier-free colloidal drug systems composed of pure active pharmaceutical ingredients (APIs) stabilized in aqueous environments. Nanocrystals offer versatile drug delivery options across various administration routes, including oral, parenteral, pulmonary, ocular, and dermal applications [3]. Nanoscale dimensions nanocrystals encase many great attributes, such as enhanced adhesion, surface area, and permeability, which finally imputes better bioavailability [4]. Among the 20 approved nanoformulations, a significant majority are nanocrystals. These formulations demonstrate high encapsulation efficiencies, improving drug transport and achieving optimal concentrations for enhanced pharmacodynamic effects [3].

Nanotechnology catches all our daily lives when we look at computer chips, biotechnology, or cosmetic research and products. Approximately 40% of drugs in the roadway show solubility problems. These scenarios are, in turn, correlated with the bioavailability issues. Although there are various methods to enhance solubility, these solutions often come with drawbacks, including limitations related to their chemistry, molecular size, conformation, and bioavailability issues. Understanding the physical and chemical properties of nanocrystals reveals increased saturation solubility and dissolution rates, enhanced adhesiveness, prolonged stability, and improved biological performance.

10.1.2 Importance and Potential of Nanocrystals in Medicine

Nanocrystals are the focus of extensive research and development due to their efficient drug loading, consistent dissolution rates, improved structural stability, and prolonged circulation times [5]. Nanocrystals are increasingly recognized as a therapeutic option for cancer and other diseases without the need for traditional carriers. Before 2021, nanocrystal drug products accounted for 60% of the total market for nanoparticle-based drug delivery systems. The convenience of nanocrystal technology, which includes straightforward formation methods, homogeneous composition, and favorable pharmacoeconomic benefits, allows it to address some of the limitations in drug development.

The clinical effectiveness of a specific formulation depends on several factors, including size, morphology, surface charge, drug loading capacity, type of excipient used, degree of dispersibility, and targeted delivery to specific sites. Nanocrystals hold significant potential for advancing therapeutic applications. Transforming off-patent drugs into innovative nanocrystal drug products provides a significant competitive advantage for companies in the market. Moreover, a novel method that combines hybrid or multifunctional nanocrystals harnesses the full benefits of nanocrystal technology, making it especially beneficial for cancer treatment and tumor diagnostics [6]. It concurrently attains bioimaging and hence gains prominent importance in theranostics. By leveraging nanocrystal technology, pharmaceutical industries can benefit from redesigning previously existing formulations [7]. This technology's commercial potential is boosted by the comparatively brief period needed for clinical approval [8]. The initially available nanocrystals were accredited with different associates under Elan's nanocrystal technology [9]. Among these, Emend (aprepitant), Triglide (fenofibrate), and Rapamune (sirolimus) have demonstrated substantial commercial prominence in tablet and capsule dosage forms. Predominantly, they are intended to treat metabolic cases such as hypertension and lipid disorders [10]. Nanocrystals, designed for targeted drug delivery, interact precisely with receptors in specific tissues [11], as targeted drug delivery requires four different strategies such as retain, evade, target, and release [11]. Targeting ligands can be enhanced on the surfaces of nanocrystals to specifically target certain tissues or organs, which improves their effectiveness. These nanocrystals interact with body tissues differently than larger particles. Additionally, their retention time in the body is longer because they can evade the reticuloendothelial system. Although the adverse effects of nanocrystals are minimal, they can still unintentionally impact blood circulation. Therefore, it is essential to have a thorough discussion about both the beneficial effects and the limited negative effects of nanocrystals, especially considering their biodegradable nature [12]. These nanocrystals are well endured in various routes of administration in relation to conventional drug products [13].

10.2 From Bench to Bedside—Advancing Nanocrystals

10.2.1 Definition of Nanocrystals According to FDA

Nanotechnology involves nanoscale materials used for various applications in healthcare. It has advanced skin regeneration and traditional wound healing techniques, though limitations remain, such as prolonged treatment durations, significant batch-to-batch variability in medications, and allergic reactions [14]. Drug nanocrystals have been the most debated topic for the past 20 years. The study of nanocrystals began in the early 1990s, with Gary Liversidge securing the first patent related to this technology. Compared to other nanoscale drug delivery systems, nanocrystals offer a distinct advantage due to their structural simplicity, often requiring minimal use of excipients [15].

Nanocrystals are aggregates of particles typically ranging in size from 10 to 400 nm, exhibiting physical properties intermediate between bulk solids and discrete molecules. This unique size-dependent behavior contributes to enhanced solubility and permeability. They can be synthesized using top-down methods, such as wet milling, or bottom-up techniques, including antisolvent precipitation, resulting in crystalline structures that may vary with the production process. These nanocrystals are often dispersed in aqueous media containing stabilizers to maintain a stable colloidal state, forming formulations commonly called nanosuspensions [15]. The term "named drug nanoparticles" refers to a specific category of nanoparticles. This raises a question regarding polymeric nanoparticles, which typically have a lower drug loading because they consist of polymeric materials that are infused with the active pharmaceutical ingredient (API). The advantages of using nanocrystals can be categorized into three main types: enhanced dissolution rate, improved solubility, and increased surface adhesion [5]. Crystallinity is the key parameter for assessing the formation of nanocrystals, as it determines the final structure of the formulation. Evaluating crystallinity is essential for verifying the successful integration of stabilizers, surface polymers, and targeted ligands.

10.2.2 Attributes of Nanocrystals

The term "nanocrystals" denotes particles predominantly in a crystalline state, representing nano-sized suspensions that may also contain partially crystalline or amorphous drug forms [2]. Due to their crystalline structure, drug nanocrystals are classified by size. They offer numerous advantages, including high drug-loading efficiency, ease of large-scale manufacturing, and relatively low preparation costs [16]. Nanocrystal formulations introduce prominent property improvements, making them more accessible and expandable utilizing prevalent technologies [17]. The type of technique used in the preparation of nanocrystal deals with many drawbacks that are directly linked to the physicochemical characteristics of the drug molecule. Synthesis of nanocrystals can be easily done using simple procedures without well-established equipment. These exhibit a high drug-loading capacity, facilitating efficient drug delivery to or within cells and achieving elevated plasma concentrations necessary for optimal pharmacodynamic effects [2]. Nanocrystals are primarily synthesized for their small size and minimal use of surface-active agents needed for stabilization [18]. It has to be taken into account that the selection of stabilizer depends on the physical stability, and it should maintain the nano-sized particle size as long as possible to interact with the cells and cell layers [19]. Beirowski et al. studied the impact of stabilizers and cryoprotectants on the stability of drug nanocrystals during freeze-drying. Their findings indicated that sufficiently high concentrations of stabilizers surrounding the nanocrystals effectively prevent particle aggregation during drying. As a result, the formation of Van der Waals interactions between particles is inhibited, enhancing the stability of the dried nanocrystal formulation [20]. Researchers have increasingly focused on nanocrystals to produce colloidal particles with customized biological properties, thereby enhancing drug delivery and targeting efficacy [21]. Nanocrystals present several benefits and limitations. The benefits include enhanced dissolution rates compared to conventional particles, prolonged systemic circulation, and reduced tissue irritation in subcutaneous or intramuscular injections. Furthermore, the ability to produce hybrid nanocrystals offers potential for both diagnostic and therapeutic applications [22–25]. Among its limitations, achieving uniform and precise dosing remains challenging, potentially leading to issues concerning physicochemical stability. Additionally, careful handling and appropriate bulk management are essential to ensure stability during transport and storage [23, 26, 27].

10.3 Laboratory Innovations in Nanocrystal Research

10.3.1 Synthesis Methods

Drugs can be nanosized through several methods, which can be categorized into top-down, bottom-up, and combination approaches as depicted in Fig. 10.1. Top-down methods use high-energy techniques for particle size reduction, including high-pressure homogenization and media milling. In contrast, bottom-up methods employ supersaturation and precipitation techniques to reduce drug sizes to the nanoscale [23]. As mentioned earlier, there are three main categories of nanosizing methods.

10.3.2 Bottom-Up Techniques

Since precipitation creates nanoscaled drug particles that can be either crystalline or amorphous, bottom-up approaches are often referred to as precipitating strategies [13]. Using this technique, the drug is either mixed with a non-solvent, evaporated from a solvent, or precipitated from a supersaturated drug solution [28]. Because of the challenges in regulating particle growth and the procedure, these methods have not taken off in the pharmaceutical sector [15]. To increase a method's efficacy,

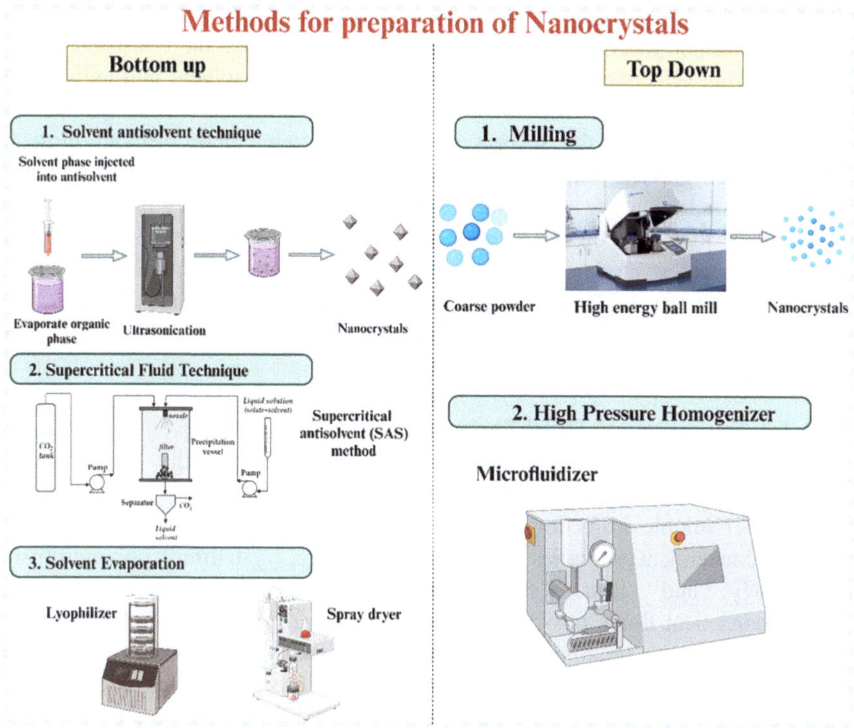

Fig. 10.1 Preparation methods of nanocrystals for drug delivery. (Self created)

bottom-up and top-down approaches can occasionally be combined. The most studied production processes include solvent-antisolvent precipitation (both conventional in bulk and with the use of microfluidics), solvent evaporation (spray dryer), temperature reduction (frozen dryer), and precipitation in supercritical fluids [13].

10.3.2.1 Precipitation by Solvent-Antisolvent Addition

The introduction of an antisolvent to a drug solution induces precipitation, offering an effective approach for nanoparticle production. This technique enables control over the morphology and particle size of the final product. The process involves dissolving the drug in a primary solvent, followed by the addition of a secondary solvent, such as water, which is miscible with the primary solvent but acts as an antisolvent for the drug [29]. As a result of the solvent diffusing into the antisolvent and the particles nucleating, the solution becomes more super-saturated. Ostwald ripening can lead to competition between particle growth and nucleation during drug precipitation; however, the process can be steered toward nucleation by employing suitable excipients and stabilizers [29, 30]. Certain key process variables significantly influence the particle size and physicochemical properties of nanosized medications. These variables include the choice of solvent and antisolvent, their volume ratio, and the order in which they are added. In the antisolvent procedure, ultrasonic sound is employed to create uniform conditions within the vessel. Sonoprecipitation generates shock waves that promote rapid and consistent nucleation, ensuring effective and homogeneous mixing of the solvent and antisolvent. The experimental setup is simple: it consists of an ultrasonic bath filled with the antisolvent, into which the solvent is atomized using compressed air pressure [31, 32]. High gravity controlled precipitation technique is mostly used to create inorganic particles that are nanoscale in size. To improve the mixing process, it has a revolving packing bed on the ground. The high gravity antisolvent precipitation (HGAP) process is an integration of the HGCP technology and the antisolvent precipitation process [33]. Using this method, cephradine drug nanocrystals with an average size of 200–400 nm were produced without significantly altering their crystalline state [34].

10.3.2.2 Supercritical Fluid Technique

Supercritical fluid (SCF) technology has been employed to modify the solid-state properties of powdered active pharmaceutical ingredients (APIs), particularly for poorly water-soluble drug compounds. Upon exposure to supercritical precipitation, key physicochemical properties, including particle size, crystallinity, polymorphism, morphology, and surface characteristics, undergo significant modification [35]. Supercritical fluids possess properties that lie in between the characteristics of gases as well as liquids, hence having low density, low viscosity, and high diffusion rate, which helps in rapid precipitation. Because of its advantageous characteristics (low cost, high safety, nonflammability, easy removal of the antisolvent by lowering the pressure, mild critical conditions, $T = 31.1$ °C, $p = 7.38$ MPa), supercritical carbon dioxide (sCO_2) is frequently used as SCF for the precipitation of micro- and nanoparticles, including drug nanocrystals and amorphous drug nanoparticles [35, 36]. This technique is further categorized into two sub-methods: rapid expansion of supercritical solution (RESS) and supercritical antisolvent (SAS). The RESS method is typically applied to drugs with high solubility in CO_2. In this process, the drug is first dissolved in supercritical CO_2 and subsequently expanded through a nozzle into a precipitation chamber, where rapid precipitation leads to the formation of nanocrystals. Key parameters influencing this process, including pressure, temperature, SCF flow rate, and the shape and diameter of the spraying nozzle, significantly affect the size and yield of the resulting nanocrystals [37]. Many medications are insoluble in supercritical carbon dioxide, which is one of the technique's main disadvantages. To address this issue, a modified procedure called rapid expansion of supercritical solution with solid cosolvent (RESS-SC) has been developed. SAS method supercritical fluids act as an antisolvent and precipitate drug from drug solution. First, the drug is dissolved in organic solvent, which is miscible with supercritical antisolvent to form a solution, and then this solution is added to supercritical antisolvent [33]. Additionally, in contrast to RESS, this can be regarded as a continual procedure, producing more consistent outcomes because the crystals develop in a constant state [37]. Variables influencing the particle size during drug precipitation include the type of solvent and antisolvent, solvent-to-antisolvent volume ratio, drug concentration, mixing intensity, antisolvent addition frequency, temperature, pressure, and nozzle diameter. Several modified versions of the supercritical antisolvent (SAS) method have been explored by researchers for producing nanosized drug crystals, including gas antisolvent (GAS) and solution-enhanced dispersion by supercritical fluids (SEDS) [38, 39].

10.3.2.3 Solvent Removal Techniques

This process transforms a fluid into hot drying gas using an atomizer. Nandiyanto et al. have thoroughly investigated how various factors—such as flow rate, atomizer diameter, fluid characteristics, and equipment design—affect the size and shape of the particles produced during spray drying [40]. A new technology called NanoCrySP® has been developed for the industrial production of nanocrystals using a spray dryer. In this process, a drug solution is sprayed with tiny particles that initiate the nucleation process, leading to the formation of drug nanocrystals within a solid dispersion of small molecules.

Another method for producing nanocrystals is freeze drying. In this technique, the drug's organic solution is mixed with an antisolvent that contains a cryoprotectant. This combination results in the formation of nanocrystals after the solutions are rapidly frozen and then dried. To optimize the size of the crystals, key parameters include the concentration of the cryoprotectant (which should be higher during crystallization in the freezing phase and lower for other processes) and the speed of freezing. The crystallization can occur either during the freezing phase or during the drying phase [33].

10.3.3 Top-Down Approach

Top-bottom approach has become increasingly recognized by the pharmaceutical industry, which involves subjecting micronized drug crystals to mechanical attrition, which reduces the particle size to a nanometer scale [41]. Media milling and high-pressure homogenization are the two primary top-down strategies.

10.3.3.1 Media Milling

Liversidge et al. first created low energy pearl milling in 1991, which paved the way for the development of NanoCrystals® technology, which reduces particle size [42]. At the moment, high energy milling is the foundation of NanoCrystal® technology, which makes this method more appealing for industrial pharmaceutical applications [43]. Low-energy ball milling is employed in wet milling processes, wherein a jar containing the milling medium is loaded with a drug dispersion in either an aqueous or non-aqueous medium and subjected to milling operations [44]. Glass beads are typically utilized as the milling media in wet ball milling, and they are moved by an electric stirrer or by moving the entire jar [15]. Special media are needed for the media milling process, which is chosen based on the agitator type and the milling chamber's inside surface. Previously, the nanosuspension was contaminated by abrasion from the milling chamber when glass or zirconium oxide beads were utilized to overcome the contamination problem. Recently, highly crosslinked polystyrene beads were introduced [45].

10.3.3.2 High Pressure Homogenization (HPH)

The High-Pressure Homogenization (HPH) method employs two distinct techniques: piston-gap homogenizers and microfluidization technology. The microfluidizer technique, developed by Skyepharma, is also known as IDD-PTM (Insoluble Drug Delivery Microparticle) technology. This method is based on jet-stream homogenization, where a homogenizing chamber subjects the drug suspension to high pressure and velocity. There are two types of chambers used in this process: "Y" type and "Z" type. In the "Z" type chamber, the direction of the suspension flow is altered, which triggers particle collisions and shear forces, ultimately reducing particle size. In contrast, the "Y" type chamber directs the suspension into two streams that collide, with shear and cavitation forces further breaking down the particles [41]. Dissocubes™ is the trademark currently held by Sky Pharma for the piston-gap HPH technology created by Muller et al. Pharmasol GmbH/Berlin is the owner of the Nanopure® technology, a second variation of piston-gap HPH that was created in 1999. The method entails homogenizing medications in a medium that has either no water or a decreased amount of water, such as a mixture of water and PEG.

10.3.3.3 Combination Techniques

Combination methods employ both top-down and bottom-up procedures sequentially. Nanocrystals are often precipitated using bottom-up techniques initially, followed by one of the top-down processes (high energy mixing, HPH, or ultrasounds) [33]. The creation of smaller nanocrystals is the main benefit of the combination of processes. The combination techniques currently in use include Nanoedge™, H69, H42, and H96.

Nanoedge™ Technology

It includes combining top-down techniques like HPH with micro-precipitation techniques.

H69 Technology

Muller and Moschwitzer created this method, which combines HPH and the microprecipitation technique. It is not the same as the Nanoedge approach since cavitation and particle production happen simultaneously.

H42 Technology

This method, which was also created by Moscwitzer, combines the HPH approach with spray-drying precipitation.

H96 Technology

The process is typically used to create nanosuspensions, involving freeze-drying as a preliminary step followed by high-pressure homogenization (HPH).

10.4 Functionalization and Surface Modification of Nanocrystals

Functionalization and surface modification of nanoparticles are essential processes in nanotechnology, as they adjust the surface properties of nanoparticles to fulfill specific application requirements. Functionalization involves attaching specific functional groups, molecules, or biomolecules to the surface of nanoparticles [46]. The primary goal is to impart new properties or enhance existing ones to suit targeted applications. Also, it has been reported that the functionalization as well as surface modification approach provides more solubility and tumor and cancer cell targeting, less elimination and side effects and more anti-cancer effects with a smaller dose. Functionalization, as well as surface modifications of nanocrystals, causes improved targeting by increasing permeability and retention, reduction in toxicity by reducing off-target effects, enhanced circulation time, overcoming drug resistance, etc. [47].

Albumin-coated nanoparticles (NCs) demonstrated superior anti-cancer efficacy compared to Abraxane in C57BL/6 mice bearing B16F10 melanoma, an aggressive cancer model. The paclitaxel (PTX) NCs were prepared in two steps: first, through crystallization in a medium containing Pluronic, followed by surface coating with albumin, resulting in the formulation known as "Cim-F-alb." The cellular uptake of the NCs was tested using murine macrophages and B16F10 melanoma cells, which represent the mononuclear phagocyte system and tumor cells that overexpress SPARC (secreted protein acidic and rich in cysteine). The results indicated that Cim-F-alb exhibited increased uptake by macrophages. This enhanced uptake was attributed to a conformational change in the surface-coated albumin. To confirm this, the researchers pretreated the macrophages with polyinosinic acid, a compound known to inhibit scavenger receptor-mediated endocytosis. Following this treatment, it was observed that the uptake of Cim-F-alb by macrophages was reduced. Conversely, the presence of albumin coating on the Cim-F-alb formulation resulted in decreased uptake by macrophages. To further investigate the effect of albumin on cellular uptake, B16F10 cells were incubated with both Cim-F (without albumin) and Cim-F-alb. The findings revealed that PTX uptake was more pronounced in the case of Cim-F-alb, indicating that albumin plays a significant role in the cellular uptake of NCs. Cim-F-alb demonstrated lower uptake by macrophages and greater uptake by B16F10 melanoma cells. In *in vivo* anti-cancer studies, it was shown that Cim-F-alb was taken up significantly by cancer cells, attributed to the albumin coating facilitating interaction with SPARC-positive B16F10 cells. Notably, by the 10th day of treatment, only one animal in the Cim-F-alb-treated group had a tumor volume exceeding 1000 mm^3, in contrast to the PBS and Abraxane-treated groups. Additionally, the tumor growth rates in the Cim-F-alb-treated groups were significantly lower compared to those in the PBS and Abraxane-treated animals [48].

10.5 Characterization of Nanocrystals for Clinical Applications

For successful nanocrystal formulation, the most crucial factor is the characterization of nanocrystals. This ensures that the parameters of the nanocrystals fall within the specified limits. Nanocrystals are characterized through four parameters.

10.5.1 Particle Characterization

Particle characterization is essential for translating nanocrystals from lab-scale to clinical applications. Key parameters such as particle size (PS), particle size distribution (PDI), and zeta potential (ZP) are crucial indicators of formulation quality and stability. These parameters are typically measured using techniques like Photon Correlation Spectroscopy (PCS), Laser Diffractometry (LD), and Electrical Low-Pressure Impactor (ELPI). Among these, PCS, based on dynamic light scattering (DLS), is the most widely used for assessing PS, PDI, and ZP. A PDI range of 0–0.500 reflects particle size distribution uniformity, with values below 0.3 indicating optimal stability, which is crucial for the clinical success of nanocrystal formulations. The Electrical Low-Pressure Impactor (ELPI) is a cascade impactor that charges particles via a unipolar corona charger and measures their aerodynamic diameter in near real-time using electrically isolated collection stages and electrometers [49]. Liangxing Tu et al. demonstrated the critical influence of particle size and zeta potential on the stability and pharmacokinetics of puerarin nanocrystals. Nanocrystals (525.8 nm) remained stable for over 4 days, whereas microcrystals (1875.6 nm) were stable for only 6 h. Low zeta potential values (<10 mV) suggested steric stabilization as the primary mechanism for maintaining nanoparticle stability. Smaller particle size significantly enhanced bioavailability, with nanocrystals showing a 7.6-fold improvement compared to raw suspension, emphasizing the importance of particle size and zeta potential in optimizing oral drug delivery [50].

10.5.2 Solid State Characterization

Solid-state characterization is crucial for nanocrystal formulation and is performed using FTIR, PXRD, and DSC techniques. FTIR analyzes molecular interactions between the drug and excipients by detecting changes in vibrational frequencies, ensuring formulation compatibility [49]. Koradia et al. used FTIR to evaluate the interaction between albendazole and PVP K30 in albendazole nanocrystals. The characteristic peaks of albendazole at 3330 cm^{-1} (N–H stretching) and 1523 cm^{-1} (C=N stretching) were observed in the albendazole/PVP K30 physical mixture, indicating that the drug's chemical structure remained intact. This suggests no significant chemical interaction [51]. P-XRD is used to confirm the crystallinity of nanocrystals. Sharp peaks in the diffractogram indicate crystallinity, while reduced or missing peaks suggest amorphization, often caused by homogenization. The

X-ray diffraction pattern serves as a unique fingerprint for each crystalline substance, providing consistent results for pure substances and distinct patterns for mixtures [52]. In the study by Sharma et al., Soluplus® was used as a stabilizer for febuxostat nanocrystals prepared by ball milling. P-XRD analysis showed that febuxostat retained its crystalline structure, while Soluplus® was amorphous. The reduced intensity of febuxostat peaks indicated partial amorphization due to milling, which enhanced saturation solubility and accelerated dissolution, resulting in significantly improved bioavailability compared to febuxostat after oral administration [53]. DSC measures thermodynamic changes by analyzing heat flow and temperature, producing endothermic and exothermic peaks. It determines the melting point of nanocrystals through a sharp endothermic peak, indicating energy consumption [49]. DSC is widely used to assess the thermal behavior, crystallinity, and interactions of drugs and excipients after nanocrystal production [52]. In the study by Tuomela et al., DSC analysis revealed that bulk brinzolamide exhibited a sharp endothermic peak at 132 °C, confirming its crystalline nature, while brinzolamide nanocrystals showed a slightly shifted and broader peak at 128 °C, indicating partial amorphization. HPMC exhibited peaks at 80 °C (moisture evaporation) and 180 °C (T_g), reflecting its amorphous nature. The DSC results confirmed that brinzolamide's crystallinity was largely retained after milling, with no signs of recrystallization [54].

10.5.3 Morphological Characterization

Morphological characterization of nanocrystals involves SEM and TEM analysis to evaluate particle size and shape. SEM images are generated by electron beam interactions with sample atoms and are crucial for monitoring changes in particle size and shape, especially after processes like spray drying or lyophilization, where agglomeration may result in larger particles [55]. TEM captures images by transmitting electrons through samples, requiring a wet sample with sufficient concentration. AFM provides three-dimensional images, offering detailed insights into nanocrystals' size, surface texture, roughness, and morphology without the need for special sample treatments like SEM and TEM [56]. Guo et al. utilized TEM to evaluate the morphology of lovastatin nanocrystals, distinguishing rod-shaped and spherical-like forms. TEM revealed that rod-shaped nanocrystals exhibited superior dissolution rates and bioavailability compared to spherical-like counterparts, emphasizing TEM's importance in optimizing nanocrystal design for enhanced therapeutic efficacy [57]. Sharma et al. highlighted the importance of stability characterization in paclitaxel nanocrystals. Stability studies revealed that formulations stabilized with Poly sodium styrene sulfonate and Tween 80 maintained consistent particle size and zeta potential over 20 weeks, with Poly sodium styrene sulfonate providing superior stability through steric and electrostatic mechanisms. Stability testing across varying temperatures, pH, and storage conditions demonstrated the critical role of stabilizers in preventing aggregation, ensuring formulation integrity, and improving therapeutic efficacy [58].

10.5.4 *In Vitro* Drug Release Study

The *in vitro* drug release study assesses drug release from nanocrystals. BCS Class drug compounds are key targets for formulation development since their dissolution rate limits bioavailability. Factors affecting this rate include effective saturation solubility surface area, drug amount, dissolution medium volume, and diffusion coefficient [59]. Ochi et al. developed a physiochemically stable meloxicam nanocrystal formulation with improved dissolution behavior to shorten the onset of pharmacological effects. Meloxicam has low solubility and poor wettability, about 0.6 mg/mL, which likely restricts its oral absorption. To improve this, nanocrystal formulations of meloxicam were developed using wet milling technology, leading to significant enhancements in dissolution behavior [60]. Li, X., and colleagues fabricated Posaconazole Nanocrystals utilizing different stabilizers, including HPMC (NS_1), HPC (NS_2), Soluplus (NS_3), and PVPK30 (NS_4) and characterized them through various solid-state characterization methods as depicted in Fig. 10.2 [61].

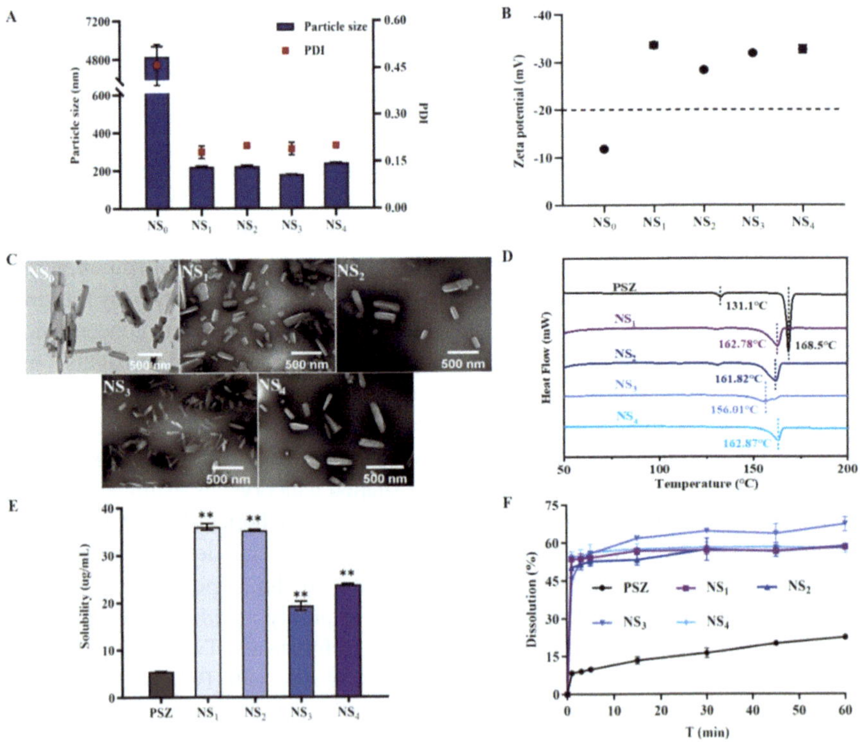

Fig. 10.2 Physicochemical characterizations of PSZ nanocrystals. (**a**) Particle size distribution (**b**) Zeta potential (**c**) TEM images scale bar = 500 nm; (**d**) DSC thermograms (**e**) Solubility $**p < 0.01$ compared to PSZ; (**f**) Dissolution profiles. (Reproduced with copyright permission from Ref. [61])

10.5.5 Preclinical Studies of Nanocrystals

The journey of a new chemical entity from discovery to market is intricate and time-consuming, often spanning several years. It begins with the synthesis of the new chemical entity and ends with the creation of an appropriate dosage form for the patient's use, which is safe and effective. This process is generally known as drug discovery and development, comprising several phases, starting with the identification of lead, and proceeding with preclinical and clinical trials [62, 63]. The drug candidate undergoes extensive toxicological evaluations and safety pharmacology studies in the preclinical phase to determine the maximum levels that can be tolerated in animal models and to assess the potential for adverse effects. The essential data gathered during this phase is needed for regulatory approval. The outcomes of these studies enable the candidate to proceed into clinical (human) trials by submitting an Investigational New Drug application to the appropriate regulatory bodies [64]. Preclinical studies generally include *in vitro*, *in vivo*, and safety toxicity evaluation. *In vitro* studies play a vital role in the drug development process, providing insights into the study of drug behavior. Primary cell culture and cell lines are used for this study. In the course of the study, researchers found the effect of drug cellular processes, DNA/RNA damage, apoptosis, and drug resistance [65]. *In vivo*, studies generally focus on the study of the pharmacokinetic and pharmacodynamic properties of the drugs. Various disease animal models are used to carry out this study. Aid in assessing the drug's safety and efficacy [66]. Preclinical toxicology evaluations include acute studies, dose-range findings, repeat-dose studies, and *in vitro* mutagenicity tests using rodent and non-rodent species. Pharmacology, the cardiovascular and respiratory systems, and the central nervous system are all evaluated for safety. Rats or rabbits are typically used to assess pharmacologic safety in conjunction with research on absorption, distribution, metabolism, and excretion [67].

10.5.6 *In Vitro* Cell Culture Studies

To ascertain how the formulation interacts with the cells, *in vitro* cell culture investigations are conducted. Cell cytotoxicity studies are used to assess the viability of cells; cellular uptake studies look at how the formulation is internalized by cells; cellular migration studies determine how the formulation affects cell movement; and cell cycle studies analyze how the formulation affects cell division and proliferation. A PTX nanocrystal (NCs) was created by Sharma et al., who also conducted an *in vitro* research on cell cytotoxicity and cell cycle arrest. The cytotoxicity of PTX nanocrystals was evaluated in breast cancer cells (MCF7 and MDA-MB) by introducing 105 cells per well in a 96-well plate and allowing them to dwell overnight at 37 °C with 5% CO2. After the cells were incubated for 36 h, the media was changed to full media containing PTX solution and PTX/NCs formulations respectively. After adding MTT (10 μL of 10 mg/mL), the mixture was incubated for 4 h. Following the crystal dissolution in 100 μL of DMSO: IPA

(50:50), the absorbance at 595 nm was measured. For cell cycle analysis, 10^6 cells were fixed in 70% alcohol after being plated and treated with 1 nM PTX or PTX/NCs for 36 h. RNase and propidium iodide were used to stain the cells, and flow cytometry was used for analysis. When compared to standard PTX solutions, PTX/NCs formulations, particularly PTX/NC-T80 (Tween 80) and PTX/NC-PSS (Polystyrene sulfonate) exhibited noticeably stronger cytotoxic effects against cancer cell lines MCF7 and MDA-MB, according to research on cell cytotoxicity and cell cycle dynamics. These formulations observed lower IC_{50} values, suggesting a significant improvement in therapeutic potency. Additionally, both formulations significantly arrested the G2/M cell cycle, but the effect of PTX/NC-T80 was the most noticeable. P-glycoprotein inhibitory action improved intracellular drug uptake and retention, which is probably the cause of this increased cytotoxicity and cell cycle modulation. All of these results point to the possibility that nanocrystal formulations could greatly increase PTX's anticancer effectiveness, making them a viable approach to cancer treatment [58].

10.5.7 *In Vivo* Animal Model Studies

Animal investigations are conducted *in vivo* to comprehend how the formulation behaves in animal models. This study incorporates many assessments, including toxicological, pharmacokinetic, and pharmacodynamic evaluations. A pharmacokinetic study looks into the nanocrystals' ADME. Blood samples are taken at different intervals following the injection of nanocrystals, and pharmacokinetic parameters such as Cmax, Tmax, T1/2, CL, etc. are calculated as part of this study. A pharmacodynamic research was conducted to investigate the drug's therapeutic efficacy and interaction with the target. To ascertain the drug's safety and possible negative effects on different organs, toxicity studies are conducted. By weighing the animals, analyzing the activities of different organs, and doing tissue histopathological analysis [68]. Shen et al. developed a quercetin nanocrystal (QT-NCs) and conducted in-vivo evaluations using male Sprague Dawley rats weighing 180 to 220 grams. After a 3-day acclimatization and a 12-h fasting period, 12 rats were divided into two groups. Blood samples (0.5 mL) were collected from the orbital venous plexus at various intervals up to 24 h following an injection of QT-NCs/GL at a dose of 50 mg/kg of quercetin. The plasma was stored at −20 °C for later analysis. For tissue distribution studies, 48 male rats were used, with three sacrificed at specific time points. After saline cardiac perfusion, the organs were cleaned, dried, and stored at −20 °C. They were then weighed and homogenized with physiological saline, and the samples were processed for quercetin extraction and analyzed by HPLC. The results demonstrated that QT-NCs/GL nanocrystals rapidly distributed to various tissues, reaching peak concentrations in the liver, likely due to glycyrrhizic acid (GL) interactions with hepatocyte receptors, indicating effective liver targeting, as shown in Fig. 10.3. Enhanced lung distribution was also observed, possibly due to GL's effect on cell membrane permeability. Overall, these findings suggest that QT-NCs/GL is a promising quercetin delivery system targeting the liver and lungs [69].

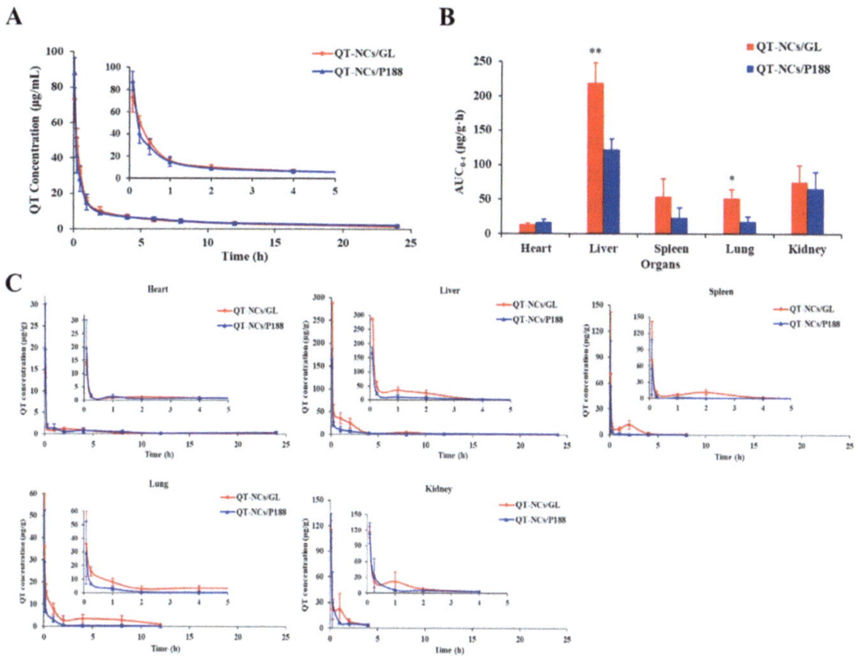

Fig. 10.3 (**a**) Plasma concentration-time curves of QT, (**b**) The $AUC_{0\text{–}t}$ of the QT distribution in various organs, (**c**) Mean QT concentration-time profiles in main organs. (Reproduced with copyright permission from Ref. [69])

10.6 Safety and Toxicity Evaluation

Safety and toxicity evaluation studies are an important part of drug development to identify the possible harmful effects of the formulation on the body or various organs. This evaluation includes the in-vivo and in-vitro pharmacological safety study and different toxicity studies, including acute, chronic toxicity, reproductive, genome toxicity, and carcinogenic risk of the formulation. Also, these studies provide insights into the required dose and tolerated dose, ensuring that the use of the drug is safe and effective in clinical practices [70]. Haroon et al. formulated aceclofenac nanocrystal (AC-N) and performed acute toxicity studies in Swiss Albino mice, to determine the median lethal dose (LD50) of unprocessed AC, a marketed formulation, and AC-N with doses ranging from 25 to 550 mg/kg. All animals in the unprocessed AC and marketed groups died at 400 mg/kg, while three mice in the AC-N group died at 250 mg/kg, which corresponded to the mortality in the unprocessed AC and marketed formulation at 350 mg/kg. As a result, the LD50 values were determined to be 350 mg/kg for unprocessed AC and the marketed drug and 250 mg/kg for AC-N. The lower LD50 of AC-N may be due to its increased saturation solubility and faster rate of dissolution [71].

10.7 Regulatory Consideration for the Development of Nanocrystals

The advancement of technological capabilities necessitates the evolution of the regulatory framework governing nanocrystal-based pharmaceuticals. In response to the fast-paced nature of innovation, global regulators are developing guidelines to ensure safety and efficacy in this emerging field. The U.S. Food and Drug Administration (FDA) has established a leadership role in formulating a regulatory framework for nanotechnology products. The FDA's definition encompasses products with sizes ranging from 1 nm to 100 nm and extends to 1000 nm for size-related properties, thereby including pharmaceutical nanocrystals [72]. Key issues for regulating such materials include material analysis, production processes, physical and chemical characteristics, and biological behavior. When reformulated in the form of nanocrystals, the product must be submitted to the FDA as a New Drug Application (NDA) to receive approval under section 505(b)(2), following section 505(b)(1). This NDA may be approved under section 505(c) of the Federal Food, Drug, and Cosmetic Act (FD&C Act) [73]. The International Council for Harmonisation (ICH) develops internationally applicable guidelines vital in developing nanocrystal formulations. These guidelines encompass critical areas such as impurities, specifications, pharmaceutical development, quality risk management, and pharmaceutical quality systems. Nanocrystals-specific regulatory considerations include characterization of particle size by using multiple analytic methods, adaptation of methods for dissolution testing, long-term stability studies, total description and validation of manufacturing processes, bioequivalence for generic products, safety aspects including the toxicological impact at nanoscale [74]. The European Medicines Agency has been actively overseeing pharmaceuticals and other medicinal products, and it regulates the approval of NMc and its effects on human health [75]. The recent proposal includes all product-specific regulations concerning safety requirements, information requirements, and uses related to nanomaterials. The EMA suggests that nanomaterials are similar to conventional chemical substances; some may be toxic, while others may not be [76]. The European Commission of the recommendation of June 10th, 2022, on the definition of nanomaterial defined it as they are considered as natural, incidental, or manufactured material, including the solid particles which are present as agglomerates or aggregates and the shapes could be elongated, rod or fiber shaped [77].

10.8 Challenges in Clinical Translations

The main challenges confronting the clinical translation of nanocrystals include the issues of scalability, reducibility, quality control, and standardization. Such challenges may be more challenging for researchers to develop and successfully commercialize treatments based on nanocrystals.

10.8.1 Scalability and Reproducibility

10.8.1.1 Scalability

The manufacturing complexity, regulatory compliance, and batch-to-batch variability are the factors that mainly influence the scalability [78]. The production of nanocrystals usually involves intricate processes that are difficult to scale up. In contrast, simple nanoparticulate systems may be manufactured in scale-up [79]. More complicated formulations necessitate the development of novel methodologies or alteration of existing manufacturing processes [7]. Complexity elevated the potential of inconsistent product quality during scale-up [80]. Additional challenges arise while adhering to the stringent regulatory requirements for good manufacturing practice (GMP) and chemistry, manufacturing, and control (CMC) [80, 81]. In addition to the requirement for validation of procedures and standardization, compliance could hamper the shift from lab-scale production to commercial-scale [81]. Maintaining a uniform quality standard throughout numerous production batches is critical in fabricating nanocrystals [82]. Variations in particle size, shape, and surface attributes can compromise nanocrystal therapeutical effectiveness and safety, making scale attempts more challenging [83].

10.8.1.2 Reproducibility

The reproducibility of the nanocrystal is crucial due to the experimental variations, challenges that occur during characterization, and biological variability. During the synthesis of nanocrystals, the reproducibility is compromised by changes in experimental variables such as solvent selection, temperature, and precursor concentration. Such variables may result in notable variations in the physicochemical properties of the finished product [84]. The implementation of appropriate techniques for the characterization is necessary to ensure that every batch complies with the standards needed for clinical use. Nonetheless, the absence of proper techniques for evaluating the characteristics of nanocrystals causes challenges in the reproducibility of nanocrystals [85]. In the case of biological systems, the interactions of nanocrystals with biological systems might differ significantly between preclinical models and human patients [86]. This uncertainty affects reproducibility across the studies by making it more challenging to translate preclinical studies into clinical outcomes [86].

10.8.2 Quality Control and Standardization

10.8.2.1 Quality Control

The intricacy of nanocrystals makes it difficult to establish reliable analytical methods. It is still essential for the nanocrystals' quality control. Analytical methods must assess all pertinent properties, including size distribution and surface charges, to guarantee the final product's safety and consistency. Nanocrystals' surface characteristics and size distribution made it possible for them to display distinct toxicological profiles [87]. A thorough safety evaluation must be performed to identify negative impacts that render challenges in QC procedures. For the safety of patients, it is critical to ensure that nanocrystal formulations are devoid of endotoxins and other impurities. Establishing reliable testing procedures for these contaminants remains an enormous obstacle for QC [88].

10.8.2.2 Standardization

The standardization approaches are mainly hampered by the lack of generally accepted guidelines for testing and evaluating nanocrystals. The specifications of different regulatory agencies differ, which might trigger ambiguities and irregularities in final product development [78]. Specialized approaches to standardization are essential for the diverse formulations of nanocrystals, such as metal-based and polymeric, which might be costly and time-consuming [89]. For approval from the regulatory agencies, establishing quantitative clinical endpoints is crucial. Clinical endpoints indicate the therapeutical efficacy of nanocrystal-based treatments. It is challenging due to the complications of measuring outcomes like increased disease regulation or persistent remission [89]. In conclusion, nanocrystals offer the potential to enhance therapeutic applications, and their efficient clinical translation is contingent upon resolving challenges with scalability, reproducibility, QC, and standardization. Researchers, manufacturers, and regulatory bodies should collaborate to deal with such challenges and establish appropriate procedures and regulations that ensure constant product efficacy and quality [90]. Different nanocrystals aimed at drug delivery that are currently undergoing clinical trials are listed in Table 10.1 [91].

10.9 Clinical Applications

Broadly, poorly soluble drugs take benefit from the nanocrystal technology. Numerous products are in the pipeline for clinical trials or commercialization [49]. Currently, most nanocrystal medications have been approved for oral administration to treat diseases other than cancer. Compared to injectables, oral administration has a considerably larger market and a more straightforward process for commercialization [5]. Some examples mentioned below provide insights into prospective subsequent products. Meloxicam nanocrystal injection (MNI) is sponsored by CSPC ZhangoQi Pharmaceutical Technology Co., Ltd. It is in clinical trial III for the evaluation of the effectiveness and safety of MNI for treating moderate to severe pain after abdominal surgery [91]. Phase II research is being conducted on Panzem Nanocrystal Colloidal Dispersion (NSD), which is intended for oral administration in conjunction with prolonged oral fixed-dose Temozolomide for the treatment of recurrent Glioblastoma Multiforme [92]. 2-Methoxyestradiol NCD combined with Sunitinib Malate is in trial II for treating Metastatic Renal Cell Carcinoma [93]. Megestrol acetate oral suspension nanocrystal dispersion formulation has completed phase II studies to treat AIDS-related weight loss [94]. As some formulations are in clinical trials, some FDA-approved products are on the market. Some other examples of FDA-approved marketed products are shown in Table 10.2 [95].

Table 10.1 Nanocrystals available in clinical trials

Trade name	Sponsor	Drug	Treatment	Route of administration	ClinicalTrials.gov ID	Clinical status
Meloxicam Nanocrystal Injection	CSPC ZhongQi Pharmaceutical Technology Co., Ltd.	Meloxicam	Pain after abdominal surgery	IV infusion	NCT06379165	III
Panzem NCD	CASI Pharmaceuticals, Inc.	Temozolomide	Recurrent Glioblastoma Multiforme	Orally	NCT00481455	II
Panzem® NCD	CASI Pharmaceuticals, Inc.	Sunitinib Malate	Metastatic Renal Cell carcinoma	Orally	NCT00444314	II
Megestrol acetate oral suspension nanocrystal dispersion	Endo Pharmaceuticals	Megestrol acetate	AIDS-related Weight Loss	Orally	NCT00637572	II
NCX 4251	Nicox Ophthalmics, Inc.	Fluticasone Propionate	Acute Exacerbations of Blepharitis	Ophthalmic Suspension	NCT03926026	II

Table 10.2 Marketed nanocrystal products

Sr. no.	Drug	Marketed product	Indication	Licensed to	Year of approval
1.	Sirolimus	Rapamune®	Immunosuppressant	Elan/Wyeth	2000
2.	Aprepitant	Emend®	Antienmetic	Elan/Merck	2003
3.	Fenofibrate	Tricor®	High cholesterol and triglyceride levels	Elan/Abbott	2004
4.	Megestrol Acetate	Megace ES®	Palliative treatment of kinds of breast and uterine cancers.	Elan/Par Pharma	2005
5.	Fenofibrate	Triglide®	High cholesterol and triglyceride levels	SkyePharma/ First Horizon Pharmaceuticals	2005
6.	Paliperidone palmitate	Invega Sustenna®	Schizophrenia	Elan/Johnson and Johnson	2009

10.10 Emerging Trends in Nanocrystal Technology

Nanocrystals offer a lot of potential for future advancements in a variety of disciplines. Notably, they are expected to play a key role in quantum technologies, photodetectors, drug delivery systems, medical imaging, and environmental applications. Nanocrystals increase the solubility and bioavailability of poorly soluble medications in drug delivery systems. As such, they can be included in a variety of systems for delivery, such as hydrogels, tablets, microneedles, and functionalized liposomes [10].

Integration of nanocrystals into tablets and capsules has become the standard approach, as this formulation is most commonly seen in commercially available products, highlighting the effectiveness of tablet-based delivery systems in harnessing the potential of nanocrystals for improved drug performance [96].

In order to improve the drug's solubility and absorption, Naguib et al. prepared sublingual tablets with fibanserin nanocrystals to treat premenopausal hypoactive sexual drive condition [97]. To prevent possible interactions or interference with the nanocrystals during compression, the excipients must be carefully chosen. Increased aggregation from improper handling could lower the tablets' effective nanocrystal concentration. Their intended therapeutic efficacy may be compromised as a result of their inability to disintegrate and redistribute [98]. Moving on to hydrogels that are filled with nanocrystals, these substances are made to improve particular properties including controlled release, adhesion, and adjustable viscosity. For use on the skin, hair follicles, eyes, nasal cavity, and through parenteral injections, this provides a great deal of adaptability [99–101]. Tomic et al. created an in situ-forming hydrogel by combining azelaic acid nanocrystals with a poloxamer and hyaluronic acid hybrid in order to increase the effectiveness and safety of acne treatments [102]. A camptothecin-loaded hyaluronic acid in situ hydrogel was created by Yongsheng

Gao et al. as a long-term, local delivery method for rheumatoid arthritis [103]. A hydrogel, which is a semi-rigid polymeric network, provides additional physical stabilization through nanocrystals to help prevent issues with recrystallization [10]. Microneedle-mediated delivery of nanocrystals involves the use of microneedles or microarray patches, which are tiny spikes designed to deliver active compounds through the skin. These microneedles create small pores in the stratum corneum, allowing the drug to penetrate into the dermal tissue effectively [104]. They are made of cross-linked biocompatible polymers that typically swell when interacting with interstitial fluids, creating a drug reservoir on-site [23]. Vora and colleagues were pioneers in the process of loading Vitamin D3 into dissolvable microneedles. They initially employed the nanoprecipitation method to create a nanosuspension. Subsequently, the nanosuspension was combined with a high molecular weight polyvinylpyrrolidone (PVP) solution to form a gel matrix, which was then cast into a laser-engineered mold to fabricate microarrays. *In vitro* studies conducted using a Franz diffusion apparatus revealed a 6.8-fold increase in the transdermal delivery of vitamin D3 compared to conventional patches lacking microneedles. Recent investigations into minimally invasive intraocular delivery using microneedles have shown promising results [105]. The next section of the application discusses nanocrystal-polymer microparticles. The integration of nanocrystals within polymeric hydrophobic microparticles or microspheres creates a beneficial biodegradable solid shell that regulates release kinetics [106]. An injectable formulation can be developed that provides long-term release of drugs or nanocrystals at the site of injection, thereby reducing the frequency of administrations. By combining the characteristics of hydrophobic polymers with nanocrystal technology, we can enhance local drug release over an extended period while ensuring good drug-loading capacity [10]. Maudens et al. conducted a study that used wet milling, a technique that uses a highly soluble substance to regenerate articular cartilage. The particles created by this method have a size of 320 nm. These particles were then freeze-dried and embedded in poly(DL-lactide), producing particles with a size range of 10–20 μm that were made using spray-drying methods. With 62% of the medication released over 3 months, the formulation showed an extended-release profile. When compared to conventional keratogenic drugs, osteoarthritic mice used in *in vivo* research demonstrated higher cartilage regeneration activity [107]. Unlike microparticle polymers, the solubility and specific surface area of nanocrystals govern the prolonged release over a number of months. We intend to implant surface-engineered and liposomal nanocrystals in situ within lipid layers in our next application. Li et al. have made great strides in creating ciprofloxacin in situ-forming nanocrystals in liposomes for oral administration [108]. The drug precipitates inside the vesicles after the freeze-thawing step. The solid-state of the drug dictates the liposome's release profile, ensuring consistent drug release under non-digestive conditions, with an increased release observed later in a simulated intestinal fluid medium. PEGylated liposomes functionalized with folic acid enclose nanocrystals produced using the wet ball milling process. The resultant liposomal formulation exhibits improved stability and increases drug loading capacity by up to 20% [10]. Nowadays, covering the outside of nanocrystals with proteins or PEG is a popular method in this field. This technique reduces the specific adsorption of proteins and the phagocytosis of nanoparticles by increasing bioavailability and biocompatibility after intravenous administration [109].

10.11 Summary of Translational Strategies

Nanocrystals have emerged as a transformative platform in the pharmaceutical industry, primarily due to their capacity to enhance the bioavailability, stability, and targeted delivery of poorly soluble drugs. However, the transition from laboratory-scale production to industrial-scale manufacturing involves addressing numerous technical and regulatory challenges. Critical considerations in scaling up nanocrystal production include selecting appropriate fabrication techniques, such as high-pressure homogenization, milling, and precipitation. These methods must be meticulously optimized to ensure compliance with quality standards and scalability demands. Key process parameters, including energy input, processing time, and formulation composition, may require refinement when adapting lab-scale protocols to industrial equipment. The application of Quality by Design (QbD) principles has become instrumental in guiding this transition. By defining and monitoring Critical Quality Attributes (CQAs)—such as particle size distribution, surface charge, crystallinity, and stability—manufacturers can systematically optimize production parameters while ensuring therapeutic efficacy and product safety. Analytical characterization and rigorous stability testing are essential to maintain nanocrystal properties within predefined specifications throughout their shelf life. Regulatory compliance remains a pivotal aspect of production scale-up, requiring strict adherence to guidelines for consistent batch quality, particle size control, and stability. Cost-effectiveness is equally critical for commercial viability, necessitating strategies to balance material consumption, process efficiency, and production expenses. Continuous manufacturing techniques are increasingly explored for their potential to reduce costs and enhance throughput, though they require significant technological investment. Innovative approaches such as microfluidics are gaining traction, offering precise control over reaction conditions and the production of uniform nanocrystals. Additionally, drying techniques like spray and freeze drying are being optimized for large-scale operations to stabilize nanocrystals and improve handling. A multidisciplinary approach integrating pharmaceutical science, engineering, regulatory frameworks, and economic analysis is essential to successfully scale up nanocrystal production. By overcoming technical, logistical, and regulatory barriers, manufacturers can ensure the development of nanocrystal-based formulations with consistent quality and therapeutic performance, paving the way for their broad clinical application. Continued advancements in scalable production technologies and supportive regulatory frameworks underscore the potential of nanocrystals to revolutionize drug delivery and expand therapeutic options across diverse disease areas.

Author Contribution

Niveditha Manne, Vishal Labhade, Vaibhav Dhumal, Pooja Khade, Sourabh Tapekar, Bhavana Gajare: Literature, Compilation of literature, drafting and writing the chapter.
Anitha Sriram, Anish Dhuri: Summarization of chapter, writing and review the manuscript.
Pankaj Kumar Singh: Final review of chapter, summarization, writing, review and editing.

References

1. Pardhi E et al (2024) Nanocrystals technologies in biomedical science: bench to beside the clinic. Drug Discov Today 29(3):103913
2. Rajan DS, Subashini R, Murugananthan G (2021) Pharmaceutical nanocrystals. In: Handbook on nanobiomaterials for therapeutics and diagnostic applications. Elsevier, San Diego, pp 409–420
3. Malamatari M et al (2018) Pharmaceutical nanocrystals: production by wet milling and applications. Drug Discov Today 23(3):534–547
4. Chary PS et al (2024) Emerging role of nanocrystals in pharmaceutical applications: a review of regulatory aspects and drug development process. Appl Mater Today 40:102334
5. Jarvis M, Krishnan V, Mitragotri S (2019) Nanocrystals: a perspective on translational research and clinical studies. Bioeng Transl Med 4(1):5–16
6. Joseph E, Singhvi G (2019) Chapter 4 – Multifunctional nanocrystals for cancer therapy: a potential nanocarrier. In: Nanomaterials for drug delivery and therapy. William Andrew Publishing, Norwich
7. Srivalli KMR, Mishra B (2016) Drug nanocrystals: a way toward scale-up. Saudi Pharm J 24(4):386–404
8. Bobo D et al (2016) Nanoparticle-based medicines: a review of FDA-approved materials and clinical trials to date. Pharm Res 33:2373–2387
9. Jacob S, Nair AB, Shah J (2020) Emerging role of nanosuspensions in drug delivery systems. Biomater Res 24(1):3
10. Rossier B et al (2024) Nanocrystals and nanosuspensions: an exploration from classic formulations to advanced drug delivery systems. Drug Deliv Transl Res 14(12):1–14
11. Strebhardt K, Ullrich A (2008) Paul Ehrlich's magic bullet concept: 100 years of progress. Nat Rev Cancer 8(6):473–480
12. Jahangir MA et al (2020) Nanocrystals: characterization overview, applications in drug delivery, and their toxicity concerns. J Pharm Innov 17:1–12
13. Junyaprasert VB, Morakul B (2015) Nanocrystals for enhancement of oral bioavailability of poorly water-soluble drugs. Asian J Pharm Sci 10(1):13–23
14. Litany RJ, Praseetha P (2022) Tiny tots for a big-league in wound repair: tools for tissue regeneration by nanotechniques of today. J Control Release 349:443–459
15. Möschwitzer JP (2013) Drug nanocrystals in the commercial pharmaceutical development process. Int J Pharm 453(1):142–156
16. Gao L et al (2012) Drug nanocrystals: *in vivo* performances. J Control Release 160(3):418–430
17. Mohammad IS et al (2019) Drug nanocrystals: fabrication methods and promising therapeutic applications. Int J Pharm 562:187–202
18. Peltonen L, Hirvonen J (2018) Drug nanocrystals–versatile option for formulation of poorly soluble materials. Int J Pharm 537(1–2):73–83
19. Tuomela A, Hirvonen J, Peltonen L (2016) Stabilizing agents for drug nanocrystals: effect on bioavailability. Pharmaceutics 8(2):16
20. Beirowski J et al (2011) Freeze-drying of nanosuspensions, 1: freezing rate versus formulation design as critical factors to preserve the original particle size distribution. J Pharm Sci 100(5): 1958–1968

21. Gigliobianco MR et al (2018) Nanocrystals of poorly soluble drugs: drug bioavailability and physicochemical stability. Pharmaceutics 10(3):134
22. Colombo M et al (2017) In situ determination of the saturation solubility of nanocrystals of poorly soluble drugs for dermal application. Int J Pharm 521(1–2):156–166
23. Al-Kassas R, Bansal M, Shaw J (2017) Nanosizing techniques for improving bioavailability of drugs. J Control Release 260:202–212
24. Chaubal MV (2004) Application of formulation technologies in lead candidate selection and optimization. Drug Discov Today 9(14):603–609
25. Zhao R et al (2011) Hybrid nanocrystals: achieving concurrent therapeutic and bioimaging functionalities toward solid tumors. Mol Pharm 8(5):1985–1991
26. Jermain SV, Brough C, Williams RO III (2018) Amorphous solid dispersions and nanocrystal technologies for poorly water-soluble drug delivery–an update. Int J Pharm 535(1–2):379–392
27. Junghanns J-UA, Müller RH (2008) Nanocrystal technology, drug delivery and clinical applications. Int J Nanomedicine 3(3):295–310
28. Salazar J et al (2012) Nanocrystals: comparison of the size reduction effectiveness of a novel combinative method with conventional top-down approaches. Eur J Pharm Biopharm 81(1): 82–90
29. Dong Y et al (2011) Controlled antisolvent precipitation of spironolactone nanoparticles by impingement mixing. Int J Pharm 410(1–2):175–179
30. Shi Y et al (2009) Recent advances in intravenous delivery of poorly water-soluble compounds. Expert Opin Drug Deliv 6(12):1261–1282
31. Thi Ngoc Vo A et al (2017) Sonication-assisted nanoprecipitation in drug delivery. Curr Drug Metab 18(2):145–156
32. Aher S et al (2010) Ultrasound assisted cocrystallization from solution (USSC) containing a non-congruently soluble cocrystal component pair: caffeine/maleic acid. Eur J Pharm Sci 41(5):597–602
33. Sinha B, Müller RH, Möschwitzer JP (2013) Bottom-up approaches for preparing drug nanocrystals: formulations and factors affecting particle size. Int J Pharm 453(1):126–141
34. Zhong J et al (2005) Preparation and characterization of uniform nanosized cephradine by combination of reactive precipitation and liquid anti-solvent precipitation under high gravity environment. Int J Pharm 301(1–2):286–293
35. Kim M-S et al (2008) Preparation, characterization and *in vivo* evaluation of amorphous atorvastatin calcium nanoparticles using supercritical antisolvent (SAS) process. Eur J Pharm Biopharm 69(2):454–465
36. Padrela L et al (2010) Screening for pharmaceutical cocrystals using the supercritical fluid enhanced atomization process. J Supercrit Fluids 53(1–3):156–164
37. Badens E et al (2018) Current situation and perspectives in drug formulation by using supercritical fluid technology. J Supercrit Fluids 134:274–283
38. Esfandiari N, Ghoreishi SM (2013) Synthesis of 5-fluorouracil nanoparticles via supercritical gas antisolvent process. J Supercrit Fluids 84:205–210
39. Tabernero A, del Valle EMM, Galán MA (2012) Precipitation of tretinoin and acetaminophen with solution enhanced dispersion by supercritical fluids (SEDS). Role of phase equilibria to optimize particle diameter. Powder Technol 217:177–188
40. Nandiyanto ABD, Okuyama K (2011) Progress in developing spray-drying methods for the production of controlled morphology particles: from the nanometer to submicrometer size ranges. Adv Powder Technol 22(1):1–19
41. Alam MA, Al-Jenoobi FI, Al-Mohizea AM (2013) Commercially bioavailable proprietary technologies and their marketed products. Drug Discov Today 18(19–20):936–949
42. Liversidge GG et al (1992) Surface modified drug nanoparticles. Google Patents
43. Merisko-Liversidge EM, Liversidge GG (2008) Drug nanoparticles: formulating poorly water-soluble compounds. Toxicol Pathol 36(1):43–48
44. Loh ZH, Samanta AK, Heng PWS (2015) Overview of milling techniques for improving the solubility of poorly water-soluble drugs. Asian J Pharm Sci 10(4):255–274

45. Merisko-Liversidge E, Liversidge GG (2011) Nanosizing for oral and parenteral drug delivery: a perspective on formulating poorly-water soluble compounds using wet media milling technology. Adv Drug Deliv Rev 63(6):427–440
46. Oh J-H et al (2015) Recent advances in chemical functionalization of nanoparticles with biomolecules for analytical applications. Anal Bioanal Chem 407:8627–8645
47. Sapsford KE et al (2013) Functionalizing nanoparticles with biological molecules: developing chemistries that facilitate nanotechnology. Chem Rev 113(3):1904–2074
48. Park J, Sun B, Yeo Y (2017) Albumin-coated nanocrystals for carrier-free delivery of paclitaxel. J Control Release 263:90–101
49. Pardhi VP et al (2018) Nanocrystals: an overview of fabrication, characterization and therapeutic applications in drug delivery. Curr Pharm Des 24(43):5129–5146
50. Tu L et al (2013) Effects of particle size on the pharmacokinetics of puerarin nanocrystals and microcrystals after oral administration to rat. Int J Pharm 458(1):135–140
51. Koradia KD, Parikh RH, Koradia HD (2018) Albendazole nanocrystals: optimization, spectroscopic, thermal and anthelmintic studies. J Drug Deliv Sci Technol 43:369–378
52. Chogale MM, Ghodake VN, Patravale VB (2016) Performance parameters and characterizations of nanocrystals: a brief review. Pharmaceutics 8(3):26
53. Sharma OP, Patel V, Mehta T (2016) Design of experiment approach in development of febuxostat nanocrystal: application of Soluplus® as stabilizer. Powder Technol 302:396–405
54. Tuomela A et al (2014) Brinzolamide nanocrystal formulations for ophthalmic delivery: reduction of elevated intraocular pressure *in vivo*. Int J Pharm 467(1–2):34–41
55. Gao L, Zhang D, Chen M (2008) Drug nanocrystals for the formulation of poorly soluble drugs and its application as a potential drug delivery system. J Nanopart Res 10:845–862
56. Du J et al (2015) Nanosuspensions of poorly water-soluble drugs prepared by bottom-up technologies. Int J Pharm 495(2):738–749
57. Guo M et al (2015) Rod shaped nanocrystals exhibit superior *in vitro* dissolution and *in vivo* bioavailability over spherical like nanocrystals: a case study of lovastatin. Colloids Surf B: Biointerfaces 128:410–418
58. Sharma S et al (2015) Development of stabilized paclitaxel nanocrystals: in-vitro and in-vivo efficacy studies. Eur J Pharm Sci 69:51–60
59. Müller RH et al (2012) Long-term stability of quercetin nanocrystals prepared by different methods. J Pharm Pharmacol 64(10):1394–1402
60. Ochi M et al (2014) Development of nanocrystal formulation of meloxicam with improved dissolution and pharmacokinetic behaviors. Int J Pharm 474(1–2):151–156
61. Li X et al (2025) Posaconazole nanocrystals dry powder inhalers for the local treatment of invasive pulmonary aspergillosis. Int J Pharm 668:124938
62. Shegokar R (2020) Preclinical testing—understanding the basics first. In: Drug delivery aspects. Elsevier, pp 19–32
63. Padmanabhan S (2014) Clinical trials in pharmacogenomics and stratified medicine. In: Handbook of pharmacogenomics and stratified medicine. Elsevier/Academic Press, London
64. Sinha S, Divya V (2018) Drug discovery and development: An overview. Pharmaceutical medicine and translational clinical research, 19–32
65. Bajpai M, Esmay JD (2002) *In vitro* studies in drug discovery and development: an analysis of study objectives and application of good laboratory practices (GLP). Drug Metab Rev 34(4): 679–689
66. Maheshwari R et al (2018) Guiding principles for human and animal research during pharmaceutical product development. In: Dosage form design parameters. Elsevier, London, pp 621–664
67. Gupta PK (2016) Chapter 14 – Preclinical toxicological investigations of pharmaceutical products. In: Gupta PK (ed) Fundamentals of toxicology. Academic Press, San Diego, pp 165–171
68. Cheng L et al (2011) *In vivo* pharmacokinetics, long-term biodistribution and toxicology study of functionalized upconversion nanoparticles in mice. Nanomedicine 6(8):1327–1340

69. Shen B et al (2024) Fabrication and *in vitro/vivo* evaluation of quercetin nanocrystals stabilized by glycyrrhizic acid for liver targeted drug delivery. Int J Pharm X 7:100246
70. Arome D, Chinedu E (2013) The importance of toxicity testing. J Pharm BioSci 4:146–148
71. Rahim H et al (2017) Aceclofenac nanocrystals with enhanced *in vitro, in vivo* performance: formulation optimization, characterization, analgesic and acute toxicity studies. Drug Des Devel Ther 11:2443–2452
72. Tyner KM et al (2015) Product quality for nanomaterials: current U.S. experience and perspective. WIREs Interdiscip Rev Nanomed Nanobiotechnol 7(5):640–654
73. Müller R, Keck C (2011) Twenty years of drug nanocrystals: where are we, and where do we go? Eur J Pharm Biopharm 80(1):1–3
74. Sivanathan G et al (2024) Pharmaceutical nanocrystals: an extensive overview. Int J Appl Pharm 16(6):1–9
75. Rauscher H, Rasmussen K, Sokull-Klüttgen B (2017) Regulatory aspects of nanomaterials in the EU. Chemie Ingenieur Technik 89(3):224–231
76. Hansen SF, Baun A (2012) European regulation affecting nanomaterials-review of limitations and future recommendations. Dose Response 10(3):364–383
77. Nielsen MB (2024) European environmental regulation of nanomaterials: inclusion of the eco-corona. Technical University of Denmark, Copenhagen
78. Hua S et al (2018) Current trends and challenges in the clinical translation of nanoparticulate nanomedicines: pathways for translational development and commercialization. Front Pharmacol 9:790
79. Shi J et al (2017) Cancer nanomedicine: progress, challenges and opportunities. Nat Rev Cancer 17(1):20–37
80. Zheng C, Li M, Ding J (2021) Challenges and opportunities of nanomedicines in clinical translation. Bio Integr 2(2):57–60
81. Metselaar JM, Lammers T (2020) Challenges in nanomedicine clinical translation. Drug Deliv Transl Res 10:721–725
82. Barz M, Luxenhofer R, Schillmeier M (2015) Quo vadis nanomedicine? Nanomedicine 10(20):3089–3091
83. Younis MA et al (2022) Clinical translation of nanomedicines: challenges, opportunities, and keys. Adv Drug Deliv Rev 181:114083
84. Lammers T (2013) Smart drug delivery systems: back to the future vs. clinical reality. Int J Pharm 454(1):527–529
85. Wicki A et al (2015) Nanomedicine in cancer therapy: challenges, opportunities, and clinical applications. J Control Release 200:138–157
86. Zhang C et al (2020) Progress, challenges, and future of nanomedicine. Nano Today 35: 101008
87. Bourquin J et al (2018) Biodistribution, clearance, and long-term fate of clinically relevant nanomaterials. Adv Mater 30(19):1704307
88. Đorđević S et al (2022) Current hurdles to the translation of nanomedicines from bench to the clinic. Drug Deliv Transl Res 12(3):1–26
89. Mundekkad D, Cho WC (2022) Nanoparticles in clinical translation for cancer therapy. Int J Mol Sci 23(3):1685
90. Fong EL et al (2012) Building bridges: leveraging interdisciplinary collaborations in the development of biomaterials to meet clinical needs. Adv Mater 24(36):4995–5013
91. Hu S et al (2024) Bioequivalence of meloxicam nanocrystal injection in healthy Chinese volunteers. Clin Pharm Drug Dev 13(12):1339–1344
92. Yenurkar D, Nayak M, Mukherjee S (2023) Recent advances of nanocrystals in cancer theranostics. Nanoscale Adv 5(16):4018–4040
93. Joshi E, Pandya M, Desai U (2023) Current drugs and their therapeutic targets for hypoxia-inducible factors in cancer. Curr Protein Pept Sci 24(6):447–464

94. Park SR et al (2024) Comparison of the pharmacokinetic characteristics and bioequivalence between two nanosuspension formulations of megestrol acetate in healthy Korean male subjects. Transl Clin Pharm 32(1):63
95. Gülsün T, Gürsoy RN, Öner L (2011) Nanocrystal technology for oral delivery of poorly watersoluble drugs. FABAD Journal of Pharmaceutical Sciences 34.1(2009):55
96. Lopez-Vidal L et al (2022) Nanocrystal-based 3D-printed tablets: semi-solid extrusion using melting solidification printing process (MESO-PP) for oral administration of poorly soluble drugs. Int J Pharm 611:121311
97. Naguib MJ, Makhlouf AI (2021) Scalable flibanserin nanocrystal-based novel sublingual platform for female hypoactive sexual desire disorder: engineering, optimization adopting the desirability function approach and *in vivo* pharmacokinetic study. Drug Deliv 28(1): 1301–1311
98. Müller RH, Gohla S, Keck CM (2011) State of the art of nanocrystals–special features, production, nanotoxicology aspects and intracellular delivery. Eur J Pharm Biopharm 78(1): 1–9
99. Li Y et al (2018) Pramipexole nanocrystals for transdermal permeation: characterization and its enhancement micro-mechanism. Eur J Pharm Sci 124:80–88
100. Zhu S et al (2021) Effects of armodafinil nanocrystal nasal hydrogel on recovery of cognitive function in sleep-deprived rats. Int J Pharm 597:120343
101. Parmar PK, Wadhawan J, Bansal AK (2021) Pharmaceutical nanocrystals: a promising approach for improved topical drug delivery. Drug Discov Today 26(10):2329–2349
102. Tomić I et al (2021) Efficacy and safety of azelaic acid nanocrystal-loaded in situ hydrogel in the treatment of acne vulgaris. Pharmaceutics 13(4):567
103. Gao Y et al (2022) Injectable hyaluronic acid hydrogels encapsulating drug nanocrystals for long-term treatment of inflammatory arthritis. Bioeng Transl Med 7(1):e10245
104. Tuan-Mahmood T-M et al (2013) Microneedles for intradermal and transdermal drug delivery. Eur J Pharm Sci 50(5):623–637
105. Thakur RRS et al (2016) Rapidly dissolving polymeric microneedles for minimally invasive intraocular drug delivery. Drug Deliv Transl Res 6:800–815
106. Wang Y et al (2021) Improving stability and oral bioavailability of hydroxycamptothecin via nanocrystals in microparticles (NCs/MPs) technology. Int J Pharm 604:120729
107. Maudens P et al (2018) Nanocrystal–polymer particles: extended delivery carriers for osteoarthritis treatment. Small 14(8):1703108
108. Li T et al (2018) Drug nanocrystallisation within liposomes. J Control Release 288:96–110
109. Park J et al (2018) A comparative *in vivo* study of albumin-coated paclitaxel nanocrystals and abraxane. Small 14(16):1703670

Regulatory Considerations of Nanocrystal in Pharmaceutical Sciences

11

Vasu Peddinti, Shyam Sudhakar Gomte, Devaraj Yallappa, Shruti Patil, and Aakanchha Jain

Abstract

Nanocrystals (NCs), an advanced nanotechnology-based drug delivery approach, gained major interest in pharmaceutical sciences because of their ability to improve the solubility and bioavailability of poorly water-soluble drugs. Regulatory considerations are essential to ensure the safety, effectiveness, and quality of NCs during the drug development and commercialization processes. These factors include characterization, processes for manufacturing, and developing relevant specifications. Because these variables substantially influence drug performance, regulatory bodies like the European Medicines Agency (EMA) and the U.S. Food and Drug Administration (FDA) demand comprehensive physicochemical characterization of NCs, including particle size, surface area, and zeta potential. To achieve consistent product quality, manufacturing processes need to be well-controlled. The production of NCs also entails solving safety issues such as toxicity and immunogenicity, which might occur due to NCs' small size and large surface area. Regulatory requirements also stress the significance of developing robust analytical techniques, bioequivalence studies, and stability testing to ensure product quality throughout its shelf life. This chapter first introduces the overview of the NCs. Further, we discussed the relevance of regulatory concerns in the design of NC-based formulations and key regulatory agency guidelines. In addition, we shed light on the clinical implications of NCs. Furthermore, we discussed the procedure for regulatory submissions as well as new developments in NC development.

V. Peddinti · S. S. Gomte · D. Yallappa · S. Patil · A. Jain (✉)
Department of Pharmaceutics, National Institute of Pharmaceutical Education and Research (NIPER)-Ahmedabad, Palaj, Gandhinagar, Gujarat, India
e-mail: aakanchha.jain@niperahm.res.in

N. K. Mehra, K. K. Bansal (eds.), *Nanocrystal Technologies in Biomedical Sciences*,
https://doi.org/10.1007/978-981-95-2968-1_11

Keywords

Nanocrystals · Regulatory agencies · Critical quality attributes

11.1 Introduction

11.1.1 Overview of Nanocrystals

Nanocrystals (NCs) are carrier-free submicron colloidal drug delivery systems with an average particle size in the nanometer range, typically 10–400 nm. They are composed of pure drugs and the minimal amount of surface active agents needed for stability. It is possible to synthesize NCs using proven methods without costly equipment. Furthermore, NCs have a higher surface area as well as apparent solubility than larger-sized materials, which increases the rate of dissolution and bioavailability of therapeutic substances that are poorly water-soluble [1]. Moreover, NCs are easily scalable, maintain pH neutrality, show long-term stability, and may be synthesized without the use of solvents [2].

Drug solubility in water is a significant difficulty in pharmaceutical sciences [3]. The fact that a sizable fraction of currently used drugs and newly developed active candidates have limitations on water solubility makes this problem even more difficult [4]. Currently, the majority of pharmaceuticals under investigation, approximately 70% belong to class II of the Biopharmaceutics Classification System (BCS), which has low solubility and slower dissolution rates that limit absorption. "Low solubility" refers to a slower rate of dissolution, which means that bioavailability and biodistribution are affected [5]. A wide range of approaches have been methodically studied to enhance the biopharmaceutical performance and absorption of drugs that are insufficient soluble drugs. Molecular-level interventions include approaches that are effective in increasing drug solubility, such as the addition of co-solvents, the development of salts, the production of prodrugs, co-crystals [6], and complexation with cyclodextrins [7]. Particle-level strategies include co-crystallization, using metastable polymorphs [8], making solid dispersions [9], developing co-amorphous systems [10], and particle size reduction techniques. The use of nanoparticles in the development of drug delivery systems has garnered significant attention over the past 20 years. One important approach for efficiently administering hydrophobic drugs is the use of lipid-based nanoparticles [11]. However, these approaches have several drawbacks, the majority are closely related to the physicochemical properties that the drug molecule naturally possesses [12]. For instance, difficulties with producing nanocarriers are often linked to problems like inadequate long-term stability, suboptimal drug loading (usually between 5% and 30% w/w), and, in some cases, the use of organic solvents [13]. Given these limitations, developing NCs appears to be an appealing and adaptable technique for delivering water-insoluble active substances [14]. This chapter focuses on the need for regulatory considerations in the designing of NC-based products and relevant guidelines by regulatory agencies. Further, we discussed the clinical considerations of NCs. In addition, we shed light on the regulatory submission process and emerging trends in NC development.

11.1.2 Importance of Regulatory Considerations

Regulatory considerations are very important in the development and marketing of NCs in pharmaceutical sciences because of their distinct characteristics and possible risks. Regulatory bodies must address particular concerns about the formulation, safety, and efficacy of NCs as their application to increase drug solubility and bioavailability of poorly soluble compounds. Currently, no globally acknowledged regulatory structure exists for NCs. Various regulatory agencies, including the US FDA (Food and Drug Administration) and European Medicines Agency (EMA), have released guidelines; however they are not consistent. Manufacturers and healthcare professionals might get confused about the safety requirements and approval procedure for products containing NCs as a result of this discrepancy [15, 16].

NCs improve the solubility and dissolution rates of poorly soluble drugs, which is important for their bioavailability. NCs are used in approximately 30% of drug product applications, demonstrating their relevance in drug development. Regulatory frameworks must guarantee that these formulations retain their improved characteristics during the production process [17]. In contrast to conventional medications, the nanoscale size of NCs provides characterization challenges. Regulatory agencies must set precise standards for evaluating critical quality attributes (CQAs) such as stability, surface area, and particle size to ensure constant product quality. Lack of established definitions and characterization techniques can impede research and development activities [18]. Robust production procedures are necessary to specify the product's characteristics, which affect its efficacy, safety, and quality. Regulations need to take into account the difficulties of increasing output while keeping control over crucial manufacturing aspects. As part of this, strict processes must be established to guard against contamination and guarantee consistency across batches [15]. Comprehensive information about the effectiveness and safety of NC-based products is necessary by regulatory bodies. Considering their distinct pharmacokinetic and biodistribution properties, conventional methods of evaluation might not be adequate. For an accurate evaluation, a weight-of-evidence strategy that incorporates quality, non-clinical, and clinical data is required [16]. Regulatory pathways need to change to allow for generics and follow-on versions while maintaining therapeutic equivalency when NCs are used in more therapeutic applications. This includes resolving potential safety issues resulting from variations in the manufacturing techniques used for manufacturing original products vs their generics. Establishing defined regulations is essential for increasing public confidence in treatments based on NCs. The lack of standardized international regulations presents problems for firms intending to market products using NCs globally [19]. In conclusion, regulatory concerns surrounding NCs are critical for promoting advancement in drug delivery systems and promising the safety and effectiveness of pharmaceuticals.

11.2 Regulatory Framework

11.2.1 Key Regulatory Agencies and Relevant Guidelines

Several regulatory bodies regulate the review and approval processes for NCs in pharmaceutical sciences to assure their safety, effectiveness, and high quality. NCs are regulated by the US FDA as part of its larger pharmaceutical regulation. It has released guidelines outlining the special qualities of nanomaterials and stressing the importance of thorough characterization and safety evaluations. The key CQAs, manufacturing procedures, and possible hazards related to NC formulations are all assessed as part of the FDA's framework [20]. Additionally, the organization carries out technical assessments that are directed by a policy that is grounded in science and product focus. Such in-depth research makes it possible to effectively combine the unique needs of a particular product class with pharmacological expertise applied in a way that is driven by cutting-edge technology. This methodology facilitates tailored approaches that take into account the distinct features of particular products or product categories, such as NCs, and integrate technological and scientific developments in the pharmaceutical sector [21]. The majority of the data about NC biopharmaceutics quality, chemistry, manufacturing, and controls (CMC) came from the US FDA's Center for Drug Evaluation and Research (CDER) based on previously submitted Abbreviated New Drug Applications (ANDAs), New Drug Applications (NDAs), and Investigational New Drugs (INDs). The US FDA guideline document states that NCs in therapeutic forms, including active pharmaceutical ingredients (APIs) and colloidal dispersions, should not be more than size of 1000 nm [22]. In the early 2000s, a major advancement in NCs was made, opening up new possibilities for developing various pharmaceutical products using micronized (APIs) that had already been approved for commercialization. Under the 505(b)(2) procedure, a candidate may submit a reformulated product request for approval to the US FDA based on previously available data, research, or information [23].

The EMA, like the US FDA, controls NCs under its medical products regulation. With an emphasis on the physicochemical characteristics and biological activity of nanomedicines, the EMA formulated specific regulations for their evaluation. This organization also stresses the necessity of a case-by-case analysis to take into consideration the intricacies related to NCs. It has also defined nanomedicine and provided a set of particular guidelines for nanomedicine, which may be found on its website. The EMA established the European Nanomedicines Expert Group in 2009 in response to the growing need for stakeholder review of nanomedicines. Regulators from different regulatory agencies, including the FDA, met with eminent academics and regulatory science professionals from the Expert Group [24]. In

India, the CDSCO has developed guidelines for reviewing nanopharmaceuticals, which include NCs. These recommendations seek to promote innovation in products that use nanotechnology while ensuring product efficacy, safety, and quality. The framework of the CDSCO is made to adjust to the distinctive features of NCs and their uses in drug delivery [25].

11.3 Quality and Safety Considerations

11.3.1 CQAs for NCs

The assessment of CQAs is one of the important phases of a study-based QbD approach. A CQA is generally referred to as, "a physical, chemical, biological or microbiological attribute or feature that must be within an appropriate limit, range or distribution to assure the desired product quality." The CQAs are affected by CPPs or CMAs. For instance, dosage strength or dosage form cannot be altered during product development. The NC development also includes different CQAs such as particle size, particle size distribution (PSD), entrapment efficiency, release patterns, etc. that might be affected during process variables or material attributes for the manufacturing of NCs [26, 27]. The additional quality attributes involving recognizing polymorphic forms of developed NCs are of prime importance. Following guidelines given by the FDA for nanomaterials include some additional quality attributes such as shape and morphology, the material's chemical composition, the product's physical and chemical stability, surface attributes, particle concentration, in vitro release, impurities, and crystal form of the developed product [20].

Several scientists have developed NCs by considering different CQAs. For instance, Ainurofiq et al. developed resveratrol-loaded NCs into mesoporous material to enhance solubility. The CQAs for the development of NCs were kept as particle size, polydispersity index (PDI), zeta potential, drug loading, supersaturation time, and area under the curve (AUC) [28]. In another study, Awad and co-workers developed ivermectin-laden NCs for improved topical delivery. The CQAs for the NCs development were kept as particle size, PDI, and zeta potential [29].

11.3.2 Manufacturing Process Control and Validation

The manufacturing process of NCs should be deeply monitored as it affects the final quality attributes of developed NCs. The main objective of gathering agency information from NDA, ANDA, and IND applications, as well as quality submissions, was to evaluate the scientific problems related to the design of NC formulations. The key parameters that affect the quality of NCs include drug substance stability, drug product stability, manufacturing, and control of final NC products. Under each separate section the challenges were observed and later categorized into developmental and characterization issues. Thus, an in-depth understanding of process variables and characterization is essential which impacts the quality of final NCs [17, 30].

Table 11.1 Overview of fabrication methods for NCs with key limitations

Sr. no.	Method of fabrication	Limitation	References
1	Wet milling	Generation of polymorphic forms Agglomeration of particles Enhanced energy consumption	[31]
2	High-pressure homogenization	Crystallization Agglomeration of particles	[32]
3	Anti-solvent precipitation	Generation of polymorphic forms Agglomeration of particles	[33]
4	Spray drying	Generation of polymorphic forms Agglomeration of particles	[34]
5	Sonoprecipitation	Polymorphic form creation Agglomeration of particles	[35]
6	Dry milling	Generation of polymorphic forms	[36]
7	Freeze drying	Generation of polymorphic forms Agglomeration of particles	[37]
8	Laser ablation	Oxidative degradation of drugs	[38]
10	Nanoedge	Costly approach Requires sophisticated instrument	[39]
11	SmartCrystal Technology	Costly	[39]

The key development consideration in NC preparation is the assessment and recognition of in-process control or tests at critical preparation time such as uniform mixing of API within the matrix, blend homogeneity following the mixing stage, final granulation, and sampling protocol for in-process assay. The additional vital step is to control over PSD of the drug and prevent its aggregation or agglomeration of NCs after the homogenization or milling stage. In terms of the fabrication of NCs employing the top-down approach like wet milling, it is difficult to understand the mechanical attrition process, the specific kind of milling, mechanism, milling medium, important leachables, extractables, or contaminants, and suitable measures to evaluate impurities produced during the milling process. The primary problem in wet milling is the formation of residues from the milling media or substance during the milling process [22]. The various fabrication techniques used in the synthesis of NCs and their limitations are listed in Table 11.1.

The milling material is mostly composed of stainless steel, ceramics, glass, and cross-linked polystyrene beads. The shear that is generated during milling creates movement of milling media and consequently results in a reduction in particle size. The grinding material is eradicated during the milling process and process-based impurities are introduced into final products. The generated material or impurities are mostly insoluble. The implementation of assessments and the product of desired criteria to regulate contaminants and impurities recognized during the production process is a significant developmental aspect for pharmaceutical formulations incorporating NCs. The contamination of milling media might be considered in the milling procedure of NC. The fabrication variables including chiller temperature and product temperature as a function of time are taken into account to attain effective control over PSD. The other parameters such as drive motor speed and settings of peristaltic pumps are investigated in terms of their impact on PSD during and after milling steps. The reproducibility of milling is largely reliant on the viscosity of the solution, shaper, aspect ratio of API, and different input variables [22].

The hot-melt method as a bottom-up approach mostly considered the drug substance that was not changed to different polymorphic forms at the time of the melting process. The main focus in the bottom-up method should be batch-to-batch consistency of drug products in terms of PSD. The critical aspects that might be raised during the hot-melt method include (1) how to ensure homogenous blending of drugs in the matrix system, (2) how to regulate the rate of addition drug substance to melt and restrict drug aggregation, (3) How to measure and manage the appropriate viscosity of molten compound. For assay determination, sampling plans are important when several batches are mixed after the milling step as the assay helps to determine the in-process quality of developed products. These all-developed products impact the final stage of product quality. The fundamental problem with the hot-melt technique is recrystallization into the micrometer range, which must be avoided in the bottom-up approach by strict management of the production process. Moreover, some of the bottom-up utilize organic solvents that might be introduced as residue or impurity in final drug products [40–42]. The manufacturing processes and the key factors influencing regulatory approval of NC formulations are included in Table 11.2.

Table 11.2 Key parameters influencing regulatory approval of NC formulations

Sr. no.	Parameter/steps during formulation	Regulatory aspects	References
1	*Top-down strategy* High-pressure homogenization Wet milling	Time-consuming and slow process Microbiological risk Residual of milling media Not uniform particle size distribution	[43, 44]
2	*Bottom-up strategy* Nanoprecipitation Hydrosol Supercritical anti-solvent High gravity-controlled precipitation	Challenging to control the final crystal size Time-consuming Requires highly sophisticated instrument Not a suitable option for thermolabile APIs Solvent residual	[45]
3	*Combination strategy* of top-down and bottom-up approach	Increased pre-treatment procedure Higher cost	[46]
4	*Nanoedge* strategy	Organic solvent usage	[17]
5	Chemical composition	To solubilize poorly soluble therapeutic agents, the introduction of organic solvent leads to the formation of solvent-based impurities.	[47]
6	Particle size distribution	Uneven particle size results in the collision of particles and tends to form an agglomerate.	[47]

11.3.3 Safety Testing and Toxicology

The addition of an inactive pharmaceutical ingredient (IPI) into nanomaterials like NCs may provide safety and efficacy for the developed finished product. The various sizes or shapes of NCs could have potential batch reproducibility issues and it might affect the safety, efficacy, and quality of the finished product [20]. Safety is a primary obstacle in pharmaceuticals and thereby toxicity evaluation is the most vital data for the registration of new drug products. For instance, safety challenges with poorly soluble therapeutic agents might be more critical. Owing to their poor solubility, a larger amount of organic solvent could be incorporated before they are developed as drug products and that will consequently result in toxicities or undesirable side effects. Moreover, the precipitation of low soluble therapeutic agents from non-aqueous drug products when it is diluted to blood is another critical issue [48–50]. The NC-based products are found to be promising in terms of their safety and have been well tolerated in wider administration routes compared to traditional products. The smaller particle size of developed NCs helps in many aspects. The smaller particle size helps in the homogenous distribution of drug products in the gastrointestinal tract (GIT) and circumvent high and long-time local drug concentrations. These smaller particles are helpful for mucosal drug delivery like pulmonary and ocular drug delivery by reducing the occurrence of local irritation. Owing to NCs' aqueous features and safe composition, the requirement for larger organic solvents and a higher pH range for enhancing the solubility of poorly soluble therapeutic agents is omitted. This gives an additional advantage in reducing the generation of solvent-related impurities in final finished products. Safer components and composition are also suitable for pulmonary delivery as the co-solvent introduction can be avoided and thereby evade irritation [51].

11.4 Analytical Methods

11.4.1 Regulatory Requirements for Method Validation and Acceptance

The establishment of clear regulations for method validation and acceptance is crucial in ensuring the efficacy, quality, and safety of pharmaceutical products containing NCs [52]. As NCs are increasingly used to improve drug bioavailability and therapeutic effectiveness, reliable analytical techniques must be developed to accurately measure their concentration [53]. Regulatory bodies like the International Conference on Harmonisation (ICH) provide guidelines that outline essential parameters for validation, including linearity, specificity, accuracy, precision, and robustness [54].

Method validation is especially important for NCs because of their distinct properties at the nanoscale, which can significantly influence drug behavior and performance. It plays a critical role in ensuring the efficacy and safety of NC drug products [55]. The effectiveness of an NC formulation relies on its physical properties, including structure, surface charge, shape, size distribution, rheological

Table 11.3 Analytical method validation for characterization of NCs

Sr. no	Characterization parameter	Analytical methods	References
1	Surface charge	Zeta potential	[56]
2	Particle size and size distribution	Dynamic light scattering (DLS)	[57]
3	Structure and morphology	Scanning electron microscopy (SEM), transmission electron microscopy (TEM)	[58, 59]
4	Solid state analysis (crystallinity)	Powder X-ray diffraction (XRD)	[60]
5	Rheological properties (for liquid suspension)	Viscometer, Rheometer	[61]

properties, and internal atomic arrangement [45]. Various analytical techniques are commonly used for characterizing and optimizing NC properties and are summarized in Table 11.3.

11.5 Clinical Considerations

11.5.1 Clinical Trial Design for NC-Based Drugs

Clinical trials for NC drugs are designed to evaluate not only their efficacy and safety but also their pharmacokinetic profiles and therapeutic outcomes. Selecting an appropriate patient population is essential for the success of such trials. Key parameters such as disease state, severity, and previous treatments must be carefully considered to ensure that the trial results are relevant and generalizable [18]. Navigating regulatory requirements is crucial when designing clinical trials for NC formulations. Regulatory bodies often demand comprehensive data on the efficacy and safety of new drug delivery systems, necessitating rigorous preclinical studies before proceeding to human trials. Trials should establish clear primary and secondary endpoints that assess both efficacy and safety. Common endpoints may include measures of bioavailability, therapeutic response, and adverse event profiles compared to conventional drug formulations [62]. When submitting an IND application, sponsors must provide comprehensive data on formulation development, preclinical findings, and proposed clinical trial designs that adhere to FDA guidelines for nanotechnology applications [22].

Despite encouraging preclinical results, the clinical translation of NC technology is impeded by various obstacles. Regulatory hurdles, such as establishing appropriate guidelines and standards for nanomaterials, can be significant. Furthermore, thorough validating of manufacturing processes is crucial to ensure consistent quality, safety, and efficacy of the NC-based formulation. These challenges necessitate a collaborative effort between researchers, regulatory bodies, and industry to overcome obstacles and realize the full potential of NC technology in clinical [18, 62].

11.5.2 Special Considerations for Pediatric and Geriatric Populations

The pediatric population is diverse, including children of different ages with varying physical characteristics [63]. This diversity poses considerable challenges in developing and dosing NC-based therapies. Factors like age and weight are essential, with pediatric patients spanning a broad weight range from premature infants around 400–500 g to older children nearing 20 kg. Moreover, variations in organ development, including differences in organ maturity, metabolic rates, and drug absorption and elimination processes, add complexity when comparing pediatric and adult populations [64]. Regulatory bodies must consider ethical challenges when conducting clinical trials on children such as obtaining informed consent while safeguarding child safety. A significant challenge in pediatric drug development is the limited availability of pharmacokinetic (PK) data specifically for children. Children have distinct physiological and developmental characteristics compared to adults, which necessitates tailored PK studies to ensure the efficacy and safety of drug delivery systems designed for pediatric use [65].

One of the biggest obstacles to pediatric medicine development is the high failure rate of studies (about 42%). Inadequate trial design and improper dosage selection are major causes of these failures. Modeling and simulation (M&S) have been acknowledged by the US FDA and EMA as a means of enhancing efficacy, bolstering trial design, and optimizing dose selection in pediatric medication development. These regulatory bodies have implemented guidelines to facilitate the use of M&S in characterizing dose-exposure relationships and in extrapolating efficacy data from adults to children [66].

Aging of humans significantly affects pharmacokinetic processes such as elimination, distribution, absorption, and metabolism. In older adults, aging affects various pharmacokinetic processes, including decreased absorption rates and altered drug metabolism due to changes in liver and kidney function [67]. While NCs hold the potential to enhance drug absorption, they must be carefully designed to address these age-specific challenges. The long-term safety of nanomaterials remains a subject of ongoing investigation, especially in older populations who may have underlying health conditions that could influence their response to these materials. Elderly individuals experience changes in gastric pH and gut motility, which can impact the distribution of nanotherapeutics [68]. Many older patients take multiple medications, known as polypharmacy, to manage chronic conditions like hypertension, diabetes, and arthritis. While NC-based therapies can improve drug bioavailability, they may also raise the risk of interactions by altering the absorption timing and concentration of other drugs. Additionally, NC-based therapies can affect liver enzymes (e.g., cytochrome P450), influencing the metabolism of other medications. The combination of multiple drugs with NCs may increase the potential for toxicity or adverse effects. Given the variability in drug responses among older patients, individualized dosing regimens are essential to optimize efficacy and reduce the risk of adverse effects [69].

11.6 Regulatory Submission Process

11.6.1 Dossier Preparation for Marketing Authorization

Marketing authorization is the approval of a new medicine to be marketed in a country or region by the competent regulatory agency of the nation or region where the product is expected to be commercialized. The marketing approval is subjected to the evaluation of the drug product's data on quality, safety, and efficacy. A dossier containing the quality, safety, and efficacy data relating to the drug product is submitted for review to the regulatory agency for approval for marketing the drug product. There exists a common format for submission of the dossier, i.e., the Common Technical Document (CTD). CTD is a well-structured format developed by the International Council for Harmonisation of Technical Requirements for Pharmaceuticals for Human Use for submission of applications in the USA, Europe, Japan, and other countries by the applicants or sponsors.

The first step to preparing a dossier is to understand the target market as module 1 of the CTD contains administrative information and prescribing information which can have different content and format according to the requirements of the relevant regulatory authorities. Module 2 covers the overview and summaries of modules 3–5, where module 3 contains data on quality, module 4 contains data on non-clinical study reports, and module 5 contains the clinical study data [70].

EMA has developed a series of four reflection papers to guide the development of both nanomedicine and nanosimilars [71]. US FDA has laid down two guidelines for the development of drug products involving the application of nanotechnology which are as follows:

1. Draft guidance for industry on liposomal drug products (2002); and
2. Guidance for industry: drug products, including biological products, that contain nanomaterials (2017) [54].

These guidelines provide recommendations on potential risk factors for products containing nanomaterials, quality, non-clinical studies for drug products, and clinical development. Since there can be changes in chemical, physical, and biological properties between nanomaterial and their larger counterparts that can affect the quality, safety, and efficacy of the drug product, the FDA recommends evaluating the product containing nanomaterial by a risk-based approach. The risk factors to be considered are the complexity of nanomaterial, adequacy of characterization, stability, route of administration, and so on [20].

11.6.2 Common Technical Document (CTD) Format

The CTD is a globally accepted framework for pharmaceutical product regulatory applications, including formulations based on NCs. The CTD was developed by the International Council for Harmonization (ICH) and unifies the submission procedure for prominent regulatory bodies including the FDA and EMA. There are five modules in it:

Module 1: Regional Administrative Information
This category contains regional-specific documentation, such as application forms and suggested labels. Any special regulatory requirements associated with nanotechnology should be highlighted for NC-based products [72].
Module 2: Summaries and Overviews
Defines the product's pharmacologic class, suggested clinical usage, and other details. A summary of the quality, nonclinical, and clinical data about the formulation of NCs should be included [73].
Module 3: Quality
Comprehensive details on the chemical, pharmacological, and biological components of the NC formulation are included. This comprises stability data, surface properties, and NC size distribution criteria [72].
Module 4: Nonclinical Study Reports
Comprises information from preclinical research on the pharmacology, pharmacokinetics, and toxicity of products based on NCs [73].
Module 5: Clinical Study Reports
Data from clinical studies are presented, together with research procedures, methodology, findings, and interpretations of NC formulations [72].

Overall, the CTD format guarantees that all relevant information for the licensing of NC-based medications is given in a systematic and easily accessible way.

11.6.3 Bridging Studies and Extrapolation for Generic Nanocrystals

Bridging studies and extrapolation are critical in the regulatory clearance of generic NC-based formulations, demonstrating their equivalency with reference-listed drugs (RLDs). Conventional bioequivalence (BE) techniques would not be adequate because of the distinct physicochemical characteristics of NCs, including particle size, surface charge, and dissolution behavior. By assessing pharmacokinetics, in vitro–in vivo correlation (IVIVC), and CQAs, bridging studies aid in creating a scientific connection between innovators and generic products [74]. These studies evaluate the potential effects of variations in formulation characteristics, excipients, or manufacturing methods on the drug's safety, effectiveness, and bioavailability. The FDA, EMA, and other regulatory bodies need comprehensive comparative characterization using electron microscopy, XRD, zeta potential measurement, and DLS to verify nanoscale comparability [25]. Extrapolation techniques are also used to extrapolate BE results to other dose forms, strengths, or patient groups. For instance, without carrying out independent clinical studies, extrapolation may enable the use of a generic intravenous nanocrystal formulation in other routes or illness indications provided it has comparable PK and safety characteristics. Extrapolation is frequently supported by PBPK (physiologically based pharmacokinetic) modeling and in vitro dissolution and permeability investigations [74]. Regulators, however, prioritize case-by-case evaluations since nanoparticles interact with biological systems, considering elements including aggregation, systemic distribution, and long-term stability. A strong bridging and extrapolation approach ensures regulatory compliance, making it easier to approve generic nanocrystal formulations that are safe, efficient, and cost-effective [18].

11.6.4 Approval Process Overview

As such there are no specific guidance that says about the approval pathways of nano-formulation. The regulatory authorities have not laid down any guidelines for the approval of nanomedicines. Eighty-two applications for NCs have been submitted to date, among which a third are for generics. This is attributed to the simplicity and translatable properties of the NC technology [22]. The approval pathways for both nanomedicine and their follow-on versions are depicted in Fig. 11.1.

The decision to apply for marketing approval of follow-on versions of nanomedicine (nanosimilars in EMA) to the regulatory authorities relies on the sameness approach. If the applicant can demonstrate the similarity between the follow-on and reference listed drug product, the application may be filled under 505 (j) in the US or 10(1) in EMA. The complexity of nanomedicine poses challenges in the evaluation of these similarities (for example characterization of the formulation) in that case 505 (b)(2) or 10(3) pathway for approval can be

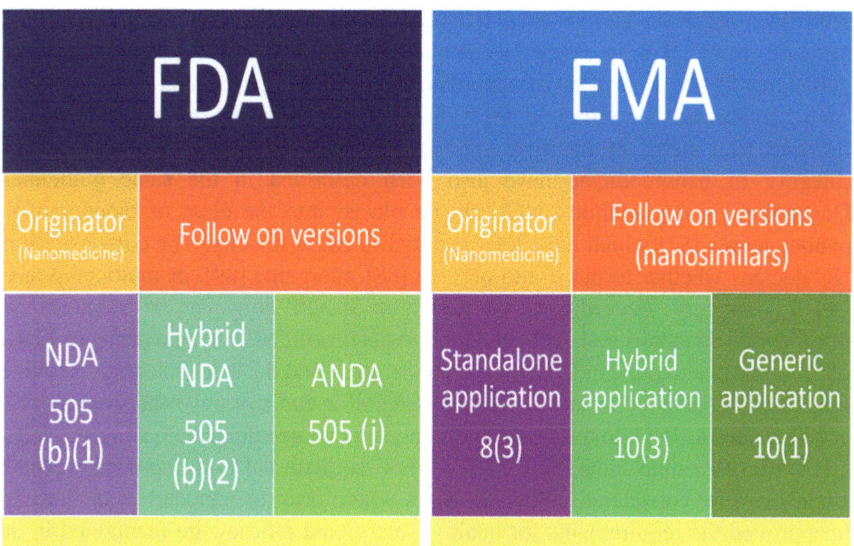

Fig. 11.1 Schematic representation of approval pathways for both nanomedicine and their generics. (Self-created)

explored. This route can also be used for reformulated drug products [15]. Nanomedicine can also be marketed through the 505 g pathway of the FD&C Act provided that the requirements are fulfilled [20]. Currently, the regulatory authorities (FDA and EMA) review and evaluate the applications case-to-case basis. Similarly, in India, the guidelines for evaluating nanopharmaceuticals state that nanomedicine's quality, safety, and efficacy are based on a case-to-case approach. CDSCO treats all nanomedicine products as Investigational New Drug (IND) and considers the physical, chemical, and biological properties of the nanotechnology-based product example general information of drug and nanomaterial, dosage form, route of administration, animal pharmacology data, etc. to review and approve the application for marketing of the product [25]. Various NC-based marketed products and their therapeutic use are outlined in Table 11.4 and a schematic illustration of the timeline from nanomedicine discovery to post-marketing surveillance is represented in Fig. 11.2.

11.6.5 Regulatory Aspects of Nanomedicines in India

CDSCO has laid down guidelines for the evaluation of nano pharmaceuticals in India. It outlines various regulatory obligations required for the approval of nanomedicine in India. Since there are no uniform guidelines for the development and evaluation of nanomedicines globally, these guidelines were framed to ensure the quality, safety, and efficacy of the developed drug product [25]. Also, the nanoregulatory task force constituted by the Nano Mission of the Government of India has published the guidelines for the safe handling of nanomaterials. These guidelines focus on various potential risk factors, safe handling procedures, and measures to be taken in case of accidental exposure to the risk factors [86]. In the guidelines for the evaluation of nanopharmaceuticals in India, nanomedicines have been categorized according to their degradability, organic or inorganic nature, function, and status of approval. The quality and safety requirements for each category of nanomedicine have also been described in the same guidelines. Depending on degradability NCs can be placed into the class of biodegradable nanomaterials [87] which have low potential to cause toxicity. The composition of NC decides whether to put it into organic [88], inorganic [89], or multicomponent [90, 91] category. According to the nanoform of API, drugs can be classified into drug-loaded nanocarrier or API converted to nanoform, NCs being a nanosized form of drug, the evaluation could be done considering it into the latter category. Based on the approval status of the drug and nanomaterial in India or any other country, NC can be treated as IND. In the case where drug and nanomedicine both are approved in a country other than India, whole data is not required and bridging or abbreviated studies can be considered. While in cases where drug or nanocarrier or both are unapproved the requirements for quality, safety, and efficacy are mentioned in the Second Schedule of the New Drugs and Clinical Trials Rules, 2019. However, due to the complex nature of the nanomaterial, the evaluation of quality, safety, and efficacy is based on a "case-by-case approach" [25].

Table 11.4 NC-based marketed products and their therapeutic use

Sr. no	Trade name	Active agent	Therapeutic application	Approved agency and year	References
1	Avinza®	Morphine sulfate	Moderate/severe pain	FDA (2002)	[75]
2	Emend®	Aprepitant	Antiemetic for chemotherapy patients	FDA (2003) and EMA (2003)	[76]
3	Focalin XR®	Dexmethylphenidate hydrochloride	Attention deficit hyperactivity disorder	FDA (2005)	[77]
4	Invega	Paliperidone palmitate	Schizophrenia	FDA (2006) and EMA (2007)	[78]
5	Naprelan	Naproxen sodium	Rheumatoid arthritis and osteoarthritis, gout	FDA (1996)	[79]
6	Rapamune	Sirolimus	Immunosuppressant	FDA (2002) and EMA (2001)	[80]
7	Ritalin LA	Methylphenidate hydrochloride	Attention deficit hyperactivity disorder	FDA (2002)	[81]
8	Tricor	Fenofibrate	Mixed dyslipidemia	FDA (2004)	[82]
9	Triglide	Fenofibrate	Mixed dyslipidemia	FDA (2005)	[83]
10	Zyprexa	Olanzapine	Schizophrenia	FDA (2009)	[84]
11	Xeplion	Paliperidone	Schizophrenia	FDA (2009) and EMA (2007)	[85]

The requirements for stability testing of nanomedicines are specified under Clause 5 of the Second Schedule of New Drugs and Clinical Trials Rules, 2019. Parameters to be tested for stability include size, PDI, drug release, drug loading, morphology, etc. [25].

11.6.6 Post-Approval Regulatory Obligations

NCs can be made using bottom-up (precipitation) or top-down (high-pressure homogenization, wet milling) methods. Variations in these procedures, however,

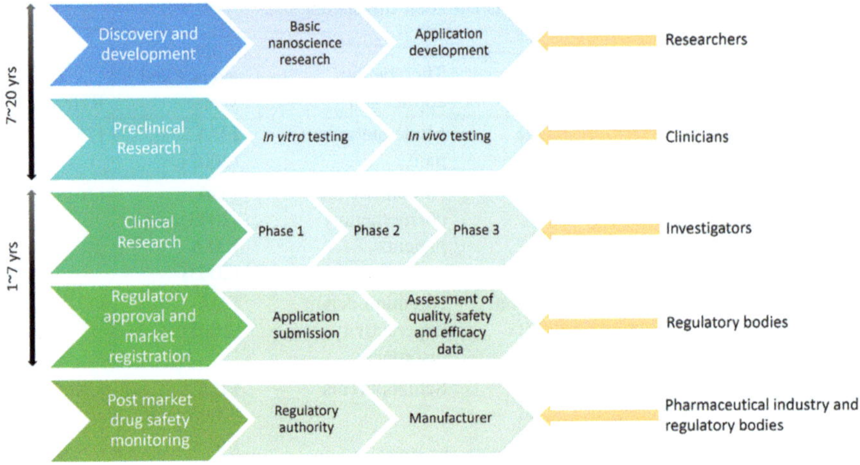

Fig. 11.2 Schematic illustration of timeline from nanomedicine discovery to post-marketing surveillance. (Self-created)

may result in inconsistent CQAs, including stability, surface charge, and particle size. The features of NCs are greatly influenced by process variables such as milling speed, duration, and stabilizer concentration. Even small changes in these characteristics can impact drug solubility, bioavailability, and therapeutic efficacy [92]. Furthermore, batch-to-batch variability is an issue that calls for stringent control measures. The process of scaling up from lab to commercial production adds even more complexity. Disparities in nanocrystal characteristics may result from variations in equipment, shear forces, and mixing dynamics. Thus, QbD methodologies and strong process analytical technology (PAT) tools are essential for ensuring uniformity and guaranteeing regulatory compliance [93].

Despite their benefits, NC-based products may raise safety issues, such as unanticipated toxicities and immunological reactions. Pharmacovigilance is essential for tracking post-market safety and guaranteeing patient well-being. Reporting adverse events is crucial for identifying infrequent but important safety indicators related to NCs. Adverse reaction data must be systematically gathered, examined, and reported to regulatory bodies like the FDA and EMA [22]. Since NCs have different pharmacokinetics than conventional formulations, their interactions with biological systems may be different, necessitating careful observation. Long-term safety issues are identified with the use of post-marketing monitoring studies, real-world data, and patient-reported outcomes. Continuous risk-benefit analyses are encouraged by regulatory frameworks, which may result in product removals or label changes as needed. Throughout their commercial life, NC-based products are guaranteed to preserve their safety profile through effective pharmacovigilance [92].

From early development to market expansion and final discontinuance, managing the life cycle of nanocrystal products entails ongoing assessment and adjustment. Formulation techniques are improved in the early stages of development to improve stability, bioavailability, and manufacturability [94]. Following approval, post-market monitoring guarantees product consistency and may offer enhancements like new medicine combinations, dosage forms, or updated formulations. Periodic safety updates and continuous stability investigations are frequently mandated by regulatory bodies. Manufacturers may submit further regulatory filings to take into account any notable breakthroughs in medication delivery or production [22]. Product life cycle plans also take into account market exclusivity and patent protection. Companies may switch to next-generation nanoformulations or reformulate pharmaceuticals based on nanocrystals as patents are about to expire to increase their market position [95].

Manufacturers have the obligation of post-marketing surveillance. They are encouraged to consult the regulatory authorities to share their understanding regarding the safety and efficacy concerns and seek recommendations from them to perform post-market monitoring. The safety data is also obtained by adverse event reporting systems. Given the complexity of the nanomedicines, it is challenging to characterize the safety, efficacy, and quality of the nano-formulation. Both regulators and manufacturers hold the responsibility of safety monitoring of nanomedicine post approval to safeguard public health [16]. In India, the CDSCO requires a pharmacovigilance plan to be submitted along with a marketing authorization application which should include potential risk factors, safety data from clinical trials, population at risk, any drug-drug or drug-food interaction, and mitigation strategies to ensure the safety of the nanopharmaceuticals [25].

11.7 Future Perspectives

Recent advances in NC research and development are opening the way for novel drug delivery systems and improved therapeutic effectiveness. These developments take advantage of the distinctive characteristics of NCs, especially when it comes to solving the problems posed by poorly soluble medications. The use of NCs for targeted therapy is growing, especially in the field of cancer. They can be designed to target both actively [96] and passively, reducing systemic exposure and enabling greater drug accumulation at the tumor region. Strategies like receptor-mediated endocytosis [97] are being investigated to improve absorption by cancer cells and selectivity. The incorporation of therapeutic and diagnostic agents into NC formulations is gaining popularity. This "theranostic" strategy increases the overall efficacy of drugs by enabling simultaneous disease therapy and monitoring. For example, imaging agents and therapeutics can be combined into NCs to deliver real-time feedback on treatment efficacy [98]. Technological developments in the fabrication process, including top-down milling and microfluidics, are making it possible to produce high-quality NCs with exact control over shape and size. These techniques improve repeatability and scalability, which are critical for industrial applications [18]. A promising technique in drug delivery is the fabrication of hybrid NCs, which mix drugs with other functional materials. These hybrids are appropriate for cutting-edge therapeutic applications because they can increase therapeutic benefits or offer multifunctional capabilities [99].

11.8 Conclusion

NCs are evolving as a promising drug delivery system in pharmaceutical sciences because of their potential to increase the solubility, bioavailability, and therapeutic efficacy of poorly soluble medications. This enhancement is attributed to NCs' distinctive characteristics. When compared to other carrier-based nanomedicines like liposomes, dendrimers, and micelles, NCs are better because of their favourable characteristics, which include easy scalability, a high drug loading capacity (up to 100% theoretically), and less toxicity. However, the regulatory landscape for NCs provides distinct problems that must be addressed to assure their safety and effective use. Regulatory bodies like the US FDA and the EMA have developed guidelines that highlight the necessity for full characterization of NCs, including their particle size, surface characteristics, stability, and dissolution behavior. These properties have a direct effect on the pharmacokinetics, pharmacodynamics, and safety of NC formulations.

Robust manufacturing procedures are essential for ensuring uniformity and repeatability from batch to batch. It is recommended to use the QbD principles to define CQAs and process variables that may have an impact on the final product's quality. Furthermore, formulations based on NCs could raise specific safety problems, such as changed toxicity profiles and biodistribution, demanding thorough preclinical and clinical testing. The pharmacokinetics, effectiveness, and safety of the formulation, which vary from conventional formulations owing to the drug's nanoscale properties, must be well documented in regulatory filings. In addition, post-marketing surveillance and risk management strategies are essential for tracking the long-term safety of formulations containing NCs. In summary, although NCs provide many benefits for drug administration, their special chemical and physical characteristics must be carefully considered for regulatory approval. The effective development and marketing of drugs based on NCs will need a proactive strategy that incorporates the most recent scientific and regulatory guidelines.

Acknowledgments All the authors are thankful to the Director of the National Institute of Pharmaceutical Education and Research (NIPER)-Ahmedabad.

References

1. Strickley RG (2004) Solubilizing excipients in oral and injectable formulations. Pharm Res 21: 201–230
2. McGuckin MB, Wang J, Ghanma R, Qin N, Palma SD, Donnelly RF, Paredes AJ (2022) Nanocrystals as a master key to deliver hydrophobic drugs via multiple administration routes. J Control Release 345:334–353
3. Chen H, Khemtong C, Yang X, Chang X, Gao J (2011) Nanonization strategies for poorly water-soluble drugs. Drug Discov Today 16:354–360
4. Möschwitzer JP (2013) Drug nanocrystals in the commercial pharmaceutical development process. Int J Pharm 453:142–156

5. Ige PP, Baria RK, Gattani SG (2013) Fabrication of fenofibrate nanocrystals by probe sonication method for enhancement of dissolution rate and oral bioavailability. Colloids Surf B Biointerfaces 108:366–373
6. Menon AM, Sidhartha NN, Shruti I, Suresh A, Meena R, Dikundwar AG, Chopra D (2024) Synthon approach in crystal engineering to modulate physicochemical properties in organic salts of chlorpropamide. Mol Pharm 21:2894–2907
7. Ainurofiq A, Putro DS, Ramadhani DA, Putra GM, Santo LDCDE (2021) A review on solubility enhancement methods for poorly water-soluble drugs. J Rep Pharm Sci 10:137–147
8. Nicoud L, Licordari F, Myerson AS (2018) Estimation of the solubility of metastable polymorphs: a critical review. Cryst Growth Des 18:7228–7237
9. Janssens S, Van den Mooter G (2009) Physical chemistry of solid dispersions. J Pharm Pharmacol 61:1571–1586
10. Knapik J, Wojnarowska Z, Grzybowska K, Jurkiewicz K, Tajber L, Paluch M (2015) Molecular dynamics and physical stability of coamorphous ezetimib and indapamide mixtures. Mol Pharm 12:3610–3619
11. Yang T, Cui F-D, Choi M-K, Lin H, Chung S-J, Shim C-K, Kim D-D (2007) Liposome formulation of paclitaxel with enhanced solubility and stability. Drug Deliv 14:301–308
12. Samia S, Chary PS, Khan O, Mehra NK (2024) Recent trends and advances in novel formulations as an armament in Bcl-2/Bax targeted breast cancer. Int J Pharm 653:123889
13. Bhalani DV, Nutan B, Kumar A, Chandel AKS (2022) Bioavailability enhancement techniques for poorly aqueous soluble drugs and therapeutics. Biomedicines 10:2055
14. Leuner C, Dressman J (2000) Improving drug solubility for oral delivery using solid dispersions. Eur J Pharm Biopharm 50:47–60
15. Mühlebach S (2018) Regulatory challenges of nanomedicines and their follow-on versions: a generic or similar approach? Adv Drug Deliv Rev 131:122–131. https://doi.org/10.1016/j.addr.2018.06.024
16. Foulkes R, Man E, Thind J, Yeung S, Joy A, Hoskins C (2020) The regulation of nanomaterials and nanomedicines for clinical application: current and future perspectives. Biomater Sci 8: 4653–4664. https://doi.org/10.1039/D0BM00558D
17. Chary PS, Shaikh S, Bhavana V, Rajana N, Vasave R, Mehra NK (2024) Emerging role of nanocrystals in pharmaceutical applications: a review of regulatory aspects and drug development process. Appl Mater Today 40:102334. https://doi.org/10.1016/j.apmt.2024.102334
18. Jarvis M, Krishnan V, Mitragotri S (2019) Nanocrystals: a perspective on translational research and clinical studies. Bioeng Transl Med 4:5–16. https://doi.org/10.1002/btm2.10122
19. Kumari R, Suman K, Karmakar S, Mishra V, Lakra SG, Saurav GK, Mahto BK (2023) Regulation and safety measures for nanotechnology-based agri-products. Front Genome Ed 5: 1–12. https://doi.org/10.3389/fgeed.2023.1200987
20. FDA (2022) Drug products, including biological products, that contain nanomaterials guidance for industry contains nonbinding recommendations, pp 1–33. https://www.fda.gov/Drugs/GuidanceComplianceRegulatoryInformation/Guidances/default.htmand/or
21. Pawar VK, Singh Y, Meher JG, Gupta S, Chourasia MK (2014) Engineered nanocrystal technology: in-vivo fate, targeting and applications in drug delivery. J Control Release 183: 51–66
22. Chen M-L, John M, Lee SL, Tyner KM (2017) Development considerations for nanocrystal drug products. AAPS J 19:642–651
23. Junghanns J-UAH, Müller RH (2008) Nanocrystal technology, drug delivery and clinical applications. Int J Nanomedicine 3:295–310
24. Pita R, Ehmann F, Papaluca M (2016) Nanomedicines in the EU-regulatory overview. AAPS J 18:1576–1582. https://doi.org/10.1208/s12248-016-9967-1
25. Gupta YK, Singht GN, Dinda AK, Pradhan AK (2019) Guidelines for evaluation of nanopharmaceuticals in India. Nanopharmaceuticals. https://dbtindia.gov.in/sites/default/files/uploadfiles/Guidelines_For_Evaluation_of_Nanopharmaceuticals_in_India_24.10.19.pdf (Accessed on 29 Sep 2024, rechecked on 29th Sep 2025)

26. Toma I, Tefas LR, Bogdan C, Tomuță I (2022) Development and characterization of loratadine liposomal gel using QbD approach. Farmacia 70:204–213
27. Jain S (2014) Quality by design (QBD): a comprehensive understanding of implementation and challenges in pharmaceuticals development. Int J Pharm Pharm Sci 6:29–35
28. Ainurofiq A, Rahayu BG, Murtadla FA, Kuncahyo I, Windarsih A, Choiri S (2024) QbD-based formulation development of resveratrol nanocrystal incorporated into soluble mesoporous material: pharmacokinetic proof of concept study. Int J Pharm 661:124459
29. Awad H, Rawas-Qalaji M, El Hosary R, Jagal J, Ahmed IS (2023) Formulation and optimization of ivermectin nanocrystals for enhanced topical delivery. Int J Pharm X 6:100210
30. Kapoor M, Lee SL, Tyner KM (2017) Liposomal drug product development and quality: current US experience and perspective. AAPS J 19:632–641
31. Hladek F, Chvíla S, Navratil O, Balouch M, Stepanek F (2022) Systematic investigation of wet-milling kinetics and colloidal stability of pharmaceutical nanocrystals. Cryst Growth Des 22:6928–6940
32. Keck CM, Müller RH (2006) Drug nanocrystals of poorly soluble drugs produced by high pressure homogenisation. Eur J Pharm Biopharm 62:3–16
33. Gigliobianco MR, Casadidio C, Censi R, Di Martino P (2018) Nanocrystals of poorly soluble drugs: drug bioavailability and physicochemical stability. Pharmaceutics 10:134
34. Arzi RS, Sosnik A (2018) Electrohydrodynamic atomization and spray-drying for the production of pure drug nanocrystals and co-crystals. Adv Drug Deliv Rev 131:79–100
35. Xia D, Gan Y, Cui F (2014) Application of precipitation methods for the production of water-insoluble drug nanocrystals: production techniques and stability of nanocrystals. Curr Pharm Des 20:408–435
36. Peltonen L, Hirvonen J (2010) Pharmaceutical nanocrystals by nanomilling: critical process parameters, particle fracturing and stabilization methods. J Pharm Pharmacol 62:1569–1579
37. Jakubowska E, Lulek J (2021) The application of freeze-drying as a production method of drug nanocrystals and solid dispersions–a review. J Drug Deliv Sci Technol 62:102357
38. Yaoting HAO, Yanhui LI, Huicong Z, Jingwen LIU, Chunlei LI (2024) Research progress of nanocrystalline drugs in different drug delivery systems. J China Pharm Univ 55:548–556
39. Bansal S, Bansal M, Kumria R (2012) Nanocrystals: current strategies and trends. Int J Res Pharm Biomed Sci 4:10
40. Shegokar R, Müller RH (2010) Nanocrystals: industrially feasible multifunctional formulation technology for poorly soluble actives. Int J Pharm 399:129–139
41. Hu J, Johnston KP, Williams RO III (2004) Nanoparticle engineering processes for enhancing the dissolution rates of poorly water soluble drugs. Drug Dev Ind Pharm 30:233–245
42. de Waard H, Frijlink HW, Hinrichs WLJ (2011) Bottom-up preparation techniques for nanocrystals of lipophilic drugs. Pharm Res 28:1220–1223
43. Patravale VB, Date AA, Kulkarni RM (2004) Nanosuspensions: a promising drug delivery strategy. J Pharm Pharmacol 56:827–840
44. Liu T, Müller RH, Möschwitzer JP (2015) Effect of drug physico-chemical properties on the efficiency of top-down process and characterization of nanosuspension. Expert Opin Drug Deliv 12:1741–1754
45. Cheng Z, Lian Y, Kamal Z, Ma X, Chen J, Zhou X, Su J, Qiu M (2018) Nanocrystals technology for pharmaceutical science. Curr Pharm Des 24:2497–2507. https://doi.org/10.2174/1381612824666180518082420
46. Salazar J, Müller RH, Möschwitzer JP (2013) Performance comparison of two novel combinative particle-size-reduction technologies. J Pharm Sci 102:1636–1649
47. Li J, Wang Z, Zhang H, Gao J, Zheng A (2021) Progress in the development of stabilization strategies for nanocrystal preparations. Drug Deliv 28:19–36. https://doi.org/10.1080/10717544.2020.1856224
48. Brazeau GA, Fung H (1990) Mechanisms of creatine kinase release from isolated rat skeletal muscles damaged by propylene glycol and ethanol. J Pharm Sci 79:393–397
49. Korttila K, Sothman A, Andersson P (1976) Polyethylene glycol as a solvent for diazepam: bioavailability and clinical effects after intramuscular administration, comparison of oral,

intramuscular and rectal administration, and precipitation from intravenous solutions. Acta Pharmacol Toxicol (Copenh) 39:104–117
50. Budden R, Kühl UG, Bahlsen J (1979) Experiments on the toxic, sedative and muscle relaxant potency of various drug solvents in mice. Pharmacol Ther 5:467–474
51. Gao L, Liu G, Ma J, Wang X, Zhou L, Li X (2012) Drug nanocrystals: in vivo performances. J Control Release 160:418–430
52. Oualikene-Gonin W, Sautou V, Ezan E, Bastos H, Bellissant E, Belgodère L, Maison P, Ankri J (2023) Regulatory assessment of nano-enabled health products in public health interest. Position of the scientific advisory board of the French National Agency for the Safety of Medicines and Health Products. Front Public Health 11:1125577. https://doi.org/10.3389/fpubh.2023.1125577
53. Patel V, Agrawal Y (2011) Nanosuspension: an approach to enhance solubility of drugs. J Adv Pharm Technol Res 2:81. https://doi.org/10.4103/2231-4040.82950
54. Bremer-Hoffmann S, Halamoda-Kenzaoui B, Borgos SE (2018) Identification of regulatory needs for nanomedicines. J Interdiscip Nanomed 3:4–15. https://doi.org/10.1002/jin2.34
55. Tadini MC, de Freitas Pinheiro AM, Carrão DB, Aguillera Forte ALS, Nikolaou S, de Oliveira ARM, Berretta AA, Marquele-Oliveira F (2017) Method validation and nanoparticle characterization assays for an innovative amphothericin B formulation to reach increased stability and safety in infectious diseases. J Pharm Biomed Anal 145:576–585. https://doi.org/10.1016/j.jpba.2017.06.034
56. Zong R, Ruan H, Zhu W, Zhang P, Feng Z, Liu C, Fan S, Liang H, Li J (2022) Curcumin nanocrystals with tunable surface zeta potential: preparation, characterization and antibacterial study. J Drug Deliv Sci Technol 76:103771. https://doi.org/10.1016/j.jddst.2022.103771
57. Rodriguez-Loya J, Lerma M, Gardea-Torresdey JL (2024) Dynamic light scattering and its application to control nanoparticle aggregation in colloidal systems: a review. Micromachines 15:24. https://doi.org/10.3390/mi15010024
58. Stevenson HP, DePonte DP, Makhov AM, Conway JF, Zeldin OB, Boutet S, Calero G, Cohen AE (2014) Transmission electron microscopy as a tool for nanocrystal characterization pre- and post-injector. Philos Trans R Soc B Biol Sci 369:1–5. https://doi.org/10.1098/rstb.2013.0322
59. Asano N, Lu J, Asahina S, Takami S (2021) Direct observation techniques using scanning electron microscope for hydrothermally synthesized nanocrystals and nanoclusters. Nanomaterials 11:908. https://doi.org/10.3390/nano11040908
60. Holder CF, Schaak RE (2019) Tutorial on powder X-ray diffraction for characterizing nanoscale materials. ACS Nano 13:7359–7365. https://doi.org/10.1021/acsnano.9b05157
61. Peng B, Tang J, Wang P, Luo J, Xiao P, Lin Y, Tam KC (2018) Rheological properties of cellulose nanocrystal-polymeric systems. Cellulose 25:3229–3240. https://doi.org/10.1007/s10570-018-1775-6
62. Junghanns JUAH, Müller RH (2008) Nanocrystal technology, drug delivery and clinical applications. Int J Nanomedicine 3:295–309
63. Wang X, Deng F, Ji T, Zhang C, Tian Y, Zhang H, Zheng A, Chen Y, He B, Dai W, Zhang H, Zhang Q, Wang X (2024) Impact of physiological characteristics on chylomicron pathway-mediated absorption of nanocrystals in the Pediatric population. ACS Nano 18:23136–23153. https://doi.org/10.1021/acsnano.4c05391
64. Marques MRC, Choo Q, Ashtikar M, Rocha TC, Bremer-Hoffmann S, Wacker MG (2019) Nanomedicines – tiny particles and big challenges. Adv Drug Deliv Rev 151–152:23–43. https://doi.org/10.1016/j.addr.2019.06.003
65. Barker CIS, Germovsek E, Hoare RL, Lestner JM, Lewis J, Standing JF (2014) Pharmacokinetic/pharmacodynamic modelling approaches in paediatric infectious diseases and immunology. Adv Drug Deliv Rev 73:127–139. https://doi.org/10.1016/j.addr.2014.01.002
66. Yellepeddi VK, Joseph A, Nance E (2019) Pharmacokinetics of nanotechnology-based formulations in pediatric populations. Adv Drug Deliv Rev 151–152:44–55. https://doi.org/10.1016/j.addr.2019.08.008

67. Chandrasekaran P, Kandasamy R (2018) Solid oral flexible formulations for paediatric and geriatric patients: age-appropriate formulation platforms. Indian J Pharm Sci 80:14–25. https://doi.org/10.4172/pharmaceutical-science.1000326
68. Xu H, Li S, Liu YS (2022) Nanoparticles in the diagnosis and treatment of vascular aging and related diseases. Signal Transduct Target Ther 7:231. https://doi.org/10.1038/s41392-022-01082-z
69. Hunt NJ, McCourt PAG, Kuncic Z, Le Couteur DG, Cogger VC (2022) Opportunities and challenges for nanotherapeutics for the aging population. Front Nanotechnol 4:1–15. https://doi.org/10.3389/fnano.2022.832524
70. The International Council for Harmonisation of Technical Requirements for Pharmaceuticals for Human Use (2021) The common technical document for the registration of pharmaceuticals for human use. In: New drug approval process. CRC Press, pp 403–437. https://doi.org/10.3109/9780203021620-25
71. P.N. Workshop (2014) EMA perspective on the development of nanomedicines
72. Jordan D (2014) An overview of the Common Technical Document (CTD) regulatory dossier. Med Writ 23:101–105. https://doi.org/10.1179/2047480614z.000000000207
73. International Council for Harmonisation (ICH) (2024) Overview of common technical document (CTD), pp 10–12. https://www.pharmagrowthhub.com/post/overview-of-common-technical-document-ctd
74. Wang T, Cao X, He Y, Chen X (2022) Innovation drug approvals based on a bridging study: from concept to practice. Transl Breast Cancer Res 3:2. https://tbcr.amegroups.org/article/view/60537
75. Kaur A, Parmar PK, Jadhav S, Bansal AK (2022) Chapter 12 – Advances in nanocrystals as drug delivery systems. In: Kesharwani P, Singh KKBT-NT (eds) Nanoparticle therapeutics. Academic Press, London, pp 413–454. https://doi.org/10.1016/B978-0-12-820757-4.00011-9
76. Müller RH, Shegokar R, Gohla S, Keck CM (2011) In: Prokop A (ed) Nanocrystals: production, cellular drug delivery, current and future products BT – intracellular delivery: fundamentals and applications. Springer Netherlands, Dordrecht, pp 411–432. https://doi.org/10.1007/978-94-007-1248-5_15
77. Childress A (2017) The safety of extended-release drug formulations for the treatment of ADHD. Expert Opin Drug Saf 16:603–615. https://doi.org/10.1080/14740338.2017.1317344
78. Lamb YN, Keating GM (2016) Paliperidone palmitate intramuscular 3-monthly formulation: a review in schizophrenia. Drugs 76:1559–1566. https://doi.org/10.1007/s40265-016-0645-5
79. Schiff MH (1996) A comparison of Naprelan and Naprosyn in the treatment of osteoarthritis of the knee. Am J Orthop (Belle Mead NJ) 25:14–20. http://europepmc.org/abstract/MED/8886213
80. Camardo J (2003) The Rapamune era of immunosuppression 2003: The journey from the laboratory to clinical transplantation. Transplant Proc 35:S18–S24. https://doi.org/10.1016/S0041-1345(03)00356-7
81. Biederman J, Quinn D, Weiss M, Markabi S, Weidenman M, Edson K, Karlsson G, Pohlmann H, Wigal S (2003) Efficacy and safety of Ritalin® LA™, a new, once daily, extended-release dosage form of methylphenidate, in children with attention deficit hyperactivity disorder. Pediatr Drugs 5:833–841. https://doi.org/10.2165/00148581-200305120-00006
82. Kaye AD, Edinoff AN, Scoon L, Youn S, Farrell KJ, Kaye AJ, Shah RJ, Cornett EM, Chami AA, Dixon BM, Alvarado MA, Viswanath O, Urits I, Calodney AK (2021) Novel interventional techniques for chronic pain with minimally invasive arthrodesis of the sacroiliac joint: (INSITE, iFuse, Tricor, Rialto, and others). Rheumatol Ther 8:1061–1072. https://doi.org/10.1007/s40744-021-00350-8
83. Tziomalos K, Athyros VG (2006) Fenofibrate: a novel formulation (Triglide™) in the treatment of lipid disorders: a review. Int J Nanomedicine 1:129–147. https://doi.org/10.2147/nano.2006.1.2.129
84. Hori H, Ueda N, Yoshimura R, Yamamoto H, Wani K, Etoh Y, Haraga K, Kitahara J, Nakamura J (2009) Olanzapine orally disintegrating tablets (Zyprexa ZydisR) rapidly improve

excitement components in the acute phase of first-episode schizophrenic patients: an open-label prospective study. World J Biol Psychiatry 10:741–745. https://doi.org/10.1080/15622970903166312
85. Talamba G, Arcabasso S, Spadafora M, Ballerini A (2013) 1360 – Long acting second generation antipsychotic therapy, from compliance to improving quality of life: the florence depot clinic preliminary report. Eur Psychiatry 28:1. https://doi.org/10.1016/S0924-9338(13)76405-8
86. Centre for Knowledge Management of Nanoscience & Technology (2016) Guidelines and best practices for safe handling of nanomaterials in research laboratories and industries, pp 1–19
87. Kumar S, Randhawa JK (2013) High melting lipid based approach for drug delivery: solid lipid nanoparticles. Mater Sci Eng C 33:1842–1852. https://doi.org/10.1016/j.msec.2013.01.037
88. Kasai H, Nakanishi H, Oikawa H (2018) Fabrication technique of organic nanocrystals and their optical properties and materialization. Elsevier B.V. https://doi.org/10.1016/B978-0-444-64110-6.00070-6
89. Palui G, Aldeek F, Wang W, Mattoussi H (2015) Strategies for interfacing inorganic nanocrystals with biological systems based on polymer-coating. Chem Soc Rev 44:193–227. https://doi.org/10.1039/c4cs00124a
90. Yin Y, Alivisatos AP (2005) Colloidal nanocrystal synthesis and the organic-inorganic interface. Nature 437:664–670. https://doi.org/10.1038/nature04165
91. Hollis CP, Dozier AK, Knutson BL, Li T (2019) Preparation and characterization of multimodal hybrid organic and inorganic nanocrystals of camptothecin and gold. Acta Pharm Sin B 9:128–134. https://doi.org/10.1016/j.apsb.2018.03.005
92. Lhaglham P, Jiramonai L, Jia Y, Huang B, Huang Y, Gao X, Zhang J, Liang X-J, Zhu M (2024) Drug nanocrystals: surface engineering and its applications in targeted delivery. iScience 27:111185. https://doi.org/10.1016/j.isci.2024.111185
93. Hutchinson N (2014) Understanding and controlling sources of process variation: risks to achieving product critical quality attributes. Bioprocess Int 12:1–15. https://www.bioprocessintl.com/bioanalytical-methods/understanding-and-controlling-sources-of-process-variation-risks-to-achieving-product-critical-quality-attributes
94. Ding Y, Zhao T, Fang J, Song J, Dong H, Liu J, Li S, Zhao M (2024) Recent developments in the use of nanocrystals to improve bioavailability of APIs. Wiley Interdiscip Rev Nanomed Nanobiotechnol 16:e1958. https://doi.org/10.1002/wnan.1958
95. Yue P, Zhou W, Huang G, Lei F, Chen Y, Ma Z, Chen L, Yang M (2022) Nanocrystals based pulmonary inhalation delivery system: advance and challenge. Drug Deliv 29:637–651
96. Anarjan FS (2019) Active targeting drug delivery nanocarriers: ligands. Nano-Struct Nano-Objects 19:100370
97. Angelopoulou A, Kolokithas-Ntoukas A, Fytas C, Avgoustakis K (2019) Folic acid-functionalized, condensed magnetic nanoparticles for targeted delivery of doxorubicin to tumor cancer cells overexpressing the folate receptor. ACS Omega 4:22214–22227
98. Yenurkar D, Nayak M, Mukherjee S (2023) Recent advances of nanocrystals in cancer theranostics. Nanoscale Adv 5(16):4018–4040
99. Kataoka T, Abe S, Tagaya M (2019) Surface-engineered design of efficient luminescent europium (III) complex-based hydroxyapatite nanocrystals for rapid HeLa cancer cell imaging. ACS Appl Mater Interfaces 11:8915–8927

MIX
Papier aus verantwortungsvollen Quellen
Paper from responsible sources
FSC® C105338

If you have any concerns about our products,
you can contact us on
ProductSafety@springernature.com

In case Publisher is established outside the EU,
the EU authorized representative is:
Springer Nature Customer Service Center GmbH
Europaplatz 3, 69115 Heidelberg, Germany

Printed by Libri Plureos GmbH
in Hamburg, Germany